Electrical Spectroscopy of Earth Materials

Electrical Spectroscopy of Earth Materials

Tsylya M. Levitskaya

Laboratory for Advanced Subsurface Imaging, University of Arizona, Tucson, AZ, United States

Ben K. Sternberg

Laboratory for Advanced Subsurface Imaging, University of Arizona, Tucson, AZ, United States

ELSEVIER

Elsevier
Radarweg 29, PO Box 211, 1000 AE Amsterdam, Netherlands
The Boulevard, Langford Lane, Kidlington, Oxford OX5 1GB, United Kingdom
50 Hampshire Street, 5th Floor, Cambridge, MA 02139, United States

Notices
Knowledge and best practice in this field are constantly changing. As new research and experience broaden
our understanding, changes in research methods, professional practices, or medical treatment may become
necessary.

Practitioners and researchers must always rely on their own experience and knowledge in evaluating and
using any information, methods, compounds, or experiments described herein. In using such information or
methods they should be mindful of their own safety and the safety of others, including parties for whom
they have a professional responsibility.

To the fullest extent of the law, neither the Publisher nor the authors, contributors, or editors, assume any
liability for any injury and/or damage to persons or property as a matter of products liability, negligence or
otherwise, or from any use or operation of any methods, products, instructions, or ideas contained in the
material herein.

British Library Cataloguing-in-Publication Data
A catalogue record for this book is available from the British Library

Library of Congress Cataloging-in-Publication Data
A catalog record for this book is available from the Library of Congress

ISBN: 978-0-12-818603-9

For Information on all Elsevier publications
visit our website at https://www.elsevier.com/books-and-journals

Publisher: Candice G. Janco
Acquisition Editor: Marisa LaFleur
Editorial Project Manager: Devlin Person
Production Project Manager: Mohanapriyan Rajendran/
                                  Prem Kumar Kaliamoorthi
Cover Designer: Miles Hitchen

Typeset by MPS Limited, Chennai, India

Dedication to Dr. James R. Wait

This book is dedicated to Dr. James R. Wait, who was affiliated with the University of Arizona during the last 18 years of his life, from 1980, when he became a Professor of Electrical and Computer Engineering Department, with a joint appointment in geosciences, at the University of Arizona, until October 1, 1998, when he died.

I had already read some of his publications when I was in Russia, and they were very helpful for my work on studying electrical properties of rocks from different locations, and various water saturation levels, in a frequency range from 50 Hz to 100 MHz. I knew that Dr. Wait was working at the University of Arizona, Tucson. When I emigrated from Russia to the United States with my relatives and learned that our destination would be Tucson, I felt lucky to have an opportunity to meet Dr. Wait in person.

We met in March 1992. I gave him copies of my Russian publications, translated into English. He was interested in geophysical research conducted in the former Soviet Union and suggested that I write a review of Russian literature on this topic. Dr. Wait introduced me to Prof. Ben K. Sternberg, Director of the Laboratory for Advanced Subsurface Imaging (LASI), University of Arizona and we collaborated on this review. We published the review in two papers in Radio Science: 1996, 31, 4, Part 1, pp. 755–779 and Part 2, pp. 781–802.

Dr. Wait was a kind and thoughtful person. Being recognized as a top specialist in electromagnetic geophysics, he offered me his consultation any time, when I needed, and he always provided his encouragement. This was very valuable to me. His readiness to explain any questions, in theory, or in the experimental work, made me more confident in my further research. Now, I miss Dr. Wait a lot.

Tsylya Levitskaya

I first met Jim Wait in 1974 during my graduate studies at the University of Wisconsin. I had set up a deep earth-probing experiment using the Navy's Sanguine antenna in Northern Wisconsin, but with a different transmitter for the frequency range 0.1−10 Hz. Other universities became involved in this large-scale experiment, and in the summer of 1974, I organized a workshop at my base camp near Clam Lake Wisconsin. Many of the deep-exploration, controlled-source, electrical methods researchers in the United States at that time attended. I immediately connected with Jim Wait. He provided important fundamental theoretical insight into what has been described as the first mega-moment source survey (K.M. Strack, Exploration with Deep Transient Electromagnetics, *Elsevier, 1992).*

I particularly enjoyed the many hours we spent in a canoe until late into the night on Clam Lake, after the meeting. Besides being a brilliant theoretician, Jim was an enthusiastic athlete.

Jim and I kept in close contact during my career at CONOCO Inc., Barringer Research, and Phoenix Geophysics. After Jim moved to the University of Arizona in 1980, Jim convinced me to apply for a faculty position at the university and Director of the LASI. I started at the University of Arizona in January 1986.

Stan Ward and I organized a series of international symposia at the University of Arizona during the 1990s. Jim played an important role in these symposia, including identifying well-known speakers and initiating exciting and rewarding discussions during the meetings. Jim was also an important sounding board during our many new technology developments in the LASI.

Ben Sternberg

Contents

Introduction

Earth materials primarily consist of rocks and soils, which are subjects of study by geologists and geophysicists. Both rock and soil are generally inhomogeneous materials. Rock and soil electrical properties depend heavily on their water content but are also related to the composition and structure of the earth materials. Rock and soil may also contain other solid, liquid, and gas components, and this affects their complicated behavior in an electromagnetic field.

We have studied electrical spectroscopy, primarily in the frequency domain, and mainly from 1 kHz to 1 GHz. In this broad frequency spectrum, various polarization processes are shown, which are linked to the material composition and structure. We have primarily used two methods: dielectric and conduction spectroscopy methods. An important part of our study has been to show how these two spectroscopy methods describe the polarization processes, and in what frequency range each of them is more informative.

Some of our previous studies are described in the following publications:

Levitskaya, T.M., Sternberg, B.K., 1996. Polarization processes in rocks: 1 Complex dielectric permittivity method. Radio Sci. 31 (4), 755−779. [A review of Soviet Literature.]
Levitskaya, T.M., Sternberg B.K., 1996. Polarization processes in rocks: 2. Complex electrical resistivity method. Radio Sci. 31 (4), 781−802. [A review of Soviet Literature.]
Sternberg, B.K., Levitskaya, T.M., 1998. Correlation between laboratory and in Situ electrical resistivity measurements of soil. J. Environ. Eng. Geophys. 63−70.
Levitskaya, T.M., Sternberg B.K., 2000. Laboratory measurement of material electrical properties: extending the application of lumped-circuit equivalent models to 1 GHz. Radio Sci. 35 (2), 371−383.
Sternberg, B.K., Levitskaya, T.M., 2001. Electrical parameters of soils in the frequency range from 1 kHz to 1 GHz, using lumped-circuit methods. Radio Sci. 36 (4), 709−719.

In this book, our previous papers are significantly expanded, and the results are shown in much greater detail.

Data on electrical properties of earth materials also provide a petrophysical background for the electrical geophysical methods, which are used for natural resources prospecting, environmental site characterization, and large-scale

Electrical Spectroscopy of Earth Materials. DOI: https://doi.org/10.1016/B978-0-12-818603-9.00001-5

geoscience studies. Magneto telluric surveys, frequency electromagnetic, and transient electromagnetic methods may operate in the range of 0.001 Hz to 10 MHz. Ground-penetrating radar methods may operate in the range of 10−1000 MHz. Electromagnetic methods are also used in well logging in the range of 1 Hz to 1000 MHz.

For field data interpretation, geophysicists need to know the electrical properties of earth materials, which depend on their water content, composition, texture, and structure over a wide frequency range. Our study on correlation between laboratory and field measurements shows the connection between lab and field measurements. A major part of our research was carried out on soils, although some rock samples were also studied.

We have used standard impedance analyzers for our measurements. Most of our measurements used the HP 4194A for the frequency range of 1 kHz to 40 MHz and the HP 4191A for the frequency range of 1 MHz to 1 GHz. Although these are not the most recent impedance analyzers, we find essentially the same impedance data with newer model impedance analyzers in this application. Furthermore, the data analysis procedures that we describe do not change with these newer instruments

Parameters describing the material behavior in an electromagnetic field

Chapter Outline

In a homogeneous, isotropic, nonmagnetic material, placed in an electromagnetic field, three phenomena may occur:

1. Transport phenomenon;
2. Polarization phenomenon; and
3. Energy dissipation.

The transport phenomenon is the movement of free charges, when an electric field is applied. In earth materials, these are usually various ions. This free-charge movement creates conduction currents in a material, which are characterized by conductivity σ (or resistivity ρ). The conductivity σ is the ability of a material to conduct an electric current; the resistivity ρ is the ability of a material to oppose the flow of current.

In a direct current (DC) field, isotropic materials usually obey Ohm's law. This law is based on experimental research performed by the German physicist George Simon Ohm, who found that the amount of electric current I through a material with resistance R is directly proportional to the voltage V across the material. Ohm's law can be stated as

$$I = \frac{V}{R} \tag{2.1}$$

In terms of conductance G, the inverse value to resistance R, the Ohm's law states

$$I = V \, G \tag{2.2}$$

Electrical Spectroscopy of Earth Materials. DOI: https://doi.org/10.1016/B978-0-12-818603-9.00002-7

The polarization phenomenon is the ability of a material to store the electrical energy, that is, to create a capacitance. This is due to restricted movement of bound charges, which can be displaced from their equilibrium positions, when an electric field is applied. As a result, electric dipole moments arrive in the material and produce an electric field opposite in sign to the applied field. This process is called polarization (Böttcher and Bordewijk, 1978; Bartnikas, 1987; Jonscher, 1983). A polarization process creates "displacement currents" in a material.

The vector of electric displacement (or electric induction) D is proportional to the applied electric field through the dielectric permittivity ε that is a measure of polarization in a material:

$$\overline{D} = \varepsilon\, \overline{E} \tag{2.3}$$

Here ε is in F/m. Usually, the relative dielectric permittivity ε_r is used: $\varepsilon_r = \varepsilon/\varepsilon_0$, where $\varepsilon_0 = 8.854 \cdot 10^{-12}$ F/m is the dielectric permittivity of vacuum or, for practical purposes, air.

In general, the behavior of a homogeneous, isotropic, nonmagnetic material in an electromagnetic field can be described by Ohm's law, which relates the current density J to the electric field intensity E through the complex conductivity σ, which includes effects from displacement currents, as follows (Wait, 1982, 1985):

$$J = \hat{\sigma}\, E = (\sigma' + j\,\omega\,\varepsilon')\, E \tag{2.4}$$

An equivalent form of Eq. (2.4) is

$$J = j\,\omega\,\hat{\varepsilon}\, E = j\,\omega \left(\varepsilon' - j\,\frac{\sigma'}{\omega} \right) E \tag{2.5}$$

Here, σ' and ε' are the real parts of complex conductivity and complex dielectric permittivity, which are interrelated as follows (Wait, 1985):

$$\hat{\sigma} = j\,\omega\,\hat{\varepsilon} \tag{2.6}$$

These complex parameters, σ and ε, which include imaginary components for energy loss occurring in a material, are

$$\hat{\sigma} = \sigma' + j\,\sigma''; \quad \hat{\varepsilon} = \varepsilon' - j\,\varepsilon'' \tag{2.7}$$

where $\sigma'' = \omega \cdot \varepsilon'$ and $\varepsilon'' = \sigma'/\omega$ are also real.

Alternatively, the complex resistivity ρ can be used which is the reciprocal of the complex conductivity (Wait, 1984):

$$\hat{\rho} = \rho' - \rho'' = \frac{1}{\hat{\sigma}} = \frac{1}{\sigma' + j\,\omega\,\varepsilon'} \tag{2.8}$$

From (2.6) and (2.8), the relation between the complex resistivity and complex permittivity is

$$\hat{\rho} = \frac{1}{j\,\omega\,\hat{\varepsilon}} \tag{2.9}$$

Both transport and polarization phenomena are accompanied by dissipation of electromagnetic energy or dielectric losses. The conductivity produces heat losses, as a result of transforming some electric energy into heat energy (Joule heat). These losses are observed not only in DC field but in alternating fields as well. The amount of such losses in DC field (in terms of power P, voltage V, current I, and resistance R) can be expressed with the experimental Joule−Lenz law, in relation to Ohm's law as follows:

$$P_{DC} = V\,I = \frac{V^2}{R} = I^2\,R \tag{2.10}$$

In alternating (sinusoidal) fields, the power of heat losses, caused by conduction current I_{real} (Fig. 2.1), can be expressed as follows (Duffin, 1990):

$$P_{AC} = V\,I_{real} = V\,I\,\cos\phi, \tag{2.11}$$

where $\cos\phi$ is known as the power factor. Because $\cos\phi \leq 1$, $P_{AC} \leq P_{DC}$.

Dielectric losses caused by slow polarization processes are also called relaxation losses. Usually, the relaxation losses are characterized by $\tan\delta$, where δ is the loss angle, which complements the phase angle ϕ between voltage and current to 90° (Fig. 2.1). Therefore, the value $\tan\delta = 1/\tan\phi$ and the phase angle ϕ are defined from measurements. As seen in Fig. 2.1, $\tan\delta = I_{real}/I_{imag}$.

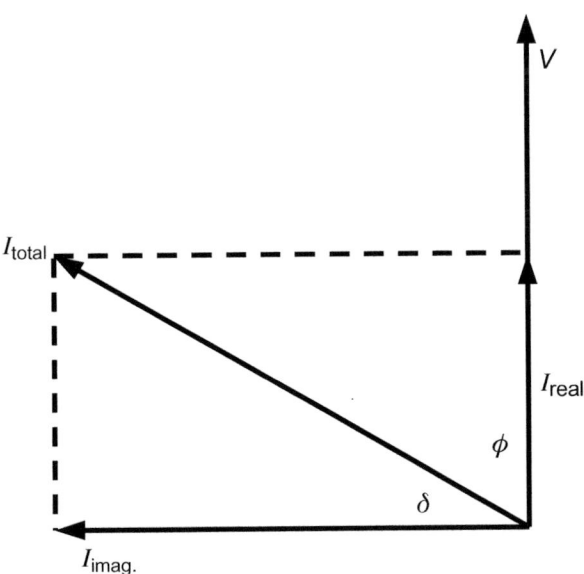

FIGURE 2.1

Schematic phasor diagram for a dielectric in alternating field.

The imaginary components for energy loss in Eqs. (2.7) and (2.8) can be expressed as tan δ and corresponding real components, as follows:

$$\varepsilon'' = \varepsilon' \tan \delta; \quad \sigma'' = \frac{\sigma'}{\tan \delta}; \quad \rho'' = \frac{\rho'}{\tan \delta}. \tag{2.12}$$

The value ε'' is called loss factor. The relationships (2.12) will be derived in Chapter 4, Measurements and analysis concept of distributed versus lumped parameters, using phasor diagrams.

2.1 Conductivity Mechanisms

The conductance of earth materials is mostly ionic in nature. Free or weakly bound ions are usually present in rocks and soils as impurities or in their salt solutions (Chernyak, 1967). Depending on the mineralogical composition and texture, electronic and mixed conductance may also exist in earth materials, if they contain electron-conductive minerals in the structure, such as ore-minerals (galena, pyrite, chalcopyrite, and magnetite) and graphite. Some rock-forming minerals, such as quartz, pyroxene, olivine, nepheline, and plagioclases, are semiconductors (Parkhomenko, 1967; Schon, 1998). Electrical conductance of all kinds results in a through (or residual) current I_{thru} (Chernyak, 1967). This current is related to the static or DC field E, where (assuming that the earth is isotropic) Ohm's law takes the form (Wait, 1982)

$$J = \sigma E \tag{2.13}$$

The constant value σ from Eq. (2.13) is often called "Ohmic conductivity" σ_{Ohmic}.

In a DC field, the conductivity σ is reciprocal to the resistivity ρ, which characterizes the ability of a material to oppose the flow of current. Then, Ohm's law can take the following form:

$$J = \frac{E}{\rho} \tag{2.14}$$

Analogously, the constant value of ρ is often called "Ohmic resistivity" ρ_{Ohmic}.

In practice, it may not be easy to measure the Ohmic values of conductivity or resistivity, because of the time-dependence of the current when a DC voltage is applied to a material. This current is called absorption current (Tareev, 1975), because it is caused by energy absorption from various polarization types, and it is actually a displacement current. Sometimes the time-dependent current is called charging current (Daniel, 1967).

A schematic diagram (Fig. 2.2) shows that the absorption current decreases with time, which may be different (rather long) depending on the type of polarization causing this displacement current. However, it usually does not take more than 1 minute after the voltage is applied. Finally, it reaches a constant residual

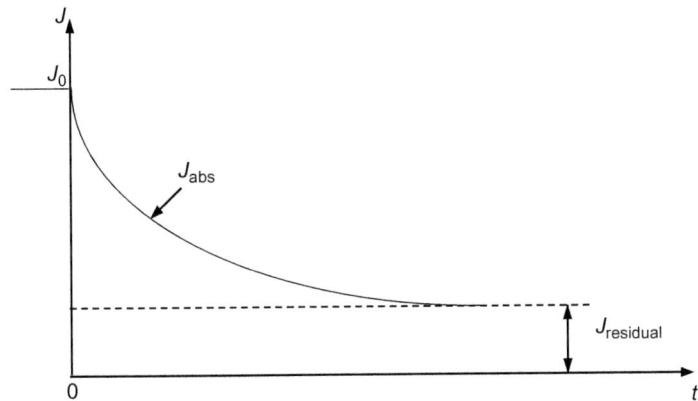

FIGURE 2.2

Current density J versus time t for a dielectric material in a DC field. *DC*, Direct current.

value equal to the DC current J_{thru}. It was shown that the transition process of decreasing the current from J_{abs} to J_{thru} obeys the exponential law as follows (Tareev, 1975; Wait, 1984):

$$J_{abs} = V \, \sigma \, \exp\left(-\frac{t}{\tau} \right) \tag{2.15}$$

Here, V is the voltage applied to the material, σ is the conductivity corresponding to the absorption current, and τ is the time during which the current I_{abs} decreases to $1/e$ of its original value.

For an alternating electromagnetic field, the conductivity σ in Ohm's law is a complex value, which includes effects from displacement currents, as follows (Wait, 1985):

$$J = \hat{\sigma} \, E = (\sigma' + j \, \omega \, \varepsilon') \, E \tag{2.16}$$

Thus, in an alternating field, the total current value is higher, than in a DC field, because in addition to I_{thru}, it contains also a displacement current I_{displ}, which, in contrast with the DC field, exists in stationary mode. In terms of current densities, it can be written as follows:

$$J_{total} = J_{thru} + J_{displ} \tag{2.17}$$

2.2 **Types and Mechanisms of Polarization**

Polarization processes occur in a material due to restricted movement of charges, which are actually bound, but can be displaced from their equilibrium positions, when an electric field is applied. There are two types of polarization with reference to the time required for the polarization process to develop:

1. Rapidly forming polarization and
2. Slowly forming polarization.

The mechanisms of these polarization types are shown in Kobranova (1989, Figure 62, page 113), and are described as follows:

1. *Rapidly forming polarization* includes various mechanisms of displacement polarization (*electron*, *atom*, and *ion* polarization). These types of polarization occur in materials containing charged and interconnected particles, which can shift under the effect of a field. Electron polarization arises when the electron orbit is displaced relative to the positively charged nucleus. It takes place in all atoms and molecules. Atom and ion polarizations arise when atoms of crystals or ions of molecules are displaced. These types of polarization may occur in minerals with valence crystals made of similar atoms (sulfur, diamond, graphite, selenium), as well as made of dissimilar atoms, and in minerals with ion crystals (quartz, corundum, calcite, halite, sylvite). The time required for these types of polarization to develop is $\tau = 10^{-1} - 10^{-15}$ seconds, which corresponds to high-frequency fields, up to optical frequencies (ultraviolet rays). For the displacement polarization, there is no irreversible dissipation of energy. The electric energy required for this polarization is completely returned to the energy source after voltage is removed. Thus, the rapidly formed types of polarization do not cause any dielectric losses.

2. *Slowly forming polarization types* are called also relaxation types of polarization. Relaxation is a delay in the response of a system to the applied electric field. When the field is removed, the restoration of a disturbed system to its equilibrium occurs within the relaxation time τ. The relaxation time, required for these types of polarization to develop, can be comparable with a half-period of applied electric field at audio and radio frequencies. These types of polarization cause an energy absorption by the material, that is, dielectric losses. There are three mechanisms of relaxation polarization:

   a. *Orientation polarization* occurs in materials containing polar molecules (H_2O), atom groups, and bonds (such as OH, C$=$O, and C$-$Cl). Before the field is applied, the dipole moments of these polar particles are oriented randomly due to the thermal molecular motion, and the resultant dipole moment of the material equals zero. Polarization is expressed in the preferable orientation of the dipoles in the field direction. The relaxation time of this polarization is $\tau = 10^{-10} - 10^{-7}$ seconds (Kobranova, 1989). Orientation polarization is observed in polar liquids contained in pores, and in crystallohydrates, clay minerals, and zeolites. Representatives of these mineral groups are cordierite, beryl, muscovite, biotite, gypsum, and talc.

   b. *Interfacial, or migration, polarization (Maxwell—Wagner effect)* takes place in heterogeneous materials, containing conducting components.

When an electric field is applied, the conduction charges migrate through the material, and some of them may accumulate on the interfaces, causing polarization in the medium. The relaxation time for this polarization $\tau = 10^{-6}-10^{-3}$ seconds. The interfacial polarization can be observed in earth materials containing ion-conductive impurities, like ore-minerals, and in wet rocks (Tareev, 1975; Lysne, 1983; Kobranova, 1989).

  c. *Induced polarization (IP)*, or *electrochemical polarization*, occurs in the electrical double layers (EDLs) on the boundaries between the solid and liquid phases (Fig. 2.3, Parkhomenko, 1971). The term "EDL" describes the arrangement of charges on the interfaces, consisting mostly of two layers. Double layers exist on any border of two phases, causing a potential difference to develop. Thus, the interface becomes electrified.

The EDLs may be formed in the following ways and are shown in Kobranova (1989, Figure 11, page 234). The ionic-layer type is formed by ion transition from the solution to the solid and back. It consists of a layer of dehydrated

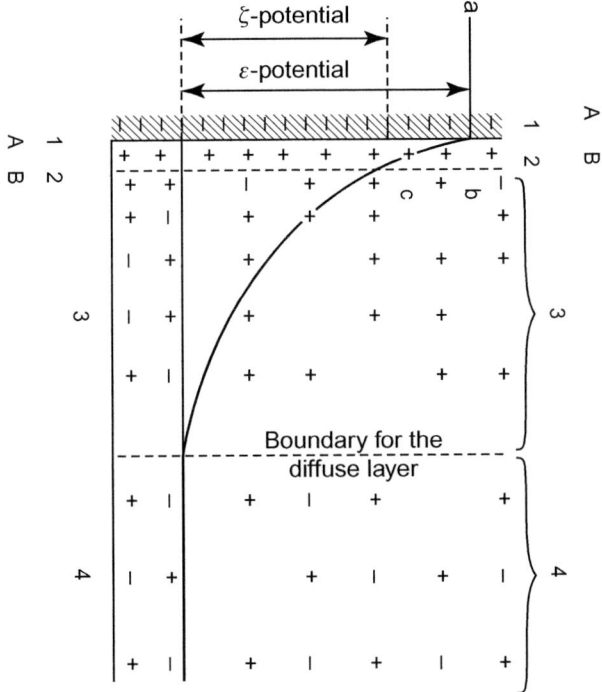

FIGURE 2.3

Induced polarization and charge distribution in the electrical double layer.

Reprinted by permission from Springer Nature, Parkhomenko, E.I., 1971, Electrification Phenomena in Rocks. Plenum Press, New York.

(unsolvated) ions in the solid phase and a layer of hydrated counter ions in the liquid phase.

The adsorption-layer type results from contact-specific adsorption of indifferent electrolyte ions on the solid uncharged surface. In different electrolytes, this means that it does not share ions with the solid phase (Bockris and Reddy, 1977). The ions adsorbed are those with greater valences and smaller hydrate radii because they have a larger electrical field and polarize solid phase atoms better. Such a charged surface may even specifically adsorb like-pole ions. The extent of the phenomenon seems to depend on the chemical nature (species) of the ion rather than on its charge (Bockris and Reddy, 1977). As described in Kobranova (1989), the structure of both ionic and absorption types of EDL looks similar; the difference is that in the absorption-layer type, the absorbed ions and counter ions are all located in the electrolyte.

The orientation type is made up of dipole water molecules. Some of them are oriented and held close to the solid phase surface, forming a layer of strongly bound water. The layer's field and residual electrostatic forces cause additional water to form a loosely bound charge-oriented layer. These two layers form an orientation type of EDL called bound water, which has special features. Strongly bound water is denser than free water (its density varies from 1.2 to 2 g/cm^3). It also displays greater viscosity, shear strength, and elasticity and lower electrical conductivity compared to distilled water. The freezing temperature of this water is much lower than $0°C$ ($-20°C$ for kaolinite, and $-193°C$ for montmorillonite), depending on the pore size and its specific surface. This water does not dissolve either salt or sugar. Loosely bound water has properties closer to those of free water, but it also has a higher viscosity, a lower freezing temperature ($-1.5°C$), and is a poor solvent (Kobranova, 1989). Depending on the properties of the solid phase and the electrolyte, one or two kinds of EDL predominate. Sometimes an EDL with all three types may occur. No matter what mechanism forms the EDL, it has the same structure and model, as shown in Kobranova (1989).

In the liquid phase, two parts may be distinguished as follows:

1. An *immobile*, or dense part, is composed of fixed ions (or oriented water dipoles), which are strongly connected to the solid surface by its electrostatic forces, although contact absorption may also occur. Ion (or water molecule) sizes determine the thickness of the dense layer, which can be imagined as an electrical capacitance. The size of the ions (or molecules) determines the distance of closest approach to the solid surface. This layer of densely packed ions and molecules, including the contact-absorbed ions, can be several tenths of a nanometer thick, depending on the degree of ion dehydration. It is called the inner Helmholtz plane. Hydrated ions are attracted to the layer of contact-adsorbed ions and they form a second layer of closest approach to the surface of the solid. This second layer of ions is called the outer Helmholtz plane.
2. The outer Helmholz plane is a boundary of approach for other hydrated ions involved in thermal motion. They form a *diffuse* layer where ion density

decreases deeper into the liquid phase. These ions are relatively mobile and may continuously exchange with other ions in the free solution. The thickness (t) of the diffuse layer (Debye length) may be considerable (up to hundreds of micrometers).

Fig. 2.3 shows the charge distribution in the EDL (Parkhomenko, 1971). The potential drop is linear within the dense part of the EDL. In the diffuse part, however, it is exponential (Fridrikhsberg and Sidorova, 1961). Potential at the boundary between the dense and the mobile, diffuse part of the EDL, which is measured by the potential level of the liquid phase, is called the zeta (ζ)-potential. The increase of concentration c forces the counter ions outward from the diffuse part into the dense layer, and the thickness of the double electric layer and the zeta-potential decrease. At high concentrations of pore electrolyte, thermal molecular motion is unable to tear ions and water dipoles away from the interface surface and diffuse them into the electrolyte. As a result, the EDL contracts. At some concentration level of the solution, all counter ions are within the dense layer, the zeta-potential is equal to zero, and the IP becomes negligible. This is the slowest type of polarization. The relaxation time τ may range from 10 ms to a few minutes (Kobranova, 1989).

The mechanisms of electrochemical polarization vary, depending on whether the solid phase is a dielectric or conductor. For a material, which consists of dielectric minerals and groundwater, a concentration-diffusion mechanism of polarization is usually adopted. This polarization appears in pores of various cross-sections, which have different solution concentration, due to different ion transference numbers in wide and narrow capillaries. Because of this imbalance, diffusion potentials appear. This causes an IP with a relaxation time $\tau > 10$ seconds.

Materials with an electron-conductive solid phase form an EDL, where the oxidation-reduction reactions occur between the solid and liquid phases, and they are in equilibrium, if no electric field is applied. The applied electric field disturbs the electrochemical balance and causes a current across the boundary. In this case, the charges have to overcome a specific energy barrier, changing from ionic to electronic conduction. Because of a very high electron conductivity of the solid phase, the charge exchange across the border in one place disturbs the potential equilibrium on its entire surface. Thus, the solid phase becomes polarized. The relaxation time τ varies from 10 ms to a few minutes, depending on the type of metal, surface condition, reaction rates, speed of ion movement, etc. The most active minerals in this respect are pyrite, chalcopyrite, pyrrhotite, and graphite, as well as oxides, such as magnetite and hematite.

The described mechanism for an electron-conductive solid phase takes place also in the case of a metallic electrode contacting an electrolyte or a wet sample (electrode polarization). The polarization of various metals increases in the following sequence: Pt (black), Pt, Ag, Cu, Al, Ni, and Au (Lopatin, 1971). The electrode polarization can cause serious errors in measurement results of electrical properties of materials. Some possible ways to eliminate the electrode polarization will be shown in Chapter 3, Methods of studying earth materials using alternating electric fields.

The extent of development of the relaxation types of polarization in a material depends on the frequency of the applied field. At low frequencies, all types of polarization occur, including the slowest, IP. That is why, the highest value of dielectric permittivity ε' of a given material is observed at the DC field (frequency = 0), and it is called the static dielectric permittivity (ε_s). At some frequency range, the dielectric permittivity remains virtually constant, and when $\tau \sim 1/\omega$ it starts to significantly decrease. At high frequencies, when $\omega\tau \gg 1$, it approaches the lowest constant value, called the optical dielectric permittivity. In this frequency range, only rapid displacement types of polarization are developed. Fig. 2.4 shows a schematic of dielectric permittivity ε' versus angular frequency ω for any type of relaxation polarization. The region of ε' decreasing with frequency is known as dielectric dispersion (by analogy with dispersion of light). If no resonance polarization takes place, the increase in frequency can only be accompanied by decreasing ε' (Tareev, 1975; Daniel, 1967). Fröhlich (1950) refers to Debye's theory of relaxation polarization (Debye, 1929), introducing the following equation that he derived. This equation describes the frequency dependence of the complex $\hat{\varepsilon}$:

$$\varepsilon' = \varepsilon_\infty + \frac{(\varepsilon_s - \varepsilon_\infty)}{1 + (\omega\tau)^2} \tag{2.18}$$

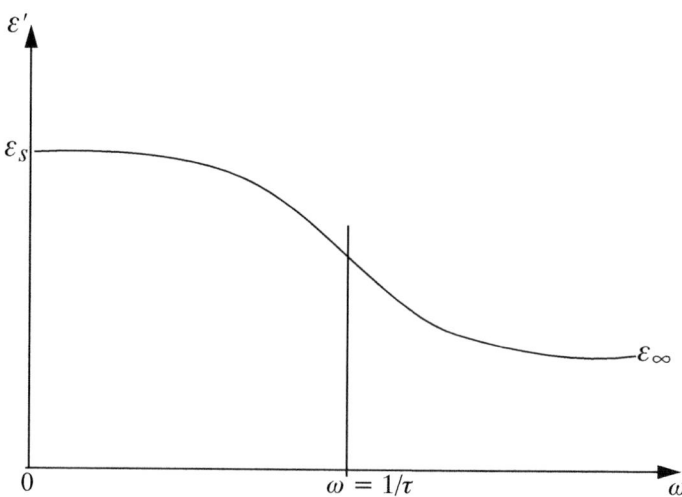

FIGURE 2.4

Schematic of dielectric permittivity versus angular frequency for a relaxation type of polarization.

2.3 Dielectric Losses

In an alternating current field, dielectric losses of a material with conductivity and relaxation polarization depend on frequency in a complicated way. Parameters that describe the dielectric losses, loss tangent tan δ, and loss factor ε'' are shown schematically versus angular frequency, in Fig. 2.5.

At low frequencies, both tan δ and ε'' decrease as ω increases. These losses, sometimes called ohmic losses, are caused by conductivity, which does not affect the ε', if no space-charge build-up occurs in the material (Von Hippel, 1954). At higher frequencies, the losses increase and are going through a

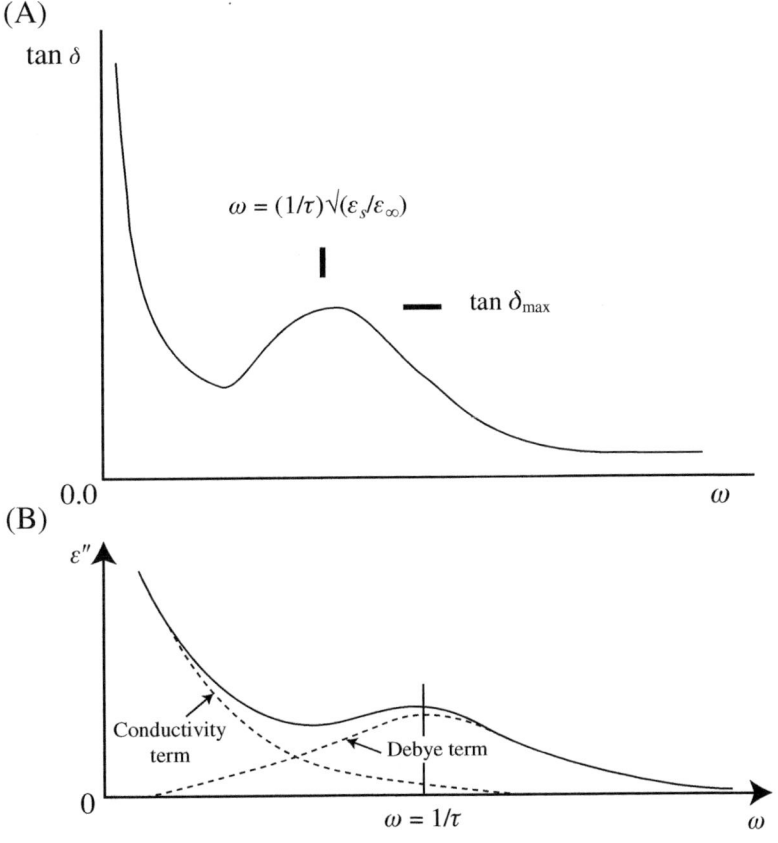

FIGURE 2.5

Schematic of loss tangent tan δ (A) and loss factor ε'' (B) versus angular frequency for a material with conductivity and relaxation losses.

Reproduced by permission from Arthur R. von Hippel, Dielectrics and Waves, Wiley & Sons, New York, 1954.

maximum and then decrease, approaching rather low values. The high-frequency part of dielectric losses occurs due to relaxation types of polarization. The relaxation loss factor is described by Debye's equation, as follows (Debye, 1929; Smyth, 1955):

$$\varepsilon''_r = \frac{(\varepsilon_s - \varepsilon_\infty)\omega\tau_0}{1 + (\omega\tau_0)^2}.$$ (2.19)

The conductivity component of the loss factor can be expressed as follows (Wait, 1985):

$$\varepsilon''_{con} = \frac{\sigma_{ohmic}}{\omega}.$$ (2.20)

The total loss factor ε''_{total} includes all the energy losses, which arise from two different mechanisms, as follows (Bartnikas, 1987; Sternberg and Levitskaya, 2001):

$$\varepsilon''_{total} = \varepsilon''_{con} + \varepsilon''_r.$$ (2.21)

Because $\varepsilon'' = \varepsilon'\tan\delta$, the expression for the relaxation component of the loss tangent $\tan\delta_r$ can be derived by dividing Eq. (2.19) by (2.18). The resulting equation is (Fröhlich, 1950; Parkhomenko, 1967; Daniel, 1967)

$$\tan\delta_r = \frac{(\varepsilon_s - \varepsilon_\infty)\omega\tau}{\varepsilon_s + \varepsilon_\infty\omega^2\tau^2}.$$ (2.22)

The conductivity component of the loss tangent can be obtained by dividing Eq. (2.20) by ε':

$$\tan\delta_{con} = \frac{\sigma_{ohmic}}{\omega\varepsilon'}.$$ (2.23)

Analogously to (2.21), the total loss tangent can be expressed as follows:

$$\tan\delta_{total} = \tan\delta_{con} + \tan\delta_r.$$ (2.24)

From Eqs. (2.19) and (2.22), it is possible to define the maximum conditions by differentiating them with respect to $\omega\tau$ and equating the derivative to zero (Appendix A). Thus, it was found that the maximum value of ε'' occurs when $\omega_{max}\tau = 1$, while for relaxation component of $\tan\delta$, the maximum condition is $\omega_{max}\tau = \sqrt{\varepsilon_s/\varepsilon_\infty}$. Inserting the expression for $\omega_{max}\tau$ into the original equations, one can find the maximum values of loss parameters, as follows (Bogoroditskiy et al., 1965; Dukhin and Shilov, 1974; Appendix A):

$$\varepsilon''_{max} = \frac{1}{2}(\varepsilon_s - \varepsilon_\infty)$$ (2.25)

$$\tan\delta_{max} = \frac{(\varepsilon_s - \varepsilon_\infty)}{2\varepsilon_s}\sqrt{\frac{\varepsilon_s}{\varepsilon_\infty}}$$ (2.26)

References

Bartnikas, R. (Ed.), 1987. Engineering Dielectrics, Volume II B, Electrical Properties of Solid Insulating Materials: Measurement Technique. ASTM STP926.

Bockris, J.O.'M., Reddy, A.K.N., 1977. Modern Electrochemistry. v.2. A Plenum/Rosetta Edition, New York.

Bogoroditskiy, N.P., Volokobinskiy, Yu.M., Vorobev, A.A., Tareev, B.M., 1965. Theory of Dielectrics. Russian, Energiya, Moscow.

Böttcher, C.J.F., Bordewijk, P., 1978. Theory of Electric Polarization, v. II, Dielectrics in Time-Dependent Fields. Elsevier, Amsterdam.

Chernyak, G.Ya, 1967. Dielectric Methods for Investigating Moist Soils. Translated from Russian, Israel Program for Scientific Translation, Jerusalem.

Daniel, V.V., 1967. Dielectric Relaxation. Academic Press, London.

Debye, P., 1929. Polar Molecules. Dover Publications, New York.

Duffin, W.J., 1990. Electricity and Magnetism, fourth ed. McGrow-Hill Book Company, London.

Dukhin, S.S., Shilov, V.N., 1974. Dielectric Phenomena and the Double Layer in Disperse Systems and Electrolytes. Halsted Press, New York.

Fridrikhsberg, D.A., Sidorova, M.P., 1961. Studying the connections between phenomenon of induced polarization and the electrokinetic properties of capillary systems. Vest Leningr. Univ., Ser. Fiz. Khim. 6 (4), 57−69.

Fröhlich, H., 1950. Theory of Dielectrics. Oxford University Press.

Jonscher, A.K., 1983. Dielectric Relaxation in Solids. Chelsea Dielectrics Press, London.

Kobranova, V.N., 1989. Petrophysics. Mir Publisher, Moscow.

Lopatin, B.A., 1971. Conductometry and Oscillometry (Translated from Russian). Israel Program for Scientific Translations, Jerusalem.

Lysne, P.C., 1983. A model for the high-frequency electrical response of wet rocks. Geophysics 48 (6), 775−786.

Parkhomenko, E.I., 1967. Electrical Properties of Rocks. Plenum Press, New York.

Parkhomenko, E.I., 1971. Electrification Phenomena in Rocks. Plenum Press, New York.

Schon, J.H., 1998. Physical properties of rocks: fundamentals and principles of petrophysics, Handbook of Geophysical Exploration, Seismic Exploration, vol.18. PERGAMON.

Smyth, C.P., 1955. Dielectric Behavior and Structure. McGraw-Hill, New York.

Sternberg, B.K., Levitskaya, T.M., 2001. Electrical parameters of soils in the frequency range from 1 kHz to 1 GHz, using lumped-circuit methods. Radio Sci. 36 (4), 709−719.

Tareev, B.M., 1975. Physics of Dielectric Materials. Mir Publishers, Moscow.

Von Hippel, A.R., 1954. Dielectrics and Waves. Wiley & Sons, New York.

Wait, J.R., 1982. Geoelectromagnetism. Academic Press, New York.

Wait, J.R., 1984. Relaxation phenomena and induced polarization. Geoexploration 22, 107−127.

Wait, J.R., 1985. Electromagnetic Wave Theory. Harper and Row, Publishers, Inc.

Further Reading

Levitskaya, T.M., Sternberg, B.K., 1996. Polarization processes in rocks, Part 1. Radio Sci. 31 (4), 755−779.

Methods of studying earth materials using alternating electric fields

3

Chapter Outline

As shown in Chapter 2, Parameters describing the material behavior in an electromagnetic field, the main parameters of a material placed in an alternating electric field, complex dielectric permittivity $\hat{\varepsilon}$, and complex conductivity $\hat{\sigma}$ (or complex resistivity $\hat{\rho}$) describe the two main processes occurring in an earth material, such as dielectric polarization and transport (conduction) phenomena. When measured over a wide frequency range, these parameters give us an entire spectrum of material behavior in an electric field (Smyth, 1955; Macdonald, 1987). The dielectric parameter $\hat{\varepsilon}$ and the transport parameters $\hat{\sigma}$ (or $\hat{\rho}$) may be more or less informative, depending on the frequency range of interest and on material composition and structure. If the dielectric processes prevail in a material, the method of "Dielectric Spectroscopy" is used. In a rock or soil with high water content, the polarization processes can be obscured by conduction phenomena. In this case, the "Conduction Spectroscopy" method is more applicable.

3.1 Dielectric Spectroscopy (in Terms of Complex Dielectric Permittivity $\hat{\varepsilon}$)

The Dielectric Spectroscopy method involves measuring the real and imaginary components of complex dielectric permittivity $\hat{\varepsilon} = \varepsilon' - j \cdot \varepsilon''$ of a material over a wide frequency range. Fröhlich (1950) refers to Debye's theory of relaxation polarization (Debye, 1929), introducing the following equation that he derived. This equation describes the frequency dependence of the complex $\hat{\varepsilon}$:

$$\hat{\varepsilon} = \varepsilon_\infty + \frac{\varepsilon_s - \varepsilon_\infty}{1 + j \cdot \omega \cdot \tau_0} \tag{3.1}$$

Electrical Spectroscopy of Earth Materials. DOI: https://doi.org/10.1016/B978-0-12-818603-9.00003-9

From (3.1), real ε' and imaginary ε'' parts of $\hat{\varepsilon}$ were separated, and the associated expressions (2.19) and (2.20) are also called Debye equations (Fröhlich, 1950; Smyth, 1955; Daniel, 1967).

If a relaxation polarization process takes place in some frequency range, it shows up as a dispersion of the real dielectric permittivity ε' and, simultaneously, a maximum of the imaginary part (or loss factor) of ε''. A schematic sketch of a real dielectric permittivity spectrum is shown in Fig. 3.1. As seen, it is possible to observe several polarization processes in a material, caused by different types of polarization (as indicated on the sketch). The maximum of the loss factor ε'' takes place at some critical frequency $\omega_c = 2\pi f_c$ that corresponds to the relaxation time of the given polarization τ_0. The classical theory for dispersion with a single relaxation time τ_0 is described by the Debye equations, shown also in Fig. 3.2A, for more clarity and convenience.

From the analysis of the Debye equations and experimental results, Cole and Cole (1941) concluded that the complex permittivity data may be represented as a semicircle, centered on the ε' axis (Fig. 3.2B, upper plot). As seen, the complex plane locus of ε'' plotted against ε' is a semicircle. Each point of ε'' corresponds to one frequency measurement. The real dielectric permittivity ε' changes from the static value ε_s to the high-frequency limit ε_∞. These limiting values of a given relaxation can be defined as intersections of the semicircle with the ε' axis. The critical frequency f_c, at which the maximum ε'' is attained, corresponds to the relaxation time τ_0 of the polarization as

$$2\pi f_c \tau_0 = 1 \tag{3.2}$$

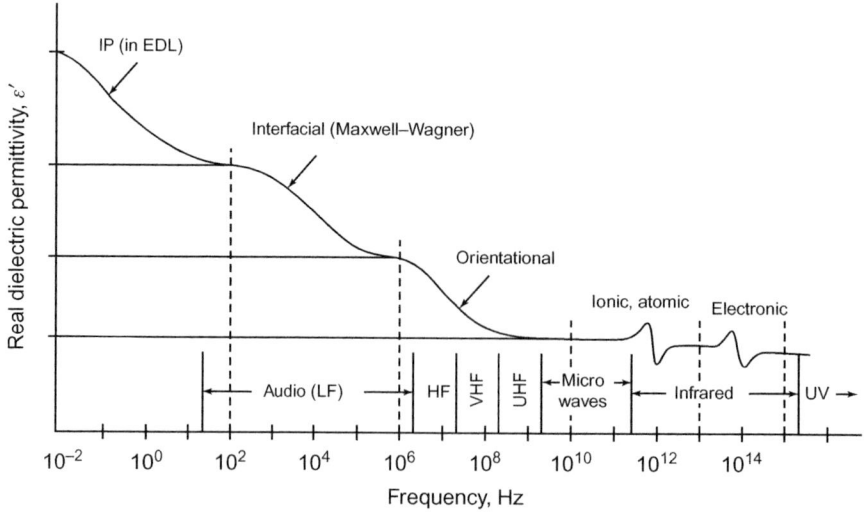

FIGURE 3.1

Schematic sketch of a dielectric permittivity spectrum.

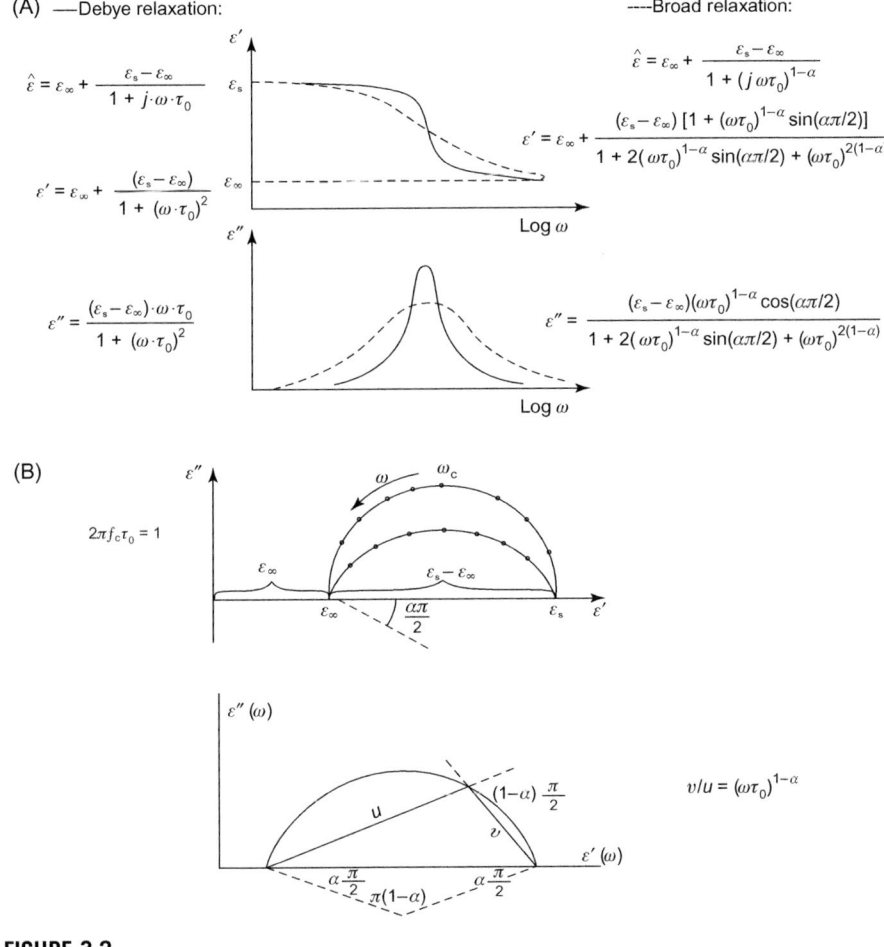

FIGURE 3.2

Relaxation plots. (A) Schematic plots of dielectric permittivity relaxation. (B) Schematic Cole–Cole diagrams in circular and arc forms.

Adapted from Daniel, V.V., 1967. Dielectric Relaxation. Academic Press, London.

This relation can be found from the derivative of the Debye equation (Fig. 3.2A) and is shown in Appendix A.1. Thus, the Debye relaxation is described with three parameters: ε_s, ε_∞, and τ_0. This type of relaxation usually occurs in homogeneous polar materials (e.g., water, alcohols, etc.), and it can be caused by the orientation polarization of the polar molecules or groups.

In heterogeneous materials (e.g., soils, porous rocks, clays, some polymers, ceramics, etc.), the relaxation polarization shows up in a broader frequency range of ε' dispersion and a smaller maximum value of ε'' than it is predicted by the Debye equations. Fig. 3.2A shows schematic plots of a broad relaxation, in comparison with a Debye relaxation. The broadening of the relaxation response may result from various relaxation agents (when a material contains two or more different dipoles). Each of these dipoles has its unique relaxation time. In this case, we are actually dealing with a set of relaxation times. Cole and Cole (1941) have found that a circular arc with a depressed center is a good approximation for such experimental results, Fig. 3.2B. The analytic expressions that describe this broad relaxation are given in Fig. 3.2A (Cole and Cole, 1941; Smyth, 1955). The distribution of the relaxation times is characterized by the parameter α, where $0 < \alpha \leq 1$. In order to define the value of the distribution parameter α for a given Cole−Cole diagram, we measure the angle between axes ε' and the radius of the arc, drawn to the point ε_∞ (Fig.3.2B), in degrees and then convert this to radians. Cole and Cole (1941) and Daniel (1967) display this angle as $\alpha \cdot \pi/2$ (Fig. 3.2B). We call this angle

$$b = \alpha \cdot \frac{\pi}{2} \tag{3.3}$$

Using (3.3), and the measured angle b on this plot, we then calculate the value α.

The value τ_0 has the physical meaning of the most probable (or major) relaxation time. When $\alpha = 0$, these equations reduce to the Debye equations, and the circular arc becomes a semicircle. The value τ_0 can be defined from the following relation, shown in Fig. 3.2B, lower plot (Cole and Cole, 1941; Smyth, 1955; Daniel, 1967):

$$\frac{v}{u} = (\omega \tau_0)^{1-\alpha} \tag{3.4}$$

The derivation of relation (3.4) is given in Appendix A.2. Thus, the Cole−Cole relaxation is described with four parameters: ε_s, ε_∞, τ_0, and α.

In homogeneous (uniform) conductive media (such as water, alcohols, crystals), which are one-phase materials (e.g., fluid or solid), there are no boundaries between different phases; therefore, no space charges occur in the material, and the real dielectric permittivity ε' is not affected by the conductivity, while the imaginary ε'' reveals conductive losses at low frequencies (Section 2.3). Our data for butanol are shown in Fig. 3.3, as an example. In addition to real ε' and imaginary ε'', we also plotted on this figure the dielectric (or relaxation) component ε''_{diel} of the total loss factor ε''_{total}, from which the conduction component ($\varepsilon''_{con} = \sigma_{Ohmic}/\omega$) was subtracted in the entire studied frequency range [Section 2.3, Eq. (2.21)]. As seen, at low frequencies (up to 2 MHz), the conduction mechanism of losses prevails over the relaxation mechanism.

In heterogeneous conductive materials, such as wet rocks and soils, the conduction processes produce space-charge build-up at the phase boundaries (mineral−grain/fluid) and thus affect the real dielectric permittivity ε', in addition to the loss parameters ε'' and tan δ, changing the shape of their plots. As shown schematically in Fig. 3.4, the higher the conductivity of the material (corresponding to

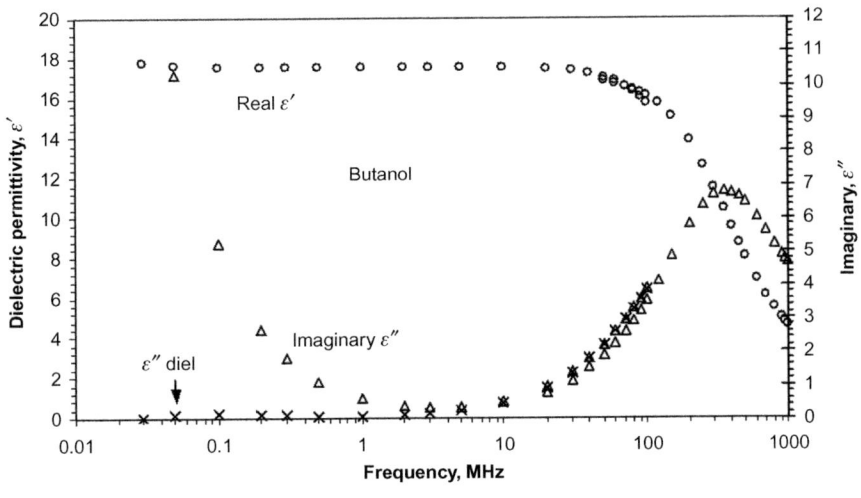

FIGURE 3.3

Dielectric permittivity versus frequency for butanol.

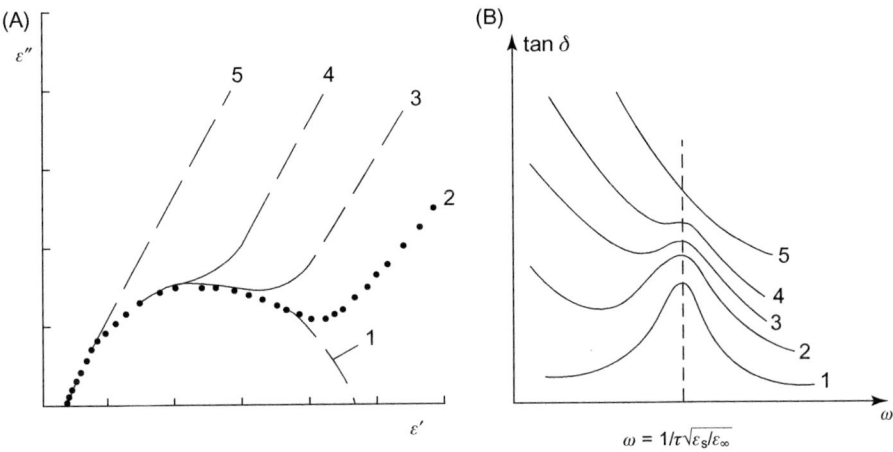

FIGURE 3.4

Schematic plots of loss factor ε'' versus ε' (A) and tan δ versus angular frequency ω (B) for materials with various conductivities. Curves labeled 1–5 are qualitative indicators of increasing water content and hence the conductivity.

Part (B) is from Skanavi, G.I., 1949. Dielectric Physics: Weak Fields Region, Gostekhteoretizdat, Leningrad

higher numbers on the curves), the flatter become the ε''- and tan δ-plots, and the Cole–Cole arc diagram may finally reduce to a straight line. This means that the conduction processes completely obscure the relaxation polarization.

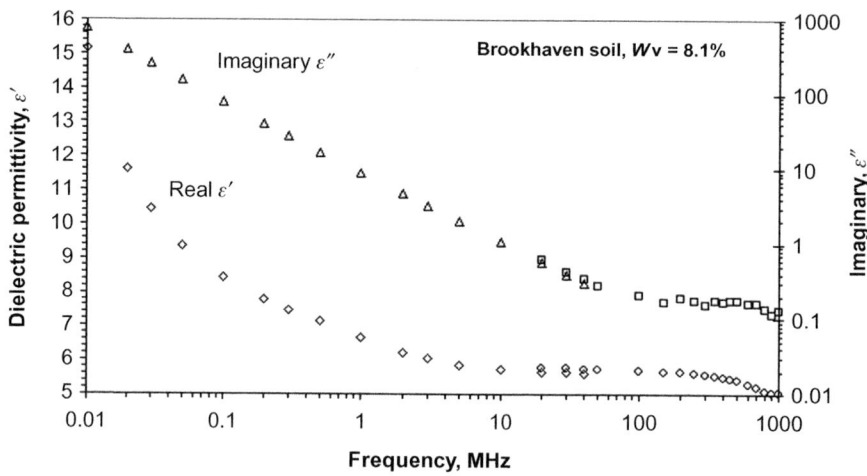

FIGURE 3.5

Dielectric permittivity versus frequency for Brookhaven soil with volume wetness $W_v = 8.1\%$.

Fig. 3.5 shows an example of dielectric permittivity dispersion for a Brookhaven soil (clean sand) with volume water content $W_v = 8.1\%$, where the effect of conduction is well pronounced for both dielectric permittivity ε' and loss factor ε'', up to 100 MHz. At higher frequencies, a small relaxation process is visible. Our data for wet soils from Avra Valley, which in addition to sand, contain some silt and clay and show that they do not reveal any relaxation in ε' or in ε'' plots (Figs. 3.6 and 3.7) (Sternberg and Levitskaya, 2001). This means that the conduction processes prevail in these soils in the entire frequency range up to 1 GHz. Although the relaxation processes are highly probable in such heterogeneous materials, the conduction phenomena obscure them. The same procedure of subtracting the conduction component from the total loss factor and plotting ε''_{diel} versus frequency, as shown for butanol, was applied to all our measured soils, thus revealing the latent relaxation polarization processes.

3.2 Conduction Spectroscopy (in Terms of Complex Resistivity $\hat{\rho}$)

As discussed above, in addition to the dielectric parameter $\hat{\varepsilon}$, which is used in the Dielectric Spectroscopy method, two parameters, complex conductivity $\hat{\sigma}$ and complex resistivity $\hat{\rho}$, characterize the material transport (or conduction) phenomena. The Conduction Spectroscopy method, in terms of either $\hat{\sigma}$ or $\hat{\rho}$, may also serve as a spectral method for studying the polarization processes in heterogeneous substances, especially with high conductivity.

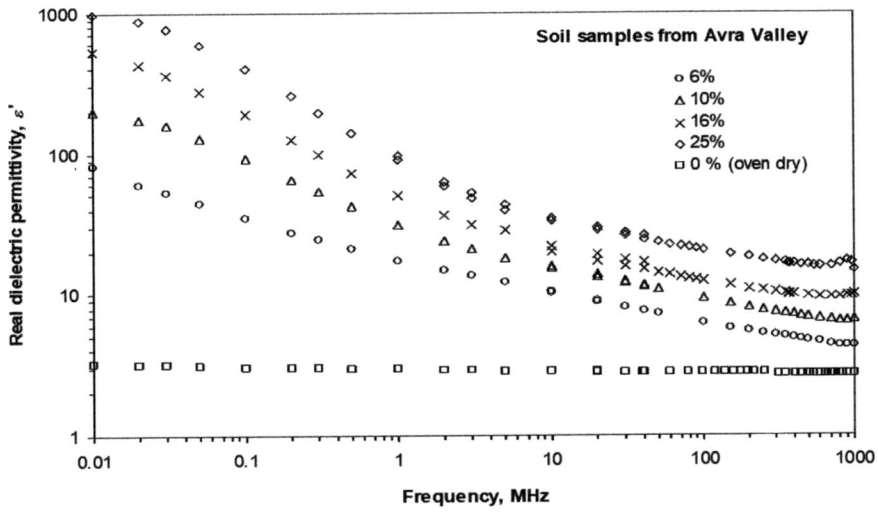

FIGURE 3.6

Real dielectric permittivity ε' versus frequency for Avra Valley samples.

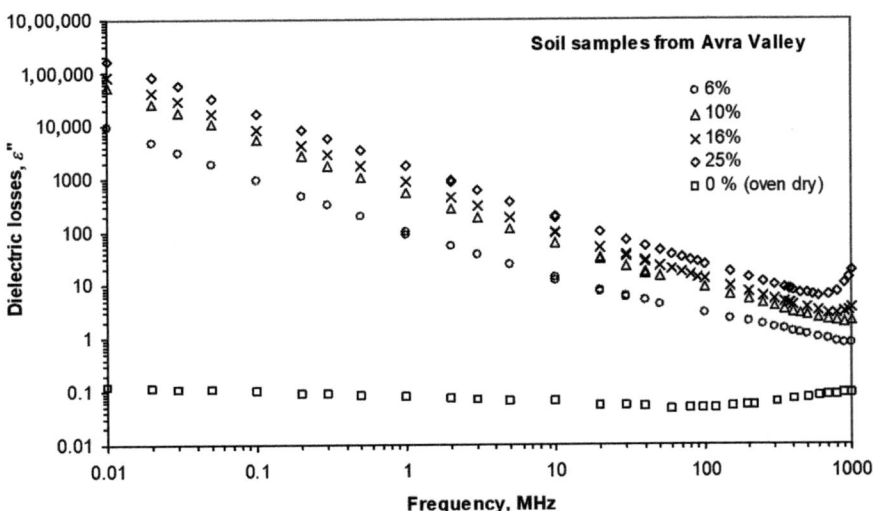

FIGURE 3.7

Dielectric losses, or loss factor ε'', versus frequency for Avra Valley samples with various wetness values.

We applied the Resistivity Spectroscopy method, which consists of measuring the complex resistivity $\hat{\rho} = \rho' - j \cdot \rho''$ in a wide range of frequencies. The complex resistivity is interrelated to the complex permittivity [Eq. (2.8)]. Therefore, the complex resistivity can, in principle, describe the same polarization processes as the complex permittivity.

In wet rocks and soils, as transport phenomena prevail, the Resistivity Spectroscopy method may be more informative. The Resistivity Spectroscopy method is especially useful at low frequencies, from 0.01 to 100 Hz for studying induced polarization (IP), when the dielectric permittivity cannot be measured in highly conductive materials. In this application, it is known as spectral IP. In complex resistivity measurements, at low frequencies for studying IP, some precautions must be taken to avoid stray phenomena occurring in the same frequency range, such as electrode polarization (EP) and contact resistance (CR). The most common methods to reduce or eliminate the EP and CR during the complex resistivity measurements are to use a four-electrode system or a black platinum two-electrode system, which are less polarizing (Section 2.2). The advantages and disadvantages of each method are described by Parkhomenko (1967), Levitskaya and Sternberg (1996a,b), and Lopatin (1971). The black (or platinized) platinum layer has a large effective surface area thus reducing current density and the polarization effect. Black platinum is formed on the flat surfaces of platinum foils by passing a direct current with the density about $J = 30$ mA/sm^2 through them while they are immersed in a 0.025 N solution of a with a small amount (0.025%) of lead acetate.

We studied NaCl solutions with concentrations C of 0.05 and 0.5 g/L, at frequencies from 5 Hz to 13 MHz, using various electrodes. An example for a salt solution of 0.5 g/L is shown in Fig. 3.8. In this frequency range, copper, silver, and bright platinum electrodes exhibit resistivity dispersion, while with the black platinum, no dispersion was observed. However, at lower frequencies, black

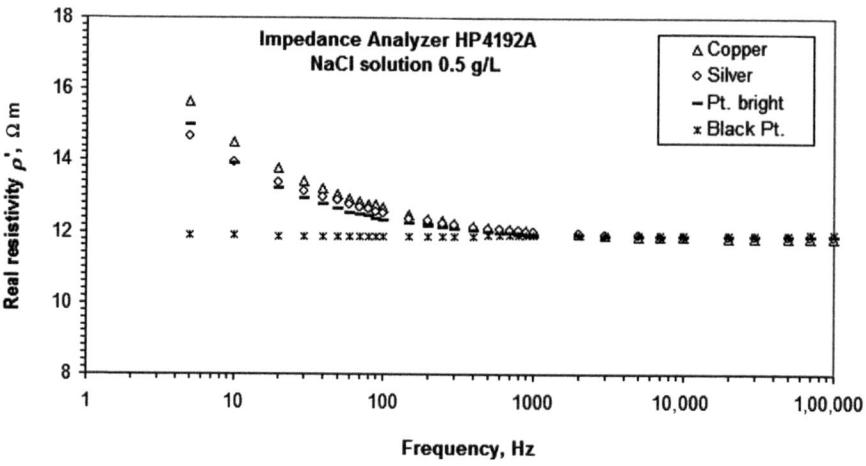

FIGURE 3.8

Real resistivity ρ' versus frequency for NaCl solution ($C = 0.5$ g/L) with various electrodes.

platinum also exhibits dispersion in the resistivity. Figs. 3.9 and 3.10 show dependencies of ρ' for black platinum, measured at frequencies of $0.01-100$ Hz, in NaCl solutions, and with $C = 0.05$ and 0.5 g/L. As seen, black platinum can be considered as nonpolarized only down to about 0.5 or 1 Hz, depending on the salt concentration, while IP may develop at lower frequencies, down to 0.001 Hz. In

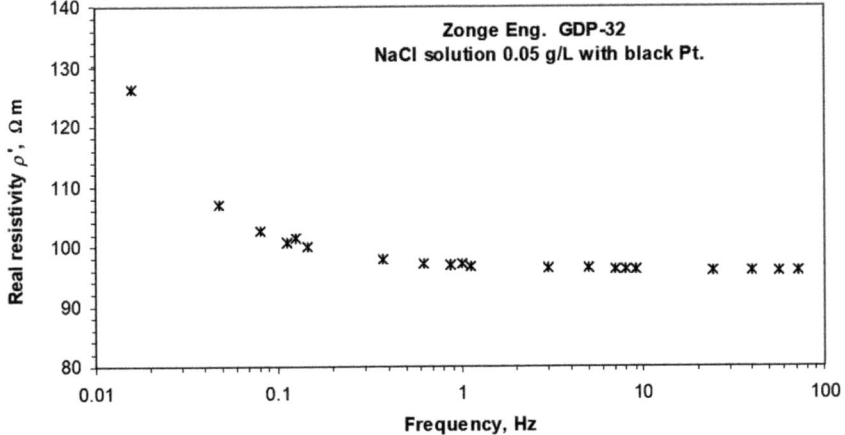

FIGURE 3.9

Real resistivity ρ' versus frequency for NaCl solution ($C = 0.05$ g/L) with black platinum electrodes.

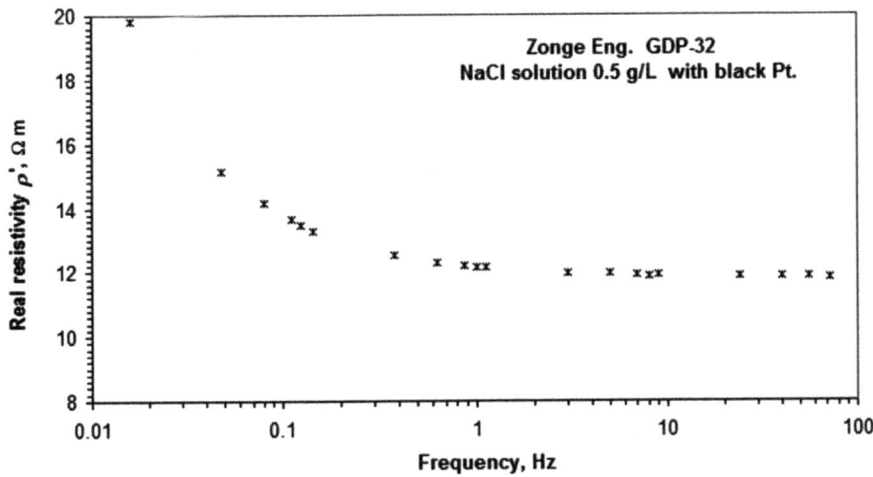

FIGURE 3.10

Real resistivity ρ' versus frequency for NaCl solution ($C = 0.5$ g/L) with black platinum electrodes.

order to eliminate the EP at frequencies less than 1 Hz, the cell resistance can be estimated throughout the frequency range with shorted electrodes. The real and imaginary components of the cell impedance can be subtracted from the corresponding values of the measured sample resistance at each frequency. Such a procedure was employed with black platinum electrodes in a cell, created in Russia. The measurements were performed in a frequency range from 0.01 Hz to 1 MHz on rock samples from Siberia (Levitskaya et al., 1992; Levitskaya and Sternberg, 1996a,b). Fig. 3.11 shows the schematic diagram of a measuring cell for rock samples. A cylindrical sample (5) is wound with adhesive polyethylene—polypropylene tape (6) so that on the upper end a volume is formed for pouring some NaCl solution that matches the concentration of the sample saturating solution. At the lower part of the sample, the tape is sealed with paraffin (7). The sample is placed in a vessel with a similar NaCl solution. The black platinum layers of the electrodes in such a cell are less likely to be damaged, because they are not squeezed. This cell gave stable and consistent results during measurements. We used a two-electrode conductivity-meter designed for electrochemical tasks at the Institute of Chemistry of Solids and Mineral Raw Materials of Russian Academy of Sciences (Novitskiy et al., 1983) for frequencies of 0.01—100 Hz. This frequency range was extended with Tesla impedance-meters BM 507 and BM 538 to 1 MHz. Some results for ρ' and ρ'' of samples, containing

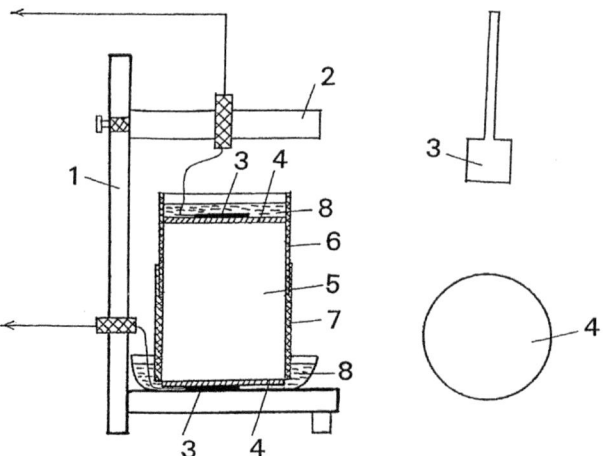

1—stand; 2—plate; black-platinum terminals; 4—black-platinum electrodes; 5—sample; 6—adhesive polyethylene tape; 7—paraffin; NaCl solution

FIGURE 3.11

Schematic of the measuring cell with black platinum electrodes.

From Levitskaya, T.M., Sternberg, B.K., 1996b. Polarization processes in rocks, Part 2. Radio Sci. 31 (4), 781–802.

ore minerals, are plotted in the frequency range from 0.01 Hz to 1 MHz (Fig. 3.12). The samples were saturated with a NaCl solution of $C = 15$ g/L. As seen, not only low- but also high-frequency dispersion domains are revealed. The latter is most pronounced in sample 15. Fig. 3.13 shows diagrams of $\rho'' = F(\rho')$ for the same samples as in Fig. 3.12. These diagrams are analogous to Cole−Cole diagrams for $[\varepsilon'' = F(\varepsilon')]$. In terms of resistivity, they are often called Argand diagrams, after the French amateur mathematician Jean Robert Argand. Argand suggested a way of representing complex numbers as points on a coordinate plane or the complex plane. The Argand plane uses the x-axis as the real axis and the

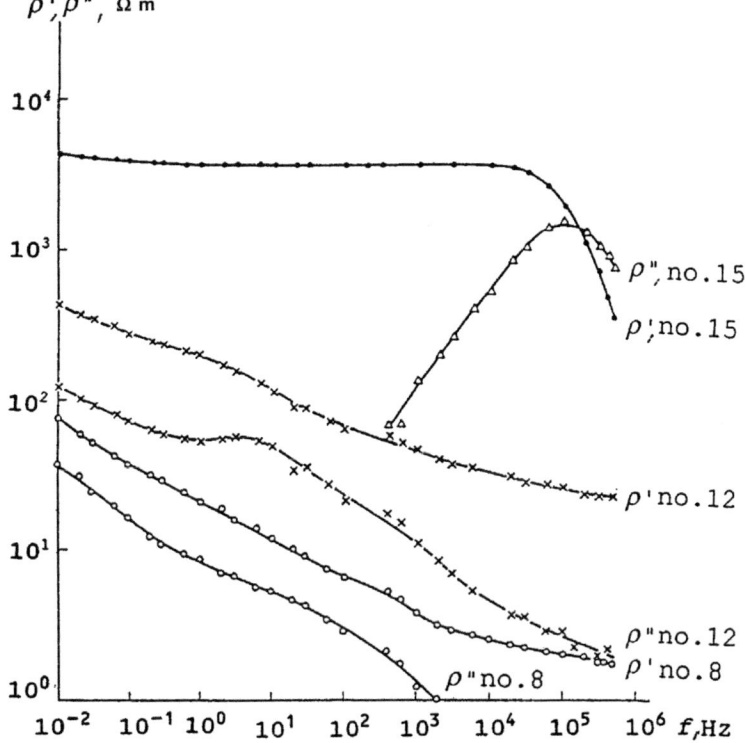

FIGURE 3.12

Real ρ' and imaginary ρ'' resistivity versus frequency for rock samples from Altai (Russia) containing ore minerals (pyrite, pyrrhotite, chlorite, galena, etc.) and saturated with a NaCl solution of concentration $C = 15$ g/L. *Sample 8*: Pyroxene with calcite and epidote veinlets. General content of ore minerals $m = 5\%-6\%$. *Sample 12*: Argillite with calcium and pyrrhotite veinlets and plant fragments, $m = 3\%-4\%$. *Sample 15*: Silty argillite with rare plant fragments, calcite and pyrrhotite veinlets, $m = 1\%-2\%$.

From Levitskaya, T.M., Sternberg, B.K., 1996b. Polarization processes in rocks, Part 2. Radio Sci. 31 (4), 781−802.

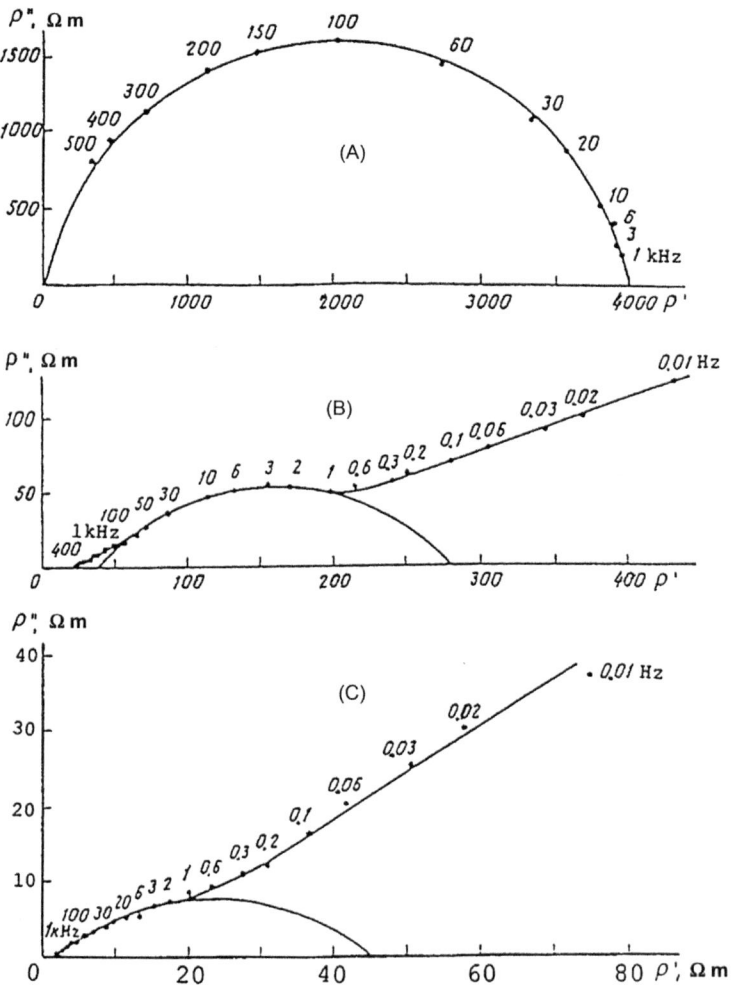

FIGURE 3.13

Argand diagrams $[\rho'' = F(\rho')]$ for the same samples as in Fig. 3.12. (A) Sample 15, (B) Sample 12, (C) Sample 8.

From Levitskaya, T.M., Sternberg, B.K., 1996b. Polarization processes in rocks, Part 2. Radio Sci. 31 (4), 781–802. From Levitskaya, T.M., Sternberg, B.K., 1996a. Polarization processes in rocks, Part 1, Radio Sci. 31 (4) 755–779.

y-axis as the imaginary axis. As seen in Fig. 3.13, the Argand diagram for sample 15 has an arc-form of a circle that apparently represents a kinetic, physical process of relaxation polarization. Two other samples exhibit a combination of straight and arc-shaped plots. This indicates the presence of both diffusion and kinetic mechanisms of polarization (Section 2.2; Olhoeft, 1985; Levitskaya et al., 1992;

Levitskaya and Sternberg, 1996a,b). The diffusion Warburg impedance is apparently formed in these samples because of their high density and small amount of conducting solution (porosities $K_p = 0.4\% - 0.9\%$). As a result, delivery of ions to the phase boundary in the electrical double layer is the slowest stage of the electrochemical process of polarization at low frequencies. This polarization process corresponds to the linear form of the Argand diagram. The arc-shaped plots at frequencies above 1 kHz probably describe the kinetic polarization induced by both electrochemical and physical processes. The kinetic electrochemical process is the limiting rate of electrochemical reactions.

In wet earth materials, which do not exhibit any dielectric polarization, even at high frequencies, the resistivity data may reveal relaxation processes not only in the IP region but also at higher frequencies, above 1 MHz. Fig. 3.14 shows an example of our impedance measurement results for an Avra Valley sample with volume water content of $W_v = 6.15\%$. A polarization process is seen at about 10 MHz. In Fig. 3.15, an Argand diagram is plotted for this sample. Argand diagrams are analogous to Cole–Cole diagrams [$\varepsilon'' = F(\varepsilon')$], which are also representing complex numbers on a complex plane. That is why Cole and Cole (1941) called them Argand diagrams. However, it is common to call the diagrams [$\varepsilon'' = F(\varepsilon')$] by the authors' names (Cole–Cole). We follow this notation, while for diagrams [$\rho'' = F(\rho')$], we will use the name Argand.

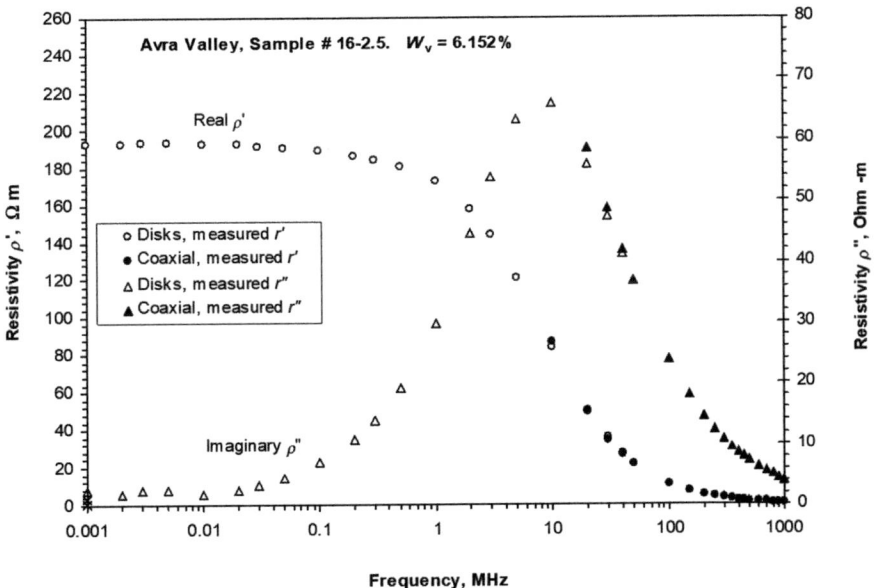

FIGURE 3.14

Complex resistivity versus frequency for Avra Valley sample with volume water content $W_v = 6.15\%$.

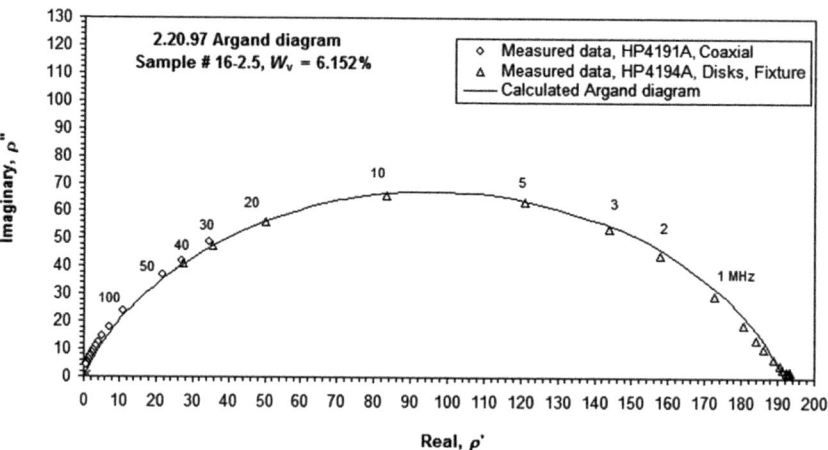

FIGURE 3.15

Argand diagram for Avra Valley sample with $W_v = 6.15\%$.

The same Cole–Cole equations can be used, where the ε-notation is substituted with ρ, as follows (Wait, 1984):

$$\hat{\rho} = \rho_\infty + \frac{\rho_0 - \rho_\infty}{1 + (j\omega\tau_0)^{1-\alpha}} \tag{3.5}$$

where ρ_0 is the static resistivity, ρ_∞ is the high-frequency limit of the given resistivity dispersion, τ_0 is the most probable relaxation time, and α is an empirical constant with values between 0 and 1 and is a measure of the distribution of the relaxation times. The τ_0 corresponds to the critical characteristic frequency f_c, at which the maximum of ρ'' is attained, analogous to the dielectric polarization [Eq. (3.2) and Appendix A.1] where $2\pi f_c \tau_0 = 1$.

Separation of the real and imaginary parts of (3.3), analogous to ε' and ε'', gives (Smyth, 1955; Sternberg and Levitskaya, 2001)

$$\rho' = \rho_\infty + \frac{(\rho_0 - \rho_\infty)[1 + (\omega\tau_0)^{1-\alpha}\sin(\alpha\pi/2)]}{1 + 2(\omega\tau_0)^{1-\alpha}\sin(\alpha\pi/2) + (\omega\tau_0)^{2(1-\alpha)}} \tag{3.6}$$

$$\rho'' = \frac{(\rho_0 - \rho_\infty)(\omega\tau_0)^{1-\alpha}\cos(\alpha\pi/2)}{1 + 2(\omega\tau_0)^{1-\alpha}\sin(\alpha\pi/2) + (\omega\tau_0)^{2(1-\alpha)}} \tag{3.7}$$

In the limit case, when $\alpha \to 0$, Eqs. (3.5)–(3.7) reduce to the Debye form, analogous to the dielectric permittivity equations, as shown in the left part of Fig. 3.2A. Pelton et al. (1978) adapted another resistivity-relaxation equation, as follows

$$\hat{\rho} = \rho_0 \left[1 - m\left(1 - \frac{1}{1 + (j\omega\tau)^c}\right)\right] \tag{3.8}$$

Eq. (3.8) was described and analyzed by Wait (1984). Here, $m = (\rho_0 - \rho_\infty)/\rho_0$ is the chargeability, and c is the dissemination parameter.

In this chapter, we considered two complex parameters that describe the polarization processes in a material (dielectric permittivity $\hat{\varepsilon}$ and resistivity $\hat{\rho}$) and showed how they were used in the Dielectric Spectroscopy and the Resistivity Spectroscopy methods, respectively.

References

Cole, K.S., Cole, R.H., 1941. Dispersion and absorption in dielectrics. J. Chem. Phys. 9, 341−351.

Daniel, V.V., 1967. Dielectric Relaxation. Academic Press, London.

Debye, P., 1929. Polar Molecules. Dover Publications, New York.

Fröhlich, H., 1950. Theory of Dielectrics. Oxford University Press.

Levitskaya, T.M., Sternberg, B.K., 1996a. Polarization processes in rocks, Part 1. Radio Sci. 31 (4), 755−779.

Levitskaya, T.M., Sternberg, B.K., 1996b. Polarization processes in rocks, Part 2. Radio Sci. 31 (4), 781−802.

Levitskaya, T.M., Palveleva, I.I., Poletaeva, N.G., Kenzin, V.I., Novitskiy, S.P., 1992. A study of rocks containing ore minerals by the method of frequency resistance dispersion, Izvestiya. Earth Phys. 28 (2).

Lopatin, B.A., 1971. Conductometry and Oscillometry (Translated from Russian). Israel Program for Scientific Translations, Jerusalem.

Macdonald, J.R. (Ed.), 1987. Impedance Spectroscopy. John Wiley, New York.

Novitskiy, S.P., Burenkov, I.I., Kenzin, V.I., Beck, R.U., 1983. Digital polarograph-impedance meters for frequency range of 10^3 to 10^5 Hz. Coll. Chem. Commun. (Prague) 48 (4), 1127−1128.

Olhoeft, G.R., 1985. Low-frequency electrical properties. Geophysics 50 (12), 2492−2503.

Parkhomenko, E.I., 1967. Electrical Properties of Rocks. Plenum Press, New York.

Pelton, W.H., Ward, S.H., Hallof, P.G., Sill, W.R., Nelson, P.H., 1978. Mineral discrimination and removal of inductive coupling with multifrequency ip. Geophysics 43 (3), 588−609. Available from: https://doi.org/10.1190/1.1440839.

Skanavi, G.I., 1949. Dielectric Physics: Weak Fields Region, Gostekhteoretizdat, Leningrad.

Smyth, C.P., 1955. Dielectric Behavior and Structure. McGraw-Hill, New York.

Sternberg, B.K., Levitskaya, T.M., 2001. Electrical parameters of soils in the frequency range from 1 kHz to 1 GHz, using lumped-circuit methods. Radio Sci. 36 (4), 709−719.

Wait, J.R., 1984. Relaxation phenomena and induced polarization. Geoexploration 22, 107−127.

Further Reading

Böttcher, C.J.F., Bordewijk, P., 1978. Theory of Electric Polarization, vol. II, Dielectrics in Time-Dependent Fields. Elsevier, Amsterdam.

Measurements and analysis: concept of distributed versus lumped parameters

Chapter Outline

When an alternating electromagnetic field is applied to an electric circuit, the voltage and current may vary from element to element of the circuit, because the alternating field is changing with time. A circuit is considered distributed or lumped, depending on the relationship between its physical dimensions and the wavelength of the current. If the circuit dimensions are a considerable fraction of a wavelength (> 0.1), the system becomes distributed. In this case, circuit parameters (resistance, capacitance, and inductance) are also varying at different points of the circuit, that is, they are distributed along the circuit. For example, this takes place in transmission lines at high frequencies, in the GHz range.

When the circuit dimensions are much smaller than the wavelength, the system is considered as lumped. In this case, the rate of field change with time is small in comparison with the speed of wave propagation. We can assume that during the time of wave propagation, the voltage and current do not change appreciably, so they are same at each point of the circuit and at each given moment. This means that the resistance, capacitance, and inductance may be considered as lumped parameters, and not distributed along the circuit. This is possible at low frequencies, when the wavelength is much larger than the sample dimensions. The circuit can be considered as a lumped system in a wide range of frequencies, up to 100 MHz or sometimes to 10 GHz, if the sizes of its elements are small enough (Bartnikas, 1987; Von Hippel, 1954; Pozar 1998; Levitskaya and Sternberg, 2000; Sternberg and Levitskaya, 2005).

Electrical Spectroscopy of Earth Materials. DOI: https://doi.org/10.1016/B978-0-12-818603-9.00004-0

4.1 Circuits With Distributed Parameters

When the field changes rapidly with time, the voltage and current change appreciably during the time of wave propagation, that is, they are not same at each point of the circuit. The wave features of the electromagnetic field prevail and cannot be neglected. In this case, the electrical parameters of a material must be considered as distributed. This takes place at high frequencies, approaching 1 GHz and above, when the wavelength becomes very much smaller than the sample dimensions. Circuits with distributed parameters are the subjects of electromagnetic wave theory that considers the wave processes in the medium in terms of reflection, absorption, and transmission of electromagnetic energy (Von Hippel, 1954; Wait, 1985; Pozar, 1998). The electromagnetic wave theory is based on Maxwell's field equations. Maxwell's field equations in differential form for sinusoidal fields show that the coupling between the electric and magnetic field vectors and their interaction with matter, are as follows:

$$\frac{\partial^2 E}{\partial x^2} = \hat{\varepsilon} \cdot \hat{\mu} \frac{\partial^2 E}{\partial t^2} \tag{4.1}$$

$$\frac{\partial^2 H}{\partial x^2} = \hat{\varepsilon} \cdot \hat{\mu} \frac{\partial^2 H}{\partial t^2} \tag{4.2}$$

where $\hat{\varepsilon}$ is the complex dielectric permittivity, and $\hat{\mu}$ is the complex magnetic permeability of the matter. From these wave equations, the electromagnetic field expressions for E and H can be derived, as follows:

$$E = E_0 e^{j\omega t - \gamma x} \tag{4.3}$$

$$H = H_0 e^{j\omega t - \gamma x} \tag{4.4}$$

where $+x$ is the direction through space; t is time; γ is the wave propagation factor.

A number of methods exist for measuring the distributed dielectric parameters of a material (Westphal, 1954; Bussey, 1967; Stuchly and Matuszewski, 1978; Rau and Wharton, 1982; Scott and Smith, 1986; Baker-Jarvis et al., 1990; Blackham and Pollard, 1997). In most of the methods, the complex parameters, intrinsic impedance Z and propagation constant γ, are determined by measuring the reflection and transmission coefficients of electromagnetic waves in a coaxial or waveguide line (Fig. 4.1). These parameters are related to the complex dielectric permittivity $\hat{\varepsilon}$ and complex magnetic permeability $\hat{\mu}$ as follows (Von Hippel, 1954; Westphal, 1954):

$$\hat{Z} = \sqrt{\frac{\hat{\mu}}{\hat{\varepsilon}}} \tag{4.5}$$

$$\hat{\gamma} = j\omega \sqrt{\hat{\varepsilon} \cdot \hat{\mu}} \tag{4.6}$$

Both $\hat{Z}$ and $\hat{\gamma}$ must be known to determine the complex dielectric permittivity $\hat{\varepsilon}$ and complex magnetic permeability $\hat{\mu}$. The complex parameters $\hat{Z}$ and $\hat{\gamma}$ are

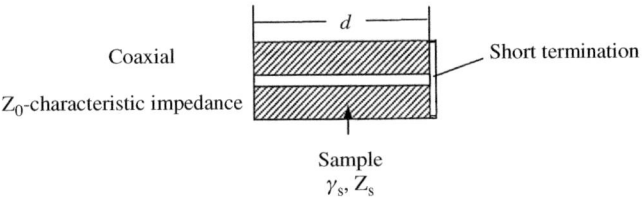

FIGURE 4.1

Coaxial sample holder used for defining distributed parameters.

related through a hyperbolic tangent function, as follows (where d is the sample thickness):

$$Z_s = Z_0 \tan h\gamma_s d \tag{4.7}$$

This hyperbolic function is transcendental and has multiple solutions (Westphal, 1954; Poley et al., 1978). Currently, a combined reflection–transmission method is widely used, which measures S parameters (scattering parameters). The scattering parameters are complex values. The procedure for defining the reflection and transmission coefficients of the material with error corrections, Γ and z, and then, $\hat{Z}$ and $\hat{\gamma}$, as well as the complex dielectric permittivity $\hat{\varepsilon}$ and complex permeability $\hat{\mu}$, from the measured S parameters is described by Nicolson and Ross (1970), Baker-Jarvis et al. (1992) and Courtney (1998). However, in this procedure, a transcendental equation must also be solved. In particular, the propagation constant γ is related to the transmission coefficient z through a natural logarithm, as follows (Nicolson and Ross, 1970; Rau and Wharton, 1982):

$$z = \exp(-\gamma \cdot d) \tag{4.8}$$

$$\gamma = \frac{1}{d} \ln\left(\frac{1}{z}\right) \tag{4.9}$$

Transcendental equations are not algebraic functions, and they can be solved only approximately, using numerical methods. Numerical solutions of these transcendental equations are inherently ambiguous, since the hyperbolic tangent function and the natural logarithm of a complex number are multivalued. This may cause an uncertainty in defining dielectric parameters. The ambiguity can be eliminated, if an approximate value of ε' is known (an initial guess) in order to initiate iterative procedures, or when measurements are made on different lengths of the sample (Von Hippel, 1954; Westphal, 1954; Nicolson and Ross, 1970; Stuchly and Matuszewski, 1978; Poley et al., 1978; Rau and Wharton, 1982; Ligthart, 1983; Baker-Jarvis et al., 1990, 1992). Measuring the group delay and comparing it to the set of calculated values for various n for the argument $(\phi + 2\pi n)$ can also help, yielding the correct value of n (Westphal, 1954; Baker-Jarvis et al., 1992; Muñoz et al., 1993). Automated systems with network analyzers and computer software may provide accurate results up to 20 GHz. However, it is useful to have an alternative method, as an independent source of information. In particular, the lumped-circuit method gives unambiguous results.

4.2 Circuits With Lumped Parameters

At low frequencies, when the wavelength is much larger than the sample dimensions, the material can be approximated with lumped elements, such as resistance R, inductance L, and capacitance C, connected in a series or a parallel circuit with wires, which have little effect on the circuit. The theory of electrical circuits is applicable in this case (Neyman and Demirchian, 1981; Duffin, 1990).

The electrical properties of a sample in a sinusoidal electric field can be obtained by measuring the magnitude Z and phase angle φ of the complex impedance $\hat{Z}$, using impedance analyzers. The sample is placed in a sample holder, which may have various electrode arrangements. Fig. 4.2 shows a schematic of the sinusoidal waveforms for voltage V and current I, which can be described with the following equations:

$$V = V_\mathrm{m} \cdot \sin(\omega t + \psi_{V_\mathrm{v}}) \quad I = I_\mathrm{m} \cdot \sin(\omega t + \psi_I) \tag{4.10}$$

where V_m and I_m are the magnitudes; ψ_v and ψ_I are the initial phase angles.

Phase angle $\varphi = \psi_\mathrm{v} - \psi_I$ characterizes the phase shift between the voltage V and current I in the circuit. When considering steady-state sinusoidal processes, one of the initial phases can be chosen arbitrarily. In particular, for a series circuit, it is convenient to choose $\psi_\mathrm{v} = \varphi$, in order to have $\psi_I = 0$. Then, the current is

$$I = I_\mathrm{m} \cdot \sin \omega t \tag{4.11}$$

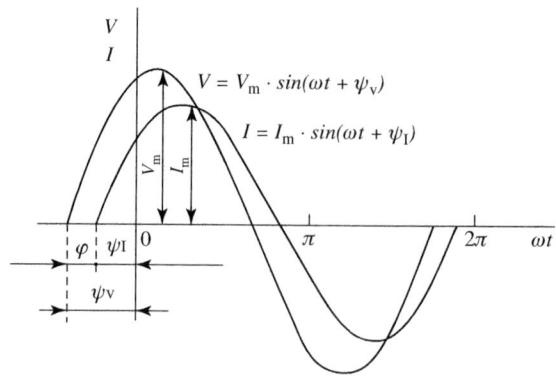

FIGURE 4.2

Schematic of sinusoidal waveforms for voltage V and current I in an AC circuit.

The sinusoidal waveform is characterized with a period T (second) and frequency $f = 1/T$ (Hz). Period T is the time occupied by one complete cycle of change. If the argument of the sinusoidal function is expressed in radians, the period is expressed by the constant value 2π (Evdokimov, 1977). The wavelength λ is the spatial period of the wave; it is the distance over which the wave's shape is repeated. Wavelength of a sine wave can be measured between any two points with the same phase (Fig. 4.2). Wavelength is inversely proportional to frequency f:

$$\lambda = \frac{v}{f} \tag{4.12}$$

where v is the speed of the wave or phase velocity. The speed of a wave depends on the medium in which it propagates. In vacuum, the electromagnetic wave travels with the speed of light, $c = 3 \times 10^8$ m/s, and is expressed using the vacuum parameters, dielectric permittivity ε_0 and magnetic permeability μ_0, as follows (Von Hippel, 1954): $c = 1/\sqrt{\varepsilon_0 \mu_0}$. The wavelength in vacuum is $\lambda_0 = c/f$. In other media, the phase velocity decreases and the wavelength is smaller, depending on the dielectric and magnetic parameters of the material (Wait, 1986; Von Hippel, 1954; Pozar, 1998). For a nonmagnetic material with a relative dielectric permittivity ε', the phase velocity is

$$v_m = \frac{c}{\sqrt{\varepsilon'}}, \quad \lambda_m = \frac{c}{f \cdot \sqrt{\varepsilon'}} \tag{4.13}$$

Fig. 4.3 represents the series circuit of R, L, and C, and its phasor diagram is shown in Fig. 4.4. In this case, the current is the same for each element of the circuit, while the voltage values are different.

1. The voltage V_R of a resistance R can be found, if the current I is known, by using Ohm's law:

$$V_R = R \cdot I = R \cdot I_m \cdot \sin \omega t \tag{4.14}$$

2. The voltage of an inductive coil L depends on the rate of current change with time (Neyman and Demirchian, 1981; Duffin, 1990). The current I produces in the surrounding space a magnetic flux

$$\Phi = L \cdot I \tag{4.15}$$

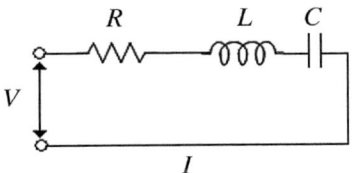

FIGURE 4.3

Series equivalent circuit for a material.

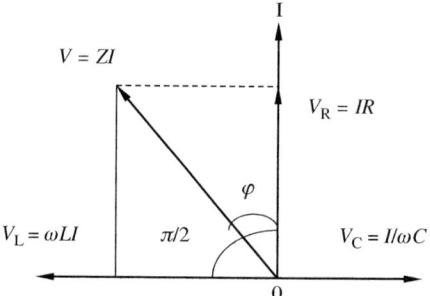

FIGURE 4.4

Phasor diagram for a series equivalent circuit.

Here, L is the inductance of the coil. Using Eq. (4.15), the electromotive force can be defined (Faraday law):

$$E = \frac{d\Phi}{dt} = -L \cdot \frac{dI}{dt} \tag{4.16}$$

The negative sign indicates that the direction of the inductive current is opposite to the external magnetic field. The voltage or the potential difference across an inductor L is (Duffin, 1990)

$$V_L = -E = L \cdot \frac{dI}{dt} \tag{4.17}$$

The voltage of a capacitor is related to its capacitance C and electric charge q as follows:

$$V_C = \frac{q}{C}; \quad \text{and the current } I = \frac{dq}{dt} = C \cdot \frac{dV_C}{dt} \tag{4.18}$$

From (4.18): $dq = Idt$, and

$$q = \int_0^t Idt + q(0) \tag{4.19}$$

From (4.18) and (4.19):

$$V_C = \frac{1}{C} \cdot \int_0^t Idt + q(0) \tag{4.20}$$

Summarizing the voltage expressions for the lumped elements of a series circuit (Fig. 4.3) gives us the differential equation for the total voltage V (Neyman and Demirchian, 1981):

$$V = V_R + V_L + V_C = R \cdot I + L \cdot \frac{dI}{dt} + \frac{1}{C} \int_0^t Idt + V_C(0) \tag{4.21}$$

After differentiation and integration, with $I = I_m \cdot \sin \omega t$, we obtain:

$$V = RI_m \sin \omega t + \omega L I_m \cos \omega t - \frac{1}{\omega C} I_m \cos \omega t \tag{4.22}$$

Using the trigonometric reduction formulas, such as $\cos \alpha = \sin(\pi/2 \pm \alpha)$, Eq. (4.22) can be rewritten as follows (Bronshtain and Semendiaev, 1971; Neyman and Demirchian, 1981):

$$V = RI_m \sin \omega t + \omega L I_m \sin\left(\omega t + \frac{\pi}{2}\right) + \frac{1}{\omega C} I_m \sin\left(\omega t - \frac{\pi}{2}\right) \tag{4.23}$$

It is seen from (4.23) that for inductance, the voltage leads the current by an angle $\pi/2$, and the vector of V_L must have a positive direction from the current vector, as shown in Fig. 4.4. For capacitance, the voltage lags the current by $\pi/2$, and hence, the vector V_C is drawn in the negative direction from the current vector in Fig. 4.4. Thus the vectors V_L and V_C are shifted from each other by π, that is, at any time they are opposite to each other.

Thus we have sinusoidal Eq. (4.11) for current and (4.23) for voltage in a series circuit with lumped elements R, L, and C. The total impedance Z of this circuit can be defined as a ratio of voltage to current. Dividing (4.19) by (4.9), we obtain

$$Z = R + \omega \cdot L - \frac{1}{\omega \cdot C} \tag{4.24}$$

In terms of complex numbers, the impedance Z consists of a real part, represented by R, and an imaginary part, reactance

$$X = \omega \cdot L - \frac{1}{\omega \cdot C} \tag{4.25}$$

The sinusoidal equations for circuit voltage V, current I (4.8), and electric intensity E are

$$V = V_m \cdot \sin(\omega t + \psi_{V_v}) \quad I = I_m \cdot \sin(\omega t + \psi_I) \quad E = E_m \cdot \sin(\omega t + \psi_e) \tag{4.26}$$

These equations can be expressed in a complex form by using Euler's formulas, as follows (Moskowitz, 2002; Neyman and Demirchian, 1981):

$$V = V_m \cdot e^{j(\omega t + \psi_v)} \quad I = I_m \cdot e^{j(\omega t + \psi_I)} \quad E = E_m \cdot e^{j(\omega t + \psi_e)} \tag{4.27}$$

Each expression in (4.27) can be represented as containing a complex magnitude, which is characterized by its original magnitude and initial phase. The corresponding complex magnitudes are

$$\hat{V}_m = V_m \cdot e^{j\psi_v} \quad \hat{I}_m = I_m \cdot e^{j\psi_I} \quad \hat{E}_m = E_m \cdot e^{j\psi_e} \tag{4.28}$$

Using the expressions for voltage and current complex amplitudes (4.28) and for X (4.25), we can find the complex impedance, as follows:

$$\hat{Z} = \frac{\hat{V}_m}{\hat{I}_m} = Z \cdot e^{j\varphi} = Z\cos\varphi + jZ\sin\varphi = R + jX = R + j\left(\omega L - \frac{1}{\omega C}\right) \tag{4.29}$$

The admittance is

$$\hat{Y} = 1\hat{Z} = Ye^{-j\varphi} = Y\cos\varphi - jY\sin\varphi = G - jB \qquad (4.30)$$

Analogously, the complex dielectric permittivity can be defined through the electric intensity E and displacement $D = \varepsilon E$, as the ratio D/E, which are also changing sinusoidally with time (Böttcher and Bordewijk, 1978; Neyman and Demirchian, 1981).

It was shown that the complex dielectric displacement $\hat{D} = D_m e^{j(\omega t - \delta)}$, when the initial phase in Eqs. (4.24) and (4.25) for the electrical intensity E is taken as zero (Böttcher and Bordewijk, 1978) is as follows:

$$\hat{\varepsilon} = \varepsilon e^{-j\delta} = \varepsilon\cos\delta - j\varepsilon\sin\delta = \varepsilon' - j\varepsilon'' \qquad (4.31)$$

where the angle δ is the difference between the initial phases of the displacement and electric intensity, which is the angle of dielectric losses. From (4.31),

$$\varepsilon' = \varepsilon \cdot \cos\delta \quad \varepsilon'' = \varepsilon \cdot \sin\delta \quad \frac{\varepsilon''}{\varepsilon'} = \tan\delta \qquad (4.32)$$

4.3 Interrelation Between the Material Electrical Parameters

We have studied nonmagnetic materials with $L = 0$; therefore their electrical equivalent models and phasor diagrams are simplified. It is conventional to represent the impedance $\hat{Z}$ with a series equivalent circuit and the admittance $\hat{Y} = 1/\hat{Z}$ with a parallel circuit. The equivalent vector diagrams (or phasor diagrams) help to calculate the lumped-circuit parameters (resistance R, conductance G, and capacitance C) from measured values of magnitude Z and phase angle φ. Fig. 4.5A shows a series equivalent model for a material with lumped parameters R and C and its phasor diagram.

As seen from Eq. (4.27), the complex impedance $\hat{Z}$, in addition to the algebraic form, can be expressed also in polar and trigonometric forms, which are related to each other by the Euler's formulas (Moskowitz, 2002; Neyman and Demirchian, 1981; Duffin, 1990):

$$\hat{Z} = Ze^{j \cdot \phi} = Z\cos\phi + jZ\sin\phi = R + jX \qquad (4.33)$$

From the measured data, we may calculate the resistance R and reactance $X = -1/\omega C_{ser}$ that are related to a series electrical model for a nonmagnetic sample [Eqs. (4.23) and (4.27) and Fig. 4.5A]. In the series circuit, the current I is the same for R and X, but there is a physical sense when we resolve the voltage V into real and imaginary parts. The impedance triangle can be obtained from the phasor diagram, and the values of resistance, reactance, and capacitance may be

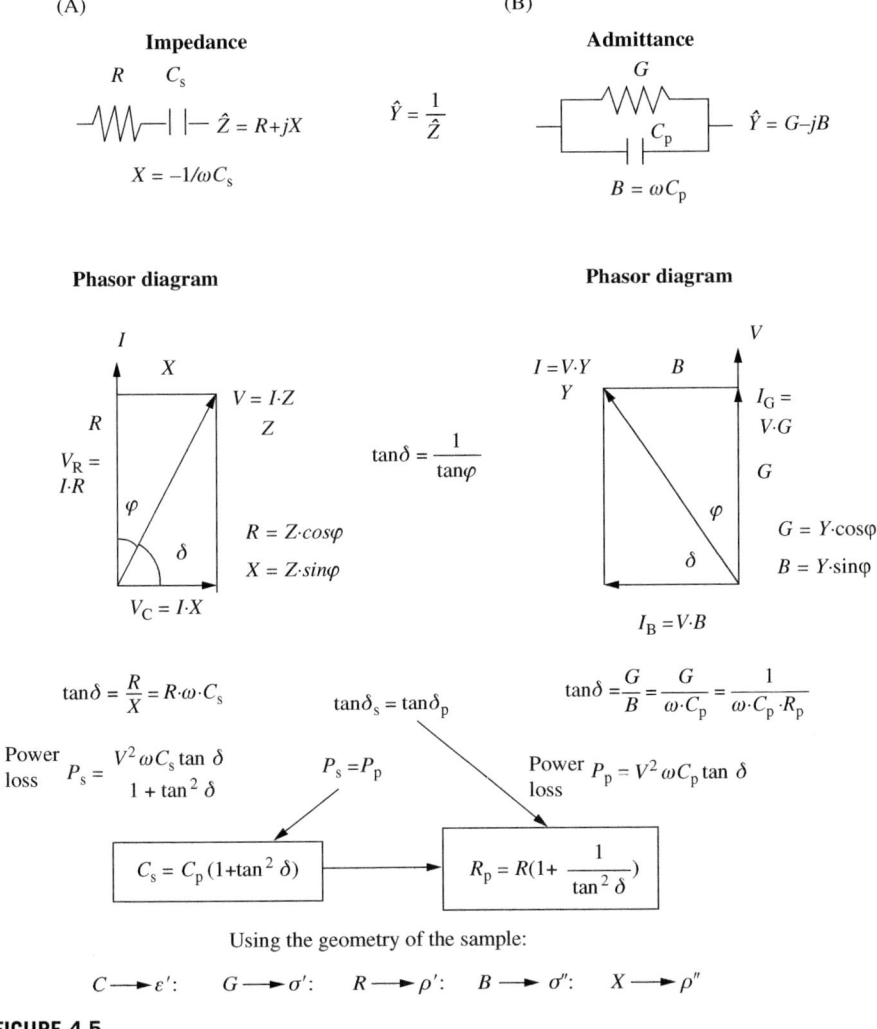

FIGURE 4.5

Electrical equivalent models and their phasor diagrams for a nonmagnetic material with lumped parameters R and C. (A) Series model and (B) parallel model.

derived in trigonometric form, as follows (Johnson, 1965; Neyman and Demirchian, 1981):

$$R = Z\cos\varphi \quad X = Z\sin\varphi \tag{4.34}$$

$$C_{ser} = -\frac{1}{\omega X} = -\frac{1}{\omega Z\sin\varphi} \tag{4.35}$$

We may also define the admittance $\hat{Y} = 1/\hat{Z}$, which can be expressed through its conductance G and susceptance $B = \omega C_{par}$, related to a parallel model (Fig. 4.5B) as follows (Neyman and Demirchian, 1981; Duffin, 1990):

$$\hat{Y} = Ye^{-j \cdot \phi} = Y\cos\phi - jY\sin\phi = G - jB \tag{4.36}$$

In the parallel circuit, the voltage V is the same for G and B, but there is a physical sense when we resolve the current I into real and imaginary parts (Tareev, 1975). In this case, it is convenient to choose the initial phase $\psi_V = 0$ in the sinusoidal equation for voltage (4.10). Then, we will have:

$$V = V_m \cdot \sin(\omega t) \quad I = I_m \cdot \sin(\omega t - \varphi) \tag{4.37}$$

The current in the capacitor $C = q/V$ can be defined as

$$I_C = \frac{dq}{dt} = \frac{d(CV)}{dt} = C \cdot \frac{dV}{dt} \tag{4.38}$$

Using the expression for voltage (4.37) in (4.38), the current in the capacitor is

$$I_C = I_B = \omega CV_m\sin\left(\omega t + \frac{\pi}{2}\right) = \omega CV \tag{4.39}$$

As indicated in Fig. 4.5B, the current $I_B = B \cdot V$, then from (4.39) $B = \omega C$.

From the parallel-circuit phasor diagram, the admittance triangle is obtained, and it's real and imaginary components G and B have the following trigonometric expressions (Neyman and Demirchian, 1981):

$$G = Y\cos\varphi, \quad B = Y\sin\varphi \tag{4.40}$$

Then the capacitance is

$$C_{par} = \frac{B}{\omega} = \frac{Y\sin\varphi}{\omega} \tag{4.41}$$

As mentioned in Chapter 2, Parameters describing the material behavior in an electromagnetic field, the dielectric losses in the material are characterized by the loss tangent $\tan\delta$. The loss angle δ complements the measured angle φ to $90°$; therefore we can define $\tan\delta$ from our measured φ values as $\tan\delta = 1/\tan\varphi$. As seen in Fig. 4.5, the values φ and δ are the same for both series and parallel equivalent circuits, and the $\tan\delta$ does not depend on the equivalent sample model. The $\tan\delta$ can also be expressed as the ratio of real to imaginary component of Z and Y, which is shown in Fig. 4.5.

In order to define the relationships between the parameters of the two models, we also need to compare the corresponding expressions for power loss P. For a series circuit

$$P_S = V_{real} \cdot I = I \cdot R_S \cdot I = I^2 \cdot R_s = \frac{I^2 \cdot \tan\delta}{\omega C_s} \tag{4.42}$$

Using the initial expressions for $I = V/Z$ and $Z^2 = R^2 + X^2$ in (4.39) and substituting I, we finally obtain the expression for Ps shown in Fig. 4.5. For a parallel circuit

$$P_\mathrm{P} = V \cdot I_\mathrm{real} = V \cdot G = \frac{V}{R_\mathrm{P}} = V^2 \cdot \omega C_\mathrm{P} \tan \delta \tag{4.43}$$

Equating the (4.42) and (4.43), as well as $\tan \delta_\mathrm{s} = \tan \delta_\mathrm{p}$, we obtain the interrelations between circuit parameters from the series and parallel models through $\tan \delta$ (Fig. 4.5). They are not always identical, depending on the value of $\tan \delta$.

   From the defined circuit parameters and the geometry of the sample, the resistivity ρ, conductivity σ, and dielectric permittivity ε of the material can be calculated. In complex form, they are

$$\hat{\rho} = \rho' + j \cdot \rho'' \quad \hat{\sigma} = \sigma' - j \cdot \sigma'' \quad \hat{\varepsilon} = \varepsilon' - j \cdot \varepsilon'' \tag{4.44}$$

   From the series and parallel phasor diagrams in Fig. 4.5, it is easy to find the relationships between the real and imaginary components of complex resistivity and conductivity and $\tan \delta$. For dielectric permittivity, this relationship is defined by Eq. (4.32). These relationships are

$$\rho'' = \frac{\rho'}{\tan \delta} \quad \sigma'' = \frac{\sigma'}{\tan \delta} \quad \varepsilon'' = \varepsilon' \cdot \tan \delta \tag{4.45}$$

   From the parallel phasor diagram (Fig. 4.5B) and using the previous relationships (4.42), we can derive the following interrelations between the material electrical parameters:

$$B = \omega C_\mathrm{p} \rightarrow \sigma'' = \omega \varepsilon' \tag{4.46}$$

$$\tan\delta = \frac{\sigma'}{\sigma''} = \frac{\sigma'}{\omega \cdot \varepsilon'_\mathrm{p}} \rightarrow \sigma' = \omega \varepsilon' \tan\delta = \omega \cdot \varepsilon'' \tag{4.47}$$

$$\hat{\rho} = \rho' - j \cdot \rho'' = 1\hat{\sigma} = \frac{1}{\sigma' + j \cdot \omega \cdot \varepsilon'} \tag{4.48}$$

$$\rho' = \frac{\sigma'}{\sigma'^2 + (\omega \cdot \varepsilon')^2}; \quad \rho'' = \frac{\omega \cdot \varepsilon'}{\sigma'^2 + (\omega \cdot \varepsilon')^2} \tag{4.49}$$

   From Ohm's law, $J = \hat{\sigma} \cdot E = (\sigma' + j\sigma'') \cdot E$ and using the relations for σ' and σ'' (4.46−4.47), we obtain

$$\hat{\sigma} = j \cdot \omega \cdot \hat{\varepsilon} \quad \text{and} \quad \hat{\rho} = \frac{1}{j \cdot \omega \cdot \hat{\varepsilon}} \tag{4.50}$$

4.4 **Experimental Procedures**

In accordance with the concept of distributed and lumped parameters, considered above, in Sections 4.1 and 4.2, there are two approaches for measurements of the electrical properties of materials, depending on the frequency range. At low frequencies, typically, below 1 GHz, and for small samples, when the wavelength is much larger than the sample dimensions, lumped-circuit methods are applied. The complex electrical parameters of a material ($\hat{\varepsilon}$, $\hat{\sigma}$, and $\hat{\rho}$) can be defined by measuring the magnitude Z and phase φ of the sample impedance $\hat{Z}$, using impedance analyzers (e.g., HP 4194A). From these measurements, the circuit lumped parameters, resistance R, conductance G, and capacitance C, can be defined (Section 4.3).

The sample is placed in a sample holder, which may have various electrode arrangements. A parallel-plate capacitor with disk electrodes is frequently used as a sample holder below 100 MHz. Fig. 4.6 shows an example of a parallel-plate sample holder that we have designed primarily for soils. Three Teflon contact clamps allow compression of the soil sample to approximately in situ conditions. A contact clamp is shown in more detail in Fig. 4.7.

The contact clamp has silver strips connected to the outside contacts. The electrodes have a diameter of $d = 5$ cm and are made from brass, about 2 mm in thickness, plated with silver. This sample holder can also be used for measuring solid samples of the same or smaller diameter than that of the electrode. For measuring fluids, thin Teflon rings (less than 1.5 mm thick) of various heights (2−9 mm) are used to support the upper electrode. The Teflon ring height determines the fluid sample thickness h. The outer diameter of the rings is 5 cm. The capacitance of the ring was calculated, included in the stray capacitance value, and thus removed from the total measured capacitance. The sample holder is connected to the HP 4194A instrument, using the HP 16047C test fixture

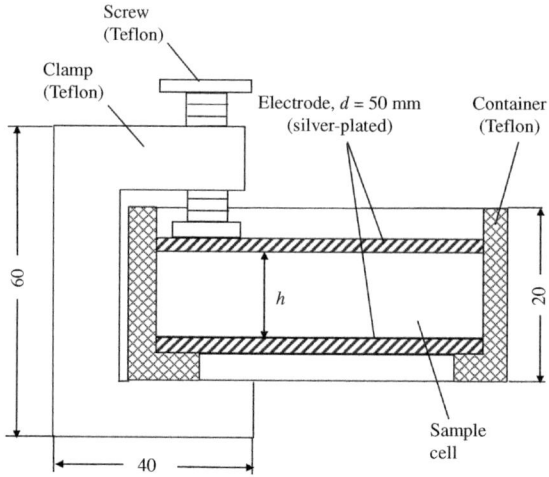

FIGURE 4.6

Schematic of the sample holder with disk electrodes and one of three hold-down clamps. Dimensions are in mm.

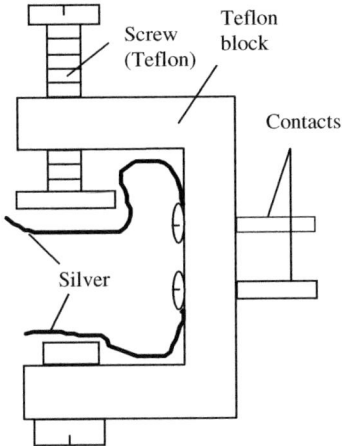

FIGURE 4.7

Schematic of the contact clamp.

($f = 0.001 - 40$ MHz), or the HP 41941A impedance Z probe ($f = 0.01 - 100$ MHz). When working with the Z probe, the sample holder must be placed in a grounded shield. Without the shield, several resonances were observed in the impedance curve between 40 and 100 MHz due to stray capacitive coupling to the sample holder's surroundings.

Basic relations between the sample dimensions, shown in Fig. 4.6, and lumped parameters R, G, and C are the following:

$$R = \rho' \frac{4h}{\pi d^2}; \quad G = \frac{\sigma' \pi d^2}{4h}; \quad C = \frac{\varepsilon \cdot A}{h} = \frac{\varepsilon_0 \varepsilon' \pi d^2}{4h} \qquad (4.51)$$

where A is the sample area in cm^2, and diameter d and thickness h are in cm.

This sample holder works up to 100 MHz. For higher frequencies, its dimensions become comparable with the wavelength. The following expression provides an estimate of the limiting dimensions for the object to be measured with a parallel-plate sample holder [adapted from Brandt (1963)]:

$$r < 0.038 \frac{\lambda}{\sqrt{\varepsilon'}} \qquad (4.52)$$

where r is the sample's radius, ε' is its real relative dielectric permittivity, and λ is the wavelength in free space (λ_0). Combining Eq. (4.52) with (4.13), we have

$$r < 0.038 \frac{c}{f \cdot \sqrt{\varepsilon'}} \qquad (4.53)$$

From (4.53) we see that at high frequencies it is easier to measure substances with low dielectric permittivity. The higher the dielectric permittivity, the lower is the frequency limit for using the lumped-parameter method, with given sample dimensions. Expression (4.53) requires very small samples at high frequencies, and it is not practical to make such a small parallel-plate sample holder. Using Eqs. (4.13) and (4.52), we find that, for example, at 100 MHz ($\lambda_0 = 300$ cm) and $\varepsilon' = 20$, the sample radius must be $r \leq 2.5$ cm, whereas for $\varepsilon' = 33$ (methanol), $r \leq 2.0$ cm. For 500 MHz ($\lambda_0 = 60$ cm) and the same values of ε', $r \leq 0.5$ cm and 0.4 cm, respectively, which are unusable sizes. From (4.51), for a material such as methanol ($\varepsilon' = 33$) and with our disk electrodes ($r = 2.5$ cm), reasonable results can be obtained only to a frequency as high as 80 MHz. This is in agreement with our measured data.

As mentioned in Section 4.1, when the sample dimensions are comparable to the wavelength, another approach may be used, based on wave-propagation theory, which considers the system distributed. However, we have extended the lumped-circuit methods up to 1 GHz by using a coaxial sample holder, as we will describe below. The coaxial sample holder does not require wires for connecting to the instrument, which would be inappropriate at high frequencies, above 100 MHz. One sample holder we use is an airline (General Radio GR 900), which is connected to the instrument by means of special adapters and thus becomes a part of the measuring system. This helps to reduce energy loss through radiation. When using the coaxial sample holder, the high-frequency limit can be extended from 100 MHz up to 1 GHz for materials with low or moderate dielectric permittivity ($\varepsilon' \leq 25$ at 1 GHz). By extending the lumped-circuit method to 1 GHz, we have unique solutions for the electrical properties of our samples at these high frequencies. This is in contrast to the distributed parameter method, which may have nonunique solutions.

The coaxial sample holder that we used was a GR 900-LZ3, with a length $h = 3$ cm (Fig. 4.8). It is connected to the instrument by means of special adapters, such as a GR-900 to N connector (GR 900-QNJ) for the HP 4194A, and a GR 900 to APC 7 connector (GR 900-QAP7) for the HP 4191A. We have also used other airlines, for example, 7 and 3.5 mm. When measuring fluid or soil samples, Teflon spacers (about 2 mm thick) were used at each end of the airline, and their capacitance was taken into account. For a coaxial sample holder, the limiting wavelength λ_{lim}, according to Brandt (1963), is related to the dielectric

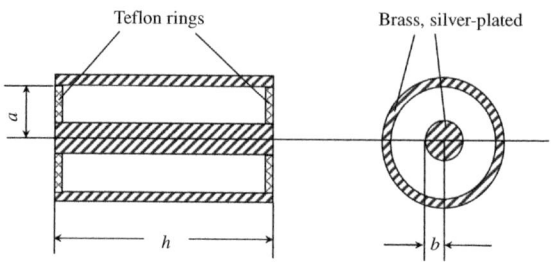

FIGURE 4.8

Schematic of the coaxial sample holder.

permittivity ε' of the substance and the average coaxial perimeter $p = \pi(b + a)$ as follows:

$$\lambda_{\text{lim}} > \pi\sqrt{\varepsilon'}(b + a) \tag{4.54}$$

where b and a are radii of the outer and inner electrodes of the coaxial, and λ_{lim} is the wavelength in measured material. The corresponding limiting frequency can be calculated, using a known relation between the frequency f and speed of light in a medium with real dielectric permittivity ε' (4.13):

$$f = \frac{c}{\lambda \cdot \sqrt{\varepsilon'}} \tag{4.55}$$

where $c = 3E10$ cm/s, the speed of light in vacuum.

For our GR900 coaxial sample holder ($a = 0.7144$ cm, $b = 0.3102$ cm, $p = 3.22$ cm), expression (4.54) indicates that the above technique may be appropriate at frequencies up to 1 GHz for materials with $\varepsilon' < 80$. For methanol ($\varepsilon' = 33$), using (4.54), the limiting wavelength $\lambda_{\text{lim}} > 18$ cm. The limiting frequency, using (4.55), $f_{\text{lim}} < 290$ MHz. In order to find out how much the limiting wavelength may exceed the coaxial dimensions, we compared these calculation results with our experimental data. Based on our measurements, the actual high-frequency limit for methanol was about 350 MHz (Section 6.2, Fig. 6.4). We can rearrange expression (4.54) in the following form:

$$\pi(b + a) < \frac{\lambda}{\sqrt{\varepsilon'}} \tag{4.56}$$

where the right-hand part is the wavelength in the given medium. From our data, using (4.55), the actual limiting wavelength in methanol is $\lambda_{\text{lim}} = 15$ cm, which is 4.7 times higher than the average perimeter of our coaxial airline. This actual limiting wavelength, according to Eq. (4.55), corresponds to the limiting frequency $f_{\text{lim}} < 340$ MHz.

The coaxial-electrode system allows us to cover the entire frequency range from 1 kHz to 1 GHz with the same sample and the same data processing procedures. The procedures of measurements with both instruments are described in Appendix B.

4.5 Preparing Soil Samples for Measurements

We have studied soil samples from Avra Valley (Arizona), Fort Huachuca (Arizona), Brookhaven (New York), and Columbia University (New York). Electrical property measurements were performed on soils with natural moisture as well as on dried samples saturated with distilled water of various contents. When we studied a sample with natural moisture, it was sieved before measurements to a fraction <0.6 mm in order to remove any stones or clumps. This was done rather quickly (in 1 minute), so the water was not lost. After performing the measurements, the sample was dried out at $T \approx 105°C$ in order to determine its moisture content. To specially prepare wet samples in the laboratory, the soil was also sieved and then dried in an oven with a temperature of $T \approx 105°C$ until the weight became constant. Then, a desired amount of distilled water ($\rho \approx 10^4 \, \Omega$ m) was added to the sample. We used distilled water rather than salt solutions because the natural salts remained in the dried sample. The measuring cell containing the sample was weighed, and the thickness of the parallel-plate sample holder between the electrodes was measured with a micrometer. The sample thickness h was determined by subtracting the known electrode thickness from the mean value of cell thickness measured at five to six points around the disk electrodes.

Subtracting the weight of the empty cell from the weight of the cell with the sample, we obtained the sample weight M with moisture. Knowing the sample's weight M (g) and geometrical volume V (cm^3), we can calculate the density of the measured sample in (g/cm^3):

$$\gamma_s = \frac{M}{V} \tag{4.57}$$

The density γ_s may serve as a measure of sample compaction in the sample holder. Also, the dry bulk density γ_b was calculated as a ratio of the sample's dry weight M_{dry} to its volume V (Hillel, 1982; Sternberg and Levitskaya, 2005).

The soil water content in the sample was calculated in percent by two ways: by weight (gravimetric, W_g) and by volume (volumetric, W_v). The difference between sample weights in the wet and dry state divided by its dry weight gives the gravimetric water content (Hillel, 1982):

$$W_g = \frac{M_{wet} - M_{dry}}{M_{dry}} \times 100\% \tag{4.58}$$

The water volume divided by the sample volume gives the volumetric water content:

$$W_v = \frac{M_{wet} - M_{dry}}{\gamma_w V} \times 100\% \tag{4.59}$$

where $\gamma_w = 1$ g/cm^3 is the water's specific weight. In our further discussion, we will use the volumetric soil water content W_v. It is easy to see from (7.2) and (7.3) that W_v and W_g are related through the dry bulk density $\gamma_b = M_{dry}/V$ (Hillel, 1982; Sternberg and Levitskaya, 2005).

References

Baker-Jarvis, J., Vanzura, E.J., Kissick, W.A., 1990. Improved technique for determining complex permittivity with the transmission/reflection method. IEEE Trans. Microwave Theory Tech. 38 (8), 1096–1103.

Baker-Jarvis, J., Geyer, R.G., Domich, P.D., 1992. A nonlinear least-squares solution with causality constraints applied to transmission line permittivity and permeability determination. IEEE Trans. Instrum. Meas. 41 (5), 646−652.

Bartnikas, R. (Ed.), Engineering dielectrics, Vol.II B, Electrical Properties of Solid Insulating Materials: Measurement Techniques, ASTM Spec. Tech. Publ., 926, 1987.

Blackham, D.V., Pollard, R.D., 1997. An improved technique for permittivity measurements using a coaxial probe. IEEE Trans. Instrum. Meas. 46 (5), 1093−1099.

Böttcher, C.J.F., Bordewijk, P., 1978. Theory of Electric Polarization, v. II, Dielectrics in Time-Dependent Fields. Elsevier, Amsterdam.

Brandt, A.A., 1963. (in Russian) Issledovanie dielektrikov na sverkhvysokykh chastotakh (Studying the Dielectrics at Very High Frequencies). State Publisher of Phys. −Math. Lit., Moscow.

Bronshtain, I.N., Semendiaev, K.A., 1971. A Guide-Book to Mathematics. H. Deutsch, Frankfurt/Main.

Bussey, H.E., 1967. Measurement of RF properties of materials. A survey. Proc. IEEE 55 (6), 1046−1053.

Courtney, C.C., 1998. Time-domain measurement of the electromagnetic properties of materials. IEEE Trans. Microwave Theory Tech. 46 (5), 517−522.

Duffin, W.J., 1990. Electricity and Magnetism, fourth ed. McGraw-Hill, New York.

Evdokimov, F.E., 1977. Fundamentals of Electricity. Mir Publisher, Moscow.

Hillel, D., 1982. Introduction to Soil Physics. Academic, San Diego, Calif.

Johnson, J.H., 1965. Introduction to Electrical Engineering. Int. Textbooks, Scranton, PA.

Levitskaya, T.M., Sternberg, B.K., 2000. Laboratory measurement of material electrical properties: extending the application of lumped-circuit equivalent models to 1 GHz. Radio Sci. 35, 371−383.

Ligthart, L.P., 1983. A fast computational technique for accurate permittivity determination using transmission line methods. IEEE Trans. Microwave Theory Tech. MTT 31 (3), 249−254.

Moskowitz, M.A., 2002. A Course in Complex Analysis in One Variable. World Scientific Publishing Co.

Muñoz, J., Rojo, M., Margineda, J., 1993. A method for measuring the permittivity without ambiguity using six-port reflectometer. IEEE Trans. Instrum. Meas. 42 (2), 222−226.

Neyman, L.R., Demirchian, K.S., 1981. (in Russian) Theoretical Foundations of Electrical Engineering, Energoizdat, St. Petersburg, Russia.

Nicolson, A.M., Ross, G.F., 1970. Measurement of the intrinsic properties of materials by time-domain techniques. IEEE Trans. Instrum. Meas. IM 19 (4), 377−382.

Poley, J.P., Nooteboom, J.J., Waal, P.J., 1978. Use of V.H.F. dielectric measurements for borehole formation analysis. Log Anal. 19 (3), 8−30.

Pozar, D.M., 1998. Microwave Engineering. John Wiley.

Rau, R.N., Wharton, R.P., 1982. Measurement of core electrical parameters at ultrahigh and microwave frequencies. J. Pet. Tech. 34 (11), 2689−2700.

Scott, W.R.J., Smith, G.S., 1986. Dielectric spectroscopy using monopole antennas of general electrical length. IEEE Trans. Antennas Propag. 34 (7), 919−929.

Sternberg, B.K., Levitskaya, T.M., 2005. Measurement of sheet-material electrical properties: extending lumped-element methods to 10 GHz. IEE Proc.-Sci. Meas. Technol. 152 (3), 123−128.

Stuchly, S.S., Matuszewski, M., 1978. A combined total reflection-transmission method in application to dielectric spectroscopy. IEEE Trans. Instrum. Meas. IM 27 (3), 285–288.

Tareev, B.M., 1975. Physics of Dielectric Materials. MIR, Moscow.

Von Hippel, A.R., 1954. Dielectrics and Waves. John Wiley, New York.

Wait, J.R., 1985. Electromagnetic Wave Theory. HarperCollins, New York.

Wait, J.R., 1986. Introduction to Antennas & Propagation. Peter Peregrinus Ltd., London.

Westphal, W.B., 1954. Dielectric measuring techniques. In: von Hippel, A.R. (Ed.), Dielectric Materials and Applications. John Wiley, New York, pp. 63–122.

Stray parameters of the measuring system and ways of defining them

5

Chapter Outline

Measured values of the sample impedance include errors due to the effects from the sample holder and its connections to the instrument. These effects, caused by the inductance, resistance, and stray capacitance of the measuring system, must be removed from the measured impedance $\hat{Z}_m$. There are also effects from the electrode polarization (EP). The complex value of stray impedance caused by resistance R_{ms} and inductance L of the measuring system is

$$\hat{Z}_L = R_{ms} + j\omega L \tag{5.1}$$

The impedance of EP is shown as

$$\hat{Z}_{el} = R_{el} + jX_{el} \tag{5.2}$$

Equivalent circuits showing these stray parameters are shown for series and parallel equivalent circuits in Figs. 5.1 and 5.2.

5.1 Stray Resistance R_{ms} and Inductance L

For disk electrodes, the resistance R_{ms} and inductance L of the system can be measured in the frequency range of interest with the sample holder electrodes shorted. For our disk electrodes sample holder (Fig. 4.6), connected to the Impedance Analyzer HP4194A with the test fixture HP16047C, the resistance of the measuring system remains almost constant at frequencies from 1 to 100 kHz

Electrical Spectroscopy of Earth Materials. DOI: https://doi.org/10.1016/B978-0-12-818603-9.00005-2

Sample impedance Z

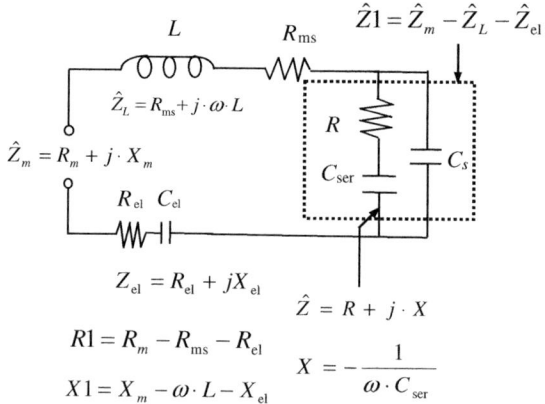

FIGURE 5.1

Series equivalent model for a material.

Sample admittance Y

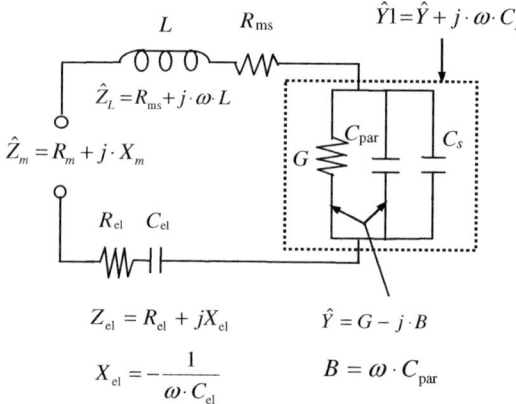

FIGURE 5.2

Parallel equivalent model for a material.

(average $R_{ms} = 0.015\,\Omega$). At higher frequencies, up to 40 MHz, it increases (because of the skin effect) and can be approximated by the power regression equation shown in Fig. 5.3. For the same sample holder and HP4194A used with the Z-Probe at frequencies from 10 kHz to 100 MHz, the resistance of the measuring system is shown in Fig. 5.4, along with the polynomial regression equation of third degree, which describes the experimental data rather well. The regression equations use f in MHz.

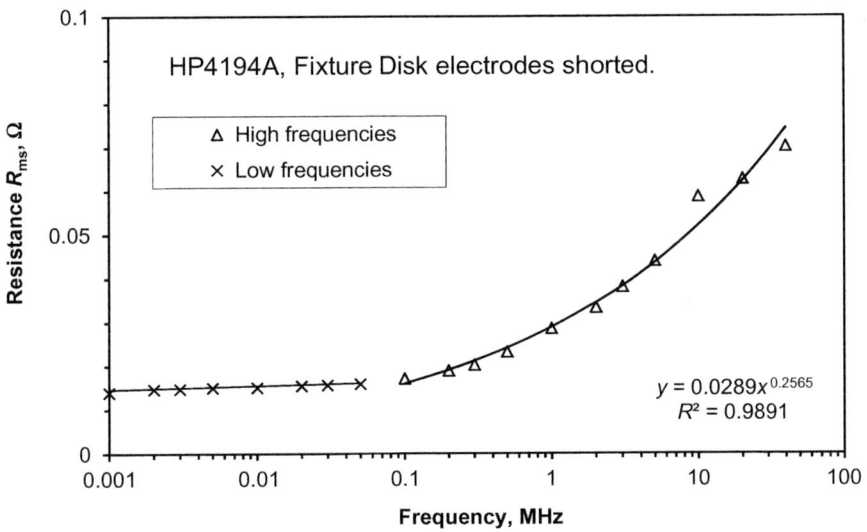

FIGURE 5.3

Resistance R_{ms} of the measuring system defined from the shorted disk electrodes with HP4194A, Fixture.

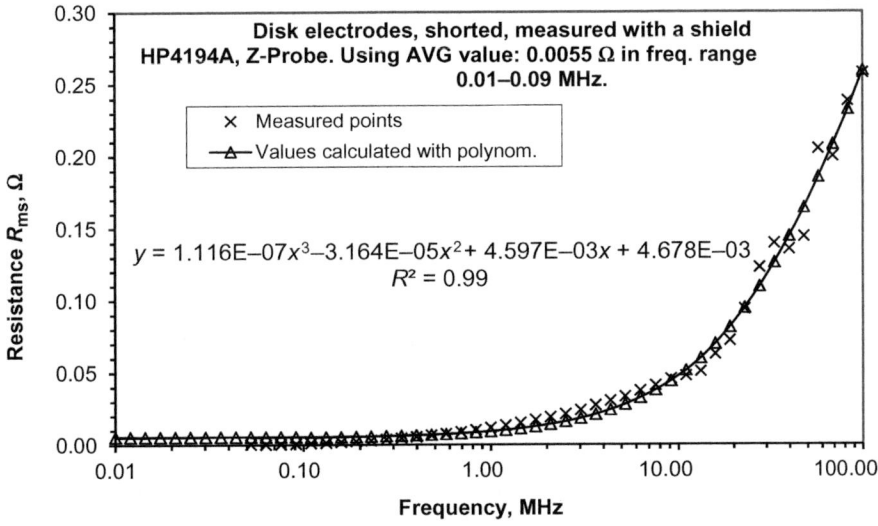

FIGURE 5.4

Resistance R_{ms} of the measuring system defined from the shorted disk electrodes with HP4194A, Z-Probe.

For our coaxial sample holder, the resistance R_{ms} was neglected for all instruments. This was based on our measurements of the resistance for the coaxial holder, shorted with various terminations, which showed very small values, close to zero.

The inductance L of the measuring system was defined for shorted sample holders with all instruments, simultaneously with the resistance R_{ms}. In addition, the stray inductance was calculated, when measuring a known material with a frequency-independent capacitance C, such as air or Teflon.

Considering, for example, a series circuit model (Fig. 5.1), we see that the sample impedance is affected by the stray impedance $\hat{Z}_L = R_{ms} + j\omega L$. The measurements actually give equivalent values of measured resistance R_m and measured capacitance C_m. Consequently, the measured impedance of the sample may be represented by the following equation (Bogdanov, 1957):

$$Z_m = (R_{ms} + R_1) + j\omega L + \frac{1}{j\omega C_L} = R_m + \frac{1 - \omega^2 L C_L}{j\omega C_L} \tag{5.3}$$

where R_1 and C_L are the sample resistance and capacitance, corrected for R_{ms} and L.

Then, equivalent values of measured resistance R_m and measured capacitance C_m are

$$R_m = R_{ms} + R_1, \quad C_m = \frac{C_L}{1 - \omega^2 L C_L} \tag{5.4}$$

Similarly, the dielectric losses will be characterized by an equivalent value of $\tan \delta_m = \omega R_m C_m$. This indicates that resistance and inductance of the measuring system affect the phase angle φ as well. Using Eq. (5.4), the sample capacitance and $\tan \delta$ can be corrected for inductance L as follows (Smith-Rose, 1935; Bogdanov, 1957):

$$C_L = \frac{C_m}{1 + \omega^2 L C_m} \tag{5.5}$$

$$\tan \delta_L = \frac{\tan \delta_m - \omega R_{ms} C_m}{1 + \omega^2 L C_m} \tag{5.6}$$

From Eq. (5.5), the inverse can be solved, defining the inductance L, when measuring a standard material with known capacitance that does not depend on frequency, using the following equation:

$$L = \frac{C_m - C_{known}}{\omega^2 C_m C_{known}} \tag{5.7}$$

In (5.7), the capacitance C_m corresponds to the highest angular frequency ω of the measured interval and for a standard material $C_L = C_{known}$. It can be calculated from the known ε' and sample dimensions, using Eq. (4.51) for C. We checked the inductance L by measuring air and Teflon. The thickness of the samples was in the range of 0.2–0.9 cm. Based on the measurement results, the inductance L for the Impedance Analyzer HP4194A with the test fixture and $f = 1$ kHz to 40 MHz can be considered as a constant value $L = 50$ nH. With the Z-Probe and $f = 10$ kHz to 100 MHz, the inductance decreases with the sample thickness h. The graph L versus h is shown in Fig. 5.5. With the following regression equation, the inductance L can be calculated for each sample thickness h:

$$L = -8.3999h + 37.205 \tag{5.8}$$

For our coaxial sample holder, the inductance L can be defined using the same procedures, such as measuring the shorted coaxial with all instruments and calculating the inductance with Eq. (5.7) from known standard material measurements. In addition to the Impedance Analyzer HP4194A with test fixture and Z-Probe, the coaxial sample holder was mostly used for high-frequency measurements from 1 MHz to 1 GHz with the HP4191A Impedance Analyzer. In order to short a coaxial, we need to use some metal connector between the outer and inner electrodes. None of the standard short terminations is appropriate, because they contain too much metal, while we do not use any metal, when measuring a sample in a coaxial. The only connector that can be used is a brass disk. It is light and thin (1.15 mm), nonmagnetic, and the error in defining the inductance L, as well as R_{ms}, is minimal. The air inside the shorted coaxial is shunted and, therefore, does not produce any significant errors.

Reasonable average L-values were obtained from shorted-coaxial measurements with the HP4194A Impedance Analyzer, with the Fixture, $L = 14.7$ nH, with the Z-Probe, $L = 4.74$ nH. For the high-frequency HP4191A, the calculated L-values are reasonable ($L = 1.54-1.55$ nH), while the shorted-coaxial measurements give too high L-values.

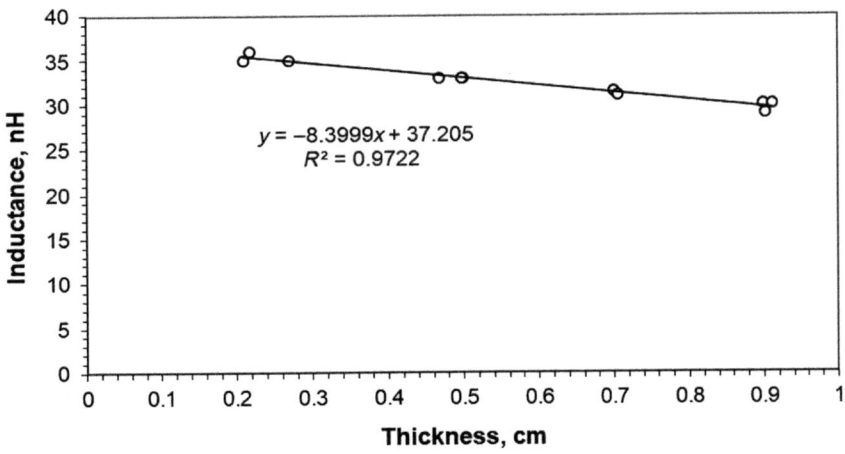

FIGURE 5.5

Stray inductance L versus sample thickness h for disk electrodes with HP4194A, Z-Probe (with a shield).

Some samples measured at high frequencies reveal a resonance, then the inductance can be calculated from the following resonance equation:

$$L = \frac{1}{C\omega_{res}^2} \tag{5.9}$$

In this case, the value L can be considered as the most reliable, because the resonance is a rigorous physical phenomenon.

5.2 Defining the Inductance L from Resonance

Resonance in an electrical circuit occurs when the input imaginary resistance or conductance is equal to zero, and there is no phase shift between the input current and voltage, that is the phase angle $\varphi = 0$. Considering, for example, the series equivalent circuit (Fig. 4.3) and its phasor diagram (Fig. 4.4), one can see that in this case, the inductive component of the reactance X is equal to its capacitive component (Neyman and Demirchian, 1981):

$$\omega L = \frac{1}{\omega C} \tag{5.10}$$

Eq. (5.10) is a condition of resonance phenomenon. From this condition, the inductance L can be expressed, as shown in Eq. (5.9). In order to calculate the inductance of the measuring system, the resonance frequency ω_{res} and capacitance C of the resonance circuit must be defined (as will be explained later in this chapter).

Let us consider the frequency characteristics of a series resonance circuit. The reactance $X = \omega L - (1/\omega C)$ versus frequency has three ultimate values. At frequencies $\omega = 0$ and infinity ∞ (the poles of the function), $X(\omega) \to \infty$. At $\omega = \omega_0$ (the zero of the function), $X(\omega) = 0$. The characteristic feature of this function is that the algebraic X-values are always rising with frequency. The impedance magnitude Z exhibits a minimum at the resonance frequency of ω_0. The phase angle φ is changing in the range from $-\pi/2$ to $+\pi/2$. At resonance, the reactance X is changing from capacitive reactance (at $\omega < \omega_0$) to inductive reactance (at $\omega > \omega_0$).

A parallel resonance circuit is characterized by G, L, and C. It is characterized by admittance, as follows [Section 4.2, Eq. (4.30)]

$$\hat{Y} = \frac{1}{\hat{Z}} = Ye^{-j\phi} = G - jB \tag{5.11}$$

where G is the real conductance and B is the imaginary conductance or susceptance.

From (5.11) and the vector diagram, the following can be derived (Neyman and Demirchian, 1981):

$$B = B_L - B_C = \frac{1}{\omega L} - \omega C \tag{5.12}$$

For a parallel circuit, the resonance occurs at the same condition as for a series one, when $\varphi = 0$. This means that $B = 0$ and

$$\frac{1}{\omega L} = \omega C \tag{5.13}$$

From (5.13), the expression for inductance at resonance is the same as (5.9) for a series circuit. The imaginary conductance B has three characteristic frequencies: two poles of the function, $\omega = 0$ and infinity, for which $B \to \infty$, and one zero of the function ω_0, when $B = 0$. At all frequencies, the conductance B is decreasing. The phase angle φ is also decreasing from $+\pi/2$ to $-\pi/2$, passing 0 at resonance. The admittance Y of the parallel circuit exhibits a minimum at the resonance frequency ω_0. The imaginary conductance B is changing its character at the moment of resonance, being inductive at $\omega < \omega_0$ ($B > 0$; $\varphi > 0$), and then it becomes capacitive at $\omega > \omega_0$ ($B < 0$; $\varphi < 0$).

From the interception of the X-function and the B-function with the ω-axis, the resonance frequency ω_0 can be defined. We used the measured imaginary resistance X versus frequency, when analyzing the resonance phenomenon in our samples.

The capacitance C for calculating the inductance L, using Eq. (5.9), can be taken from the horizontal part of the curve showing measured capacitance C_{meas} versus frequency in the beginning of the frequency range by analogy with any equivalent resonance circuit. The initial capacitance value at $f = 10$ or 20 MHz can be used for most standard materials, such as distilled water and alcohols. Fig. 5.6 shows an example of the frequency dependence of the measured

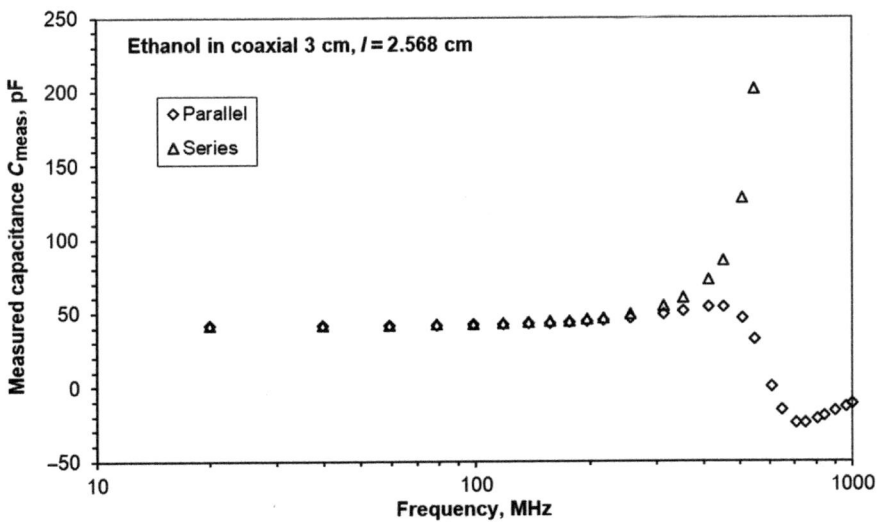

FIGURE 5.6

Measured capacitance versus frequency for ethanol.

capacitance for an ethanol sample. The capacitance defined at 20 MHz is $C = 41.82$ pF. If there is no horizontal part of the curve, and if C is decreasing with frequency (as for most wet soils), then the minimum value C_{min} before the resonance is used for the L-calculation.

Fig. 5.7 shows an example of C_{meas} versus f for a soil sample with water content 25%. The minimum value is $C_{min} = 33.18$ pF. Resonance frequencies for the same samples were defined from the frequency dependencies of reactance X, shown in Figs. 5.8 and 5.9, as follows: for ethanol $f_{res} = 608$ MHz and the calculated $L = 1.64$ nH; for the soil sample $f_{res} = 740$ MHz and $L = 1.39$ nH.

5.3 Stray Capacitance C_s

In addition to resistance and inductance, the measuring system also has a stray capacitance C_s. It may consist of two principal components, such as edge capacitance C_e of the sample holder and its cell capacitance C_c. The edge capacitance C_e is shown in Fig. 5.10 for a parallel-plate sample holder with disk electrodes and arises at the edges of electrodes as a fringing effect that results from the bending of the electric field lines (flux lines) (Bartnikas, 1987). The value C_e depends on the electrode arrangement and on the sample and electrode thicknesses. It can be either calculated or calibrated together with the cell capacitance.

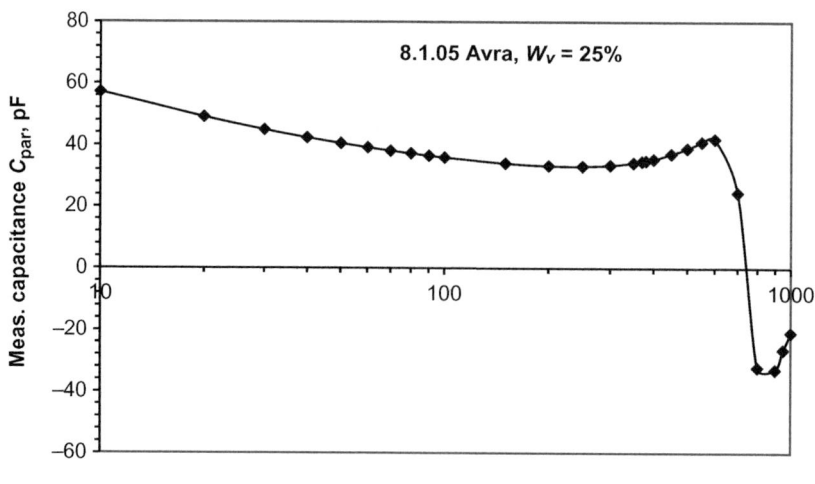

FIGURE 5.7

Measured capacitance versus frequency for a soil sample with volume water content 25%.

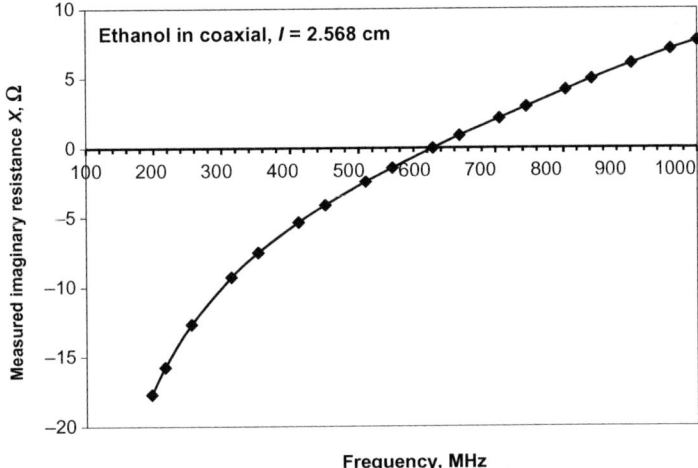

FIGURE 5.8

Measured reactance X versus frequency for ethanol.

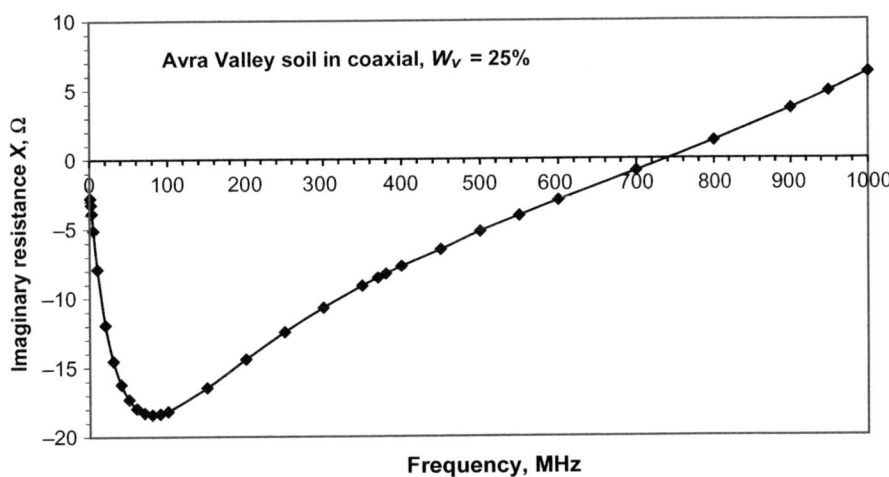

FIGURE 5.9

Measured reactance X versus frequency for a soil sample in a coaxial.

In a cell with disk electrodes, when the diameters D are equal, and the electrodes are much thinner than the sample ($t \ll h$), the edge capacitance can be calculated, using the following empirical equation (Bartnikas, 1987):

$$C_e = \pi D(0.029 - 0.058 \log h) \qquad (5.14)$$

Edge capacitance C_e

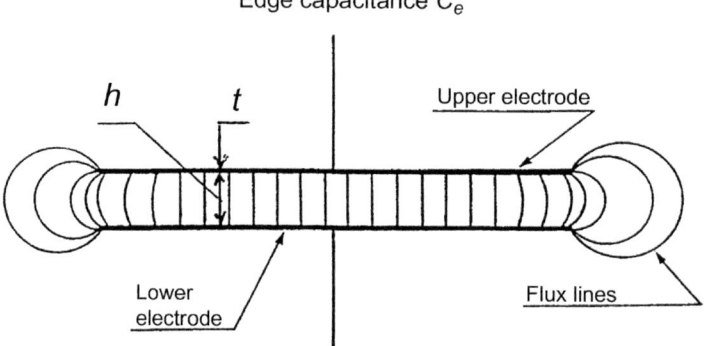

h　　t　　　　　　　　Upper electrode

Lower electrode　　　　　　　Flux lines

FIGURE 5.10

Schematic for a parallel-plate sample holder with edge capacitance C_e.

Reprinted, with permission, from Bartnikas, R. (Ed.), 1987. Engineering Dielectrics. Electrical Properties of Solid Insulating Materials: Measurement Techniques, vol. II B. ASTM Special Technical Publication 926, West Conshohocken, PA. Copyright ASTM International.

For example, Eq. (5.14) can be applied to the high-frequency measuring cell, described in Levitskaya and Sternberg (1996, Part 1) and used for measuring rock samples with thin foil electrodes. When the thickness of the electrodes t is comparable to that of the sample h (Fig. 4.6), the empirical Kirchhoff equation is applicable (Scott and Curtis, 1939):

$$C_e = \frac{1.113D}{8\pi}\left[\ln\frac{8\pi D}{h} - 3 + z\right] \tag{5.15}$$

where $z = (1 + (t/h))\ln(1 + (t/h)) - (t/h)\ln(t/h)$.

In our sample holder, the electrode thickness $t = 0.238$ cm, and the range of sample thicknesses is $h = 0.2-0.9$ cm (Section 4.4). Using Eq. (5.15) for calculating the edge capacitance at various sample thicknesses and plotting the graph $C_e(h)$, we obtained the following regression equation:

$$C_e = 0.5544h^{-0.438} \tag{5.16}$$

From (5.16), the edge capacitance C_e can be defined for any sample thickness h of the range.

5.3.1 Parallel-Plate Sample Holder

For our parallel-plate sample holder, other components of the stray capacitance, in addition to C_e, are shown in Fig. 5.11. They are parts of the cell capacitance C_c. The main part is the capacitance of the electrodes and connecting leads to ground, as well as of the connecting leads between themselves and to electrodes. The cell capacitance C_c depends on the thickness h of the sample and on its

Various components of stray capacitance C_s

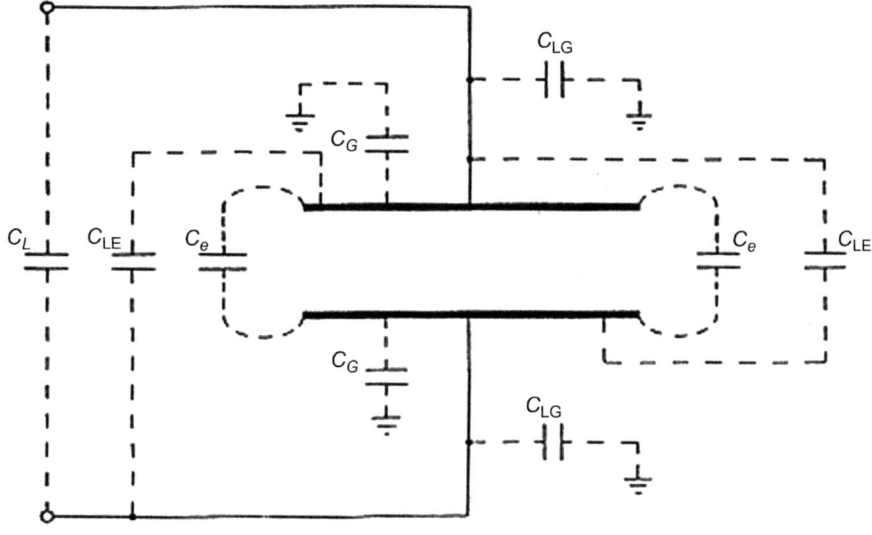

FIGURE 5.11

Schematic for a parallel-plate sample holder with various components of stray capacitance C_s.

Reprinted, with permission, from Bartnikas, R. (Ed.), 1987. Engineering Dielectrics. Electrical Properties of
Solid Insulating Materials: Measurement Techniques, vol. II B. ASTM Special Technical Publication 926,
West Conshohocken, PA. Copyright ASTM International.

distance to surrounding grounds, and C_c can be estimated by an empirical formula (Von Hippel, 1954). However, it is difficult to define the sample distance to ground, because it may change with its thickness h. The total value of the stray capacitance $C_s = C_e + C_c$ can be determined from measurements of a known standard material, such as air or Teflon.

For our parallel-plate sample holder with disk electrodes, a set of air samples was constructed, using Teflon rings of narrow width (~ 1.5 mm) with various heights (from 0.2 to 0.9 cm). Based on the assumption that the stray capacitance is the capacitance of the measuring system, excluding the sample capacitance itself, we defined the total stray capacitance C_s by subtracting the calculated capacitance between the electrodes (the air sample, including the Teflon ring) from the measured parallel capacitance C_L, corrected for inductance L with Eq. (5.5). We used the mean value of the C_L capacitance for the entire frequency range. Fig. 5.12 shows the graphs of C_s versus sample thickness defined from air measurements with an HP4194A Impedance Analyzer, using the fixture and the Z-Probe. From the regression equations, in Fig. 5.12, the stray capacitance C_s can be calculated for each sample thickness in any arrangement. When measuring

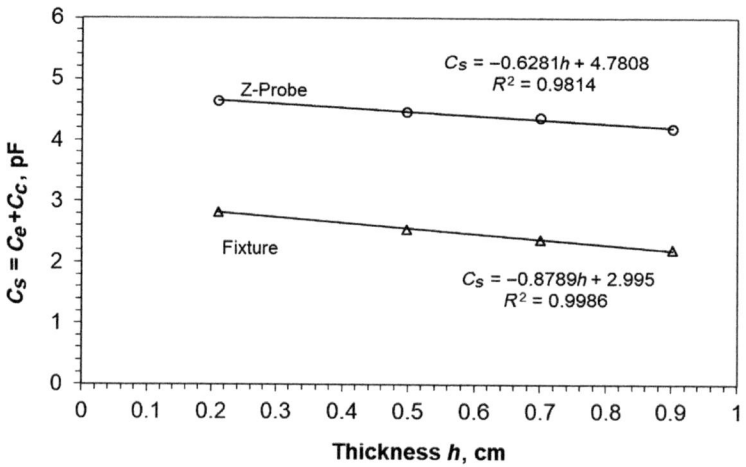

FIGURE 5.12

Stray capacitance C_s versus sample thickness defined from measuring air samples with HP4194A, Fixture and Z-Probe.

fluids with disk electrodes, we use the same Teflon rings, as for air samples. Then the capacitance of the ring must be added to the stray capacitance.

5.3.2 Coaxial Sample Holder

For our coaxial sample holder, the stray capacitance was also defined from measurements of a standard material, such as air. When measuring fluids or soils (as well as air), the Teflon spacers (rings), shown in Fig. 4.8, are used and their capacitance must be included in the stray capacitance C_s and thus removed from the measured sample capacitance. We measured air samples in the coaxial with various Teflon spacers and using Impedance Analyzers (HP4194A with Fixture and Z-Probe and HP4191A), covering the entire frequency range from 100 Hz to 1 GHz. The thickness of the two Teflon spacers is used and determined the length of the measured sample. Subtracting the calculated value of the air sample from the mean value of the measured capacitance corrects for the stray capacitance C_s. We defined the total stray capacitance C_s, for the coaxial air line for each experimental arrangement, which includes the cell capacitance C_c, with the effects from the contacts, the edge capacitance C_e, and the capacitance of the Teflon spacers C_t. For our samples, we used the stray capacitance that is defined from air measurements with the same Teflon spacers as in the given sample. In Table 5.1, some data are listed for the calculated stray capacitance with various instruments and Teflon spacers.

For some materials with high dielectric permittivity (methanol, water, etc.), we measured shorter samples, in order to increase the limiting frequency, because

Table 5.1 Stray Capacitance C_s Defined From Air Measurements in a Coaxial With Various Teflon Spacers and Instruments

| | | Stray Capacitance, C_s in pF | | |
| | | HP4194A | | HP4191A |
Teflon Spacers h, cm	Sample Length h, cm	Fixture	Z-Probe	
0.322	2.678	0.5246		
0.324	2.676	0.5285		0.6713
0.4135	2.5865	0.5861	0.5024	0.5564
0.427	2.573		0.4559	0.5498
0.432	2.568	0.4364		0.5459
0.4322	2.5678	0.4451	0.4596	

the inductance L of the measuring system decreases in this case. The capacitance of the air space in the coaxial, which is measured together with the sample capacitance, must be calculated and added to the stray capacitance.

5.4 Stray Electrode Processes. Defining the Impedance of Electrode Polarization

In addition to stray resistance R_{ms}, inductance L, and capacitance C_s of the measuring system, described above, there exist some stray electrode processes on the contact surface with the sample, in particular, when measuring wet soils. The EP occurs in the electrical double layer (EDL) on the boundary between the electrode and wet sample. In addition, a contact resistance can affect the measured data. The wet sample itself may develop a polarization at low frequencies in its own EDL on the boundary of grains and the saturating fluid. It is called induced polarization (IP) effect.

Fig. 5.13 illustrates schematically the total possible spectrum of a material conduction phenomenon in terms of resistivity ρ', in a frequency range from 10^{-3} to 10^9 Hz. On this curve, we indicate the IP effect, where the resistivity is decreasing at low frequencies. In addition to IP, this part of the curve may include the EP polarization and the effect of contact resistance. It is difficult to distinguish between the EP and IP effects because they both take place at low frequencies and decrease with frequency. As seen in Fig. 5.13, $\rho'_{ohmic} = \rho'_{dc}$ before the IP polarization. The residual resistivity, after the IP has developed, between 1 kHz and 1 MHz, is often called the "static" resistivity and is designated by both ρ_s and ρ_0 in the literature.

All the processes on electrodes are decreasing with frequency, at least up to 1 kHz, and conditions at the electrodes become less important as the frequency increases (Chang and Jaffe, 1952; Schwan, 1963, 1966). However, our data show that in some cases, the electrode processes may affect the experimental results up

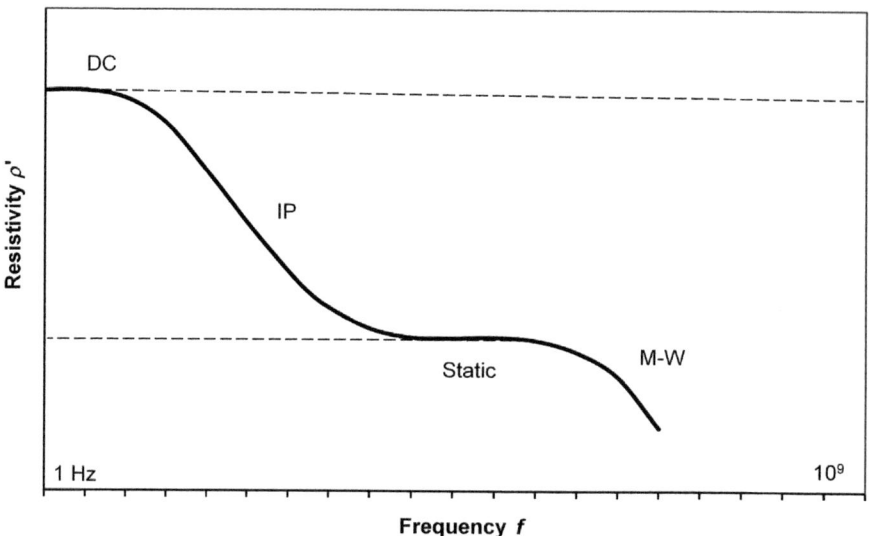

FIGURE 5.13

Schematic of a total possible spectrum of a material resistivity ρ.

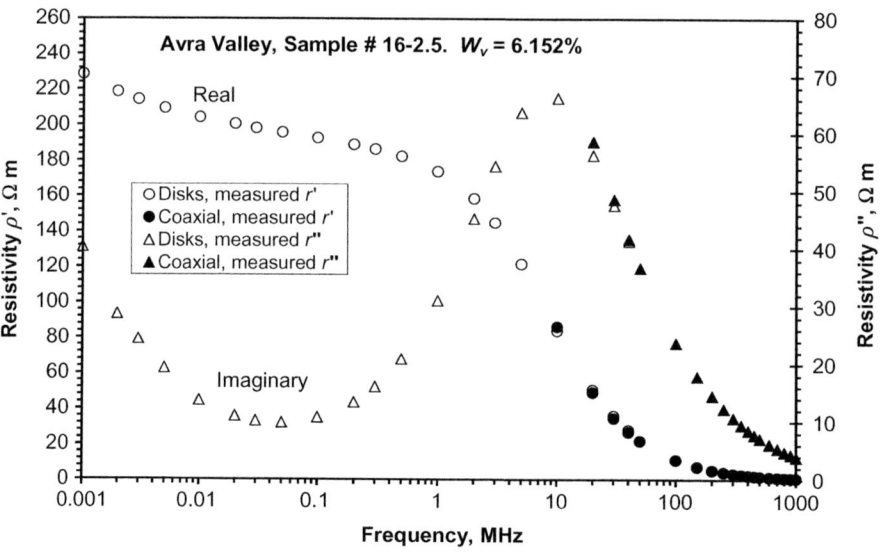

FIGURE 5.14

Complex resistivity versus frequency for a soil sample with volume wetness 6%.

to 100 kHz−1 MHz, and the static resistivity can't be defined from the graph. In Fig. 5.14, we show an example for a soil sample from Avra Valley, AZ (with volume wetness $W_v = 6.15\%$) before any corrections. From Fig. 5.14, it can be seen that, at frequencies above 1 MHz, wet soils with various water contents exhibit a relaxation process, called Maxwell—Wagner polarization.

The real resistivity ρ' measured at low frequencies may contain the components of IP polarization and EP polarization, as well as the contact resistance of the electrodes. The complex impedance, which characterizes all processes at the electrodes, is shown in Fig. 5.2 and can be described with Eq. (5.2), containing real and imaginary parts (Ferris, 1974, 1978; Schwan, 1963, 1992; Bard and Faulkner, 2001). Schwan, as well as Ferris (1974), refers to Fricke (1932), who found that the components of alternating-current EP R_s and C_s vary as a power function of frequency (ω^{-m}), that is, Fricke's law. Analyzing the available experimental results, obtained with various electrolytes and electrode materials (Pt, Ni, Ag, Au), and comparing with theoretical equations, Fricke (1932) showed that the power parameter m is, in some cases, independent of frequency, but generally it may increase with frequency in the range $0.1 < m < 0.6$. Many other experimental results contained similar dependencies, mostly with $m = 0.5$. In particular, it was found by Jones and Christian (1935), Hill et al. (1959), Schwan (1963, 1966), and Lopatin (1971) that the polarization resistance R_s and polarization capacitance C_s are decreasing with frequency inverse proportional to $\sqrt{\omega}$, as follows

$$R_s = \frac{\eta}{\sqrt{\omega}}; \quad C_s = \frac{1}{\eta\sqrt{\omega}} \tag{5.17}$$

where η is a constant. Bard and Faulkner (2001) obtained similar results in a generalized form.

Lopatin (1971) suggested a procedure using the real and imaginary components of the electrode impedance $\hat{Z}_{el}$ (5.2) and subtracting them from the measurement results. The method of subtracting the EP may also lead to extracting the IP effect and the contact resistance. Therefore, it is appropriate to first use this procedure on measurement results for some electrolytes, because they do not have an IP effect, and the contact resistance is negligible.

Lopatin analyzed data, obtained by Jaffe and Rider (1952) with nickel electrodes for KCl solutions of three concentrations (10^{-2} $N = 0.745$ g/L, 10^{-3} $N = 0.0745$ g/L, and 10^{-4} $N = 0.00745$ g/L). He found that the polarization resistance R_s became negligible at frequencies above 1 kHz. Using the same data from Jaffe and Rider (1952), we have drawn, after Lopatin, graphs of measured resistance R versus $\eta/\omega^{0.5}$ with the constant $\eta = 10^3$. As seen in Fig. 5.15, the regression equations are linear functions. For each solution, the polarization resistance R_s can be estimated as the difference between the measured resistance at a given frequency and the limit value at infinite frequency ($\eta/\omega^{0.5} = 0$).

We studied NaCl solutions with concentrations of 0.05 and 0.5 g/L, using an HP4192A Impedance Analyzer ($f = 5$ Hz—13 MHz) and a Solartron 1170 Frequency Response Analyzer ($f = 0.01$ Hz—1 kHz). First, we measured the solutions with various electrodes, such as silver, copper, bright platinum, and black platinum. Fig. 5.16 shows the results for a salt solution (0.5 g/L) obtained in a

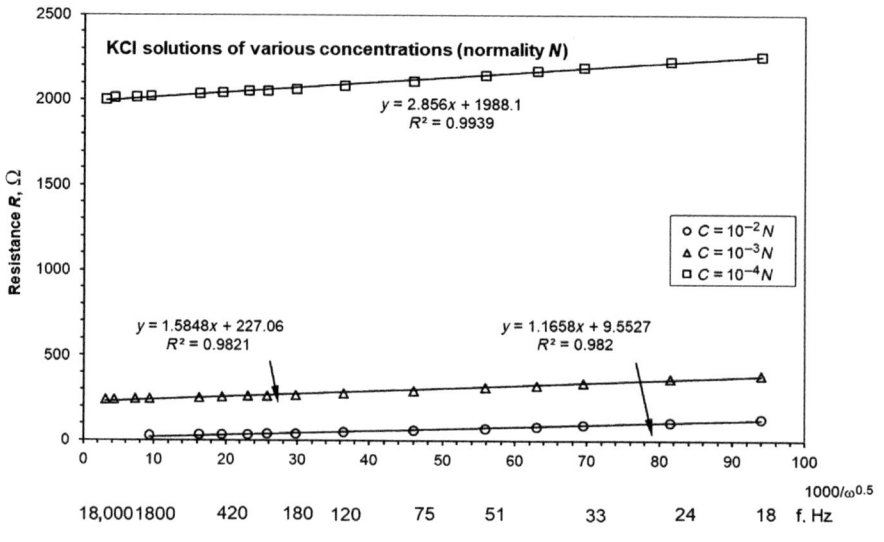

FIGURE 5.15

Resistance R versus $1000/\omega^{0.5}$ for a KCl solution with various concentrations C.

Data from Jaffe, G., Rider, J.A., 1952. Polarization in electrolytic solutions. Part II. Measurements. J. Chem. Phys. 20(7), 1077–1087.

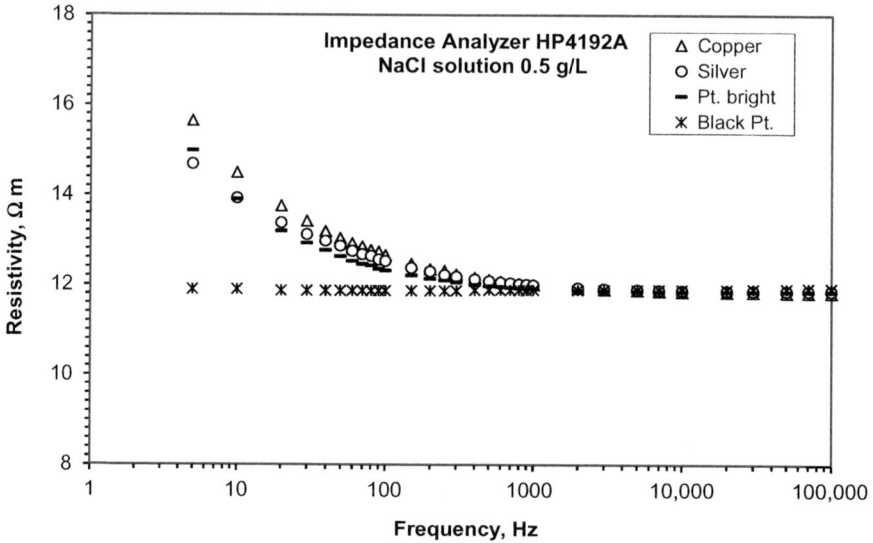

FIGURE 5.16

Resistivity ρ' versus frequency for a NaCl solution (0.5 g/L) with various electrodes.

glass with the electrode diameter equal to 3 cm and a distance between electrodes equal to 6 cm, using the HP4192A Impedance Analyzer. The resistivity versus frequency measured with Cu, Ag, and bright Pt electrodes shows a dispersion that is attributed to EP effect. With black Pt electrodes, no dispersion was observed.

By plotting the measured resistance R against $1/\sqrt{\omega}$, we obtain a graph, which is linear in the low-frequency range. An example is shown in Fig. 5.17 for a salt solution (0.5 g/L) and Ag electrodes. The line intercepts the ordinate at $R = R_{st}$ (or R_{dc}), when $\omega \to \infty$, and the electrode resistance $R_s = 0$. From the linear regression equation, $R_s = Ax + C$, and the intercept C defines the static value R_{st}, while the electrode resistance at any other frequency is $R_{sf} = 100\,A/\sqrt{\omega}$. The regression equation for a measured salt solution is shown in Fig. 5.17.

The difference between the measured resistance R and R_{sf}, at each frequency, gives the true data for the salt solution, free of EP. The true resistivity obtained from $R - R_{sf}$ is almost frequency independent and is very close to values obtained with the four-electrode approach and the Solartron (Fig. 5.18). As seen in Fig. 5.18, the EP effect is slightly higher in the parallel-plate sample holder with silver-plated electrodes, compared to the plastic box with silver-foil electrodes. The brighter and flatter the surface, the higher is the EP. This confirms Lopatin's conclusion that the surface quality of electrodes affects the polarization. The plastic box (made from Plexiglas) is shown schematically in Fig. 5.18.

Although the polarization resistance R_s produces larger errors in conductance measurements than the capacitance C_s, which affects only the phase (Lopatin, 1971), a correction for C_s is also desirable for obtaining precise values of dielectric permittivity ε. We applied the described procedure for defining the EP impedance in our moist soil samples. For each sample, in addition to resistance

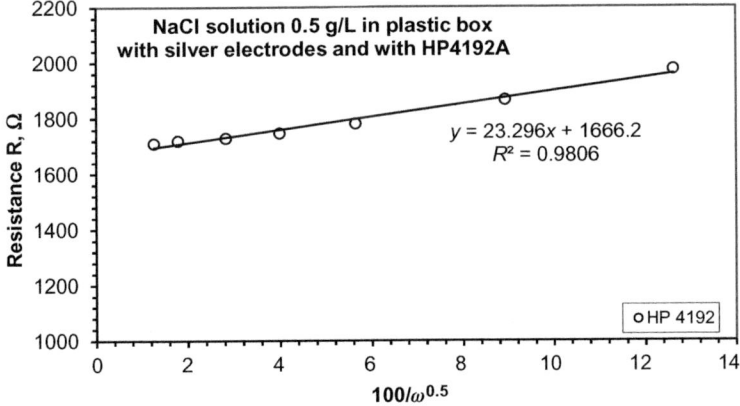

FIGURE 5.17

Resistance R versus $100/\omega^{0.5}$ for a NaCl solution (0.5 g/L), measured with silver electrodes.

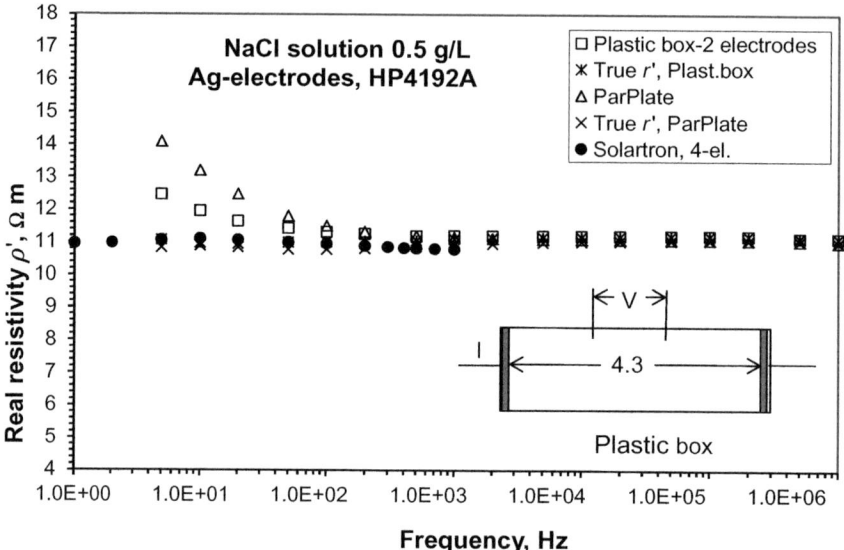

FIGURE 5.18

Resistivity ρ' versus frequency for a NaCl solution (0.5 g/L) measured with silver electrodes in various sample holders. Dimensions are in cm.

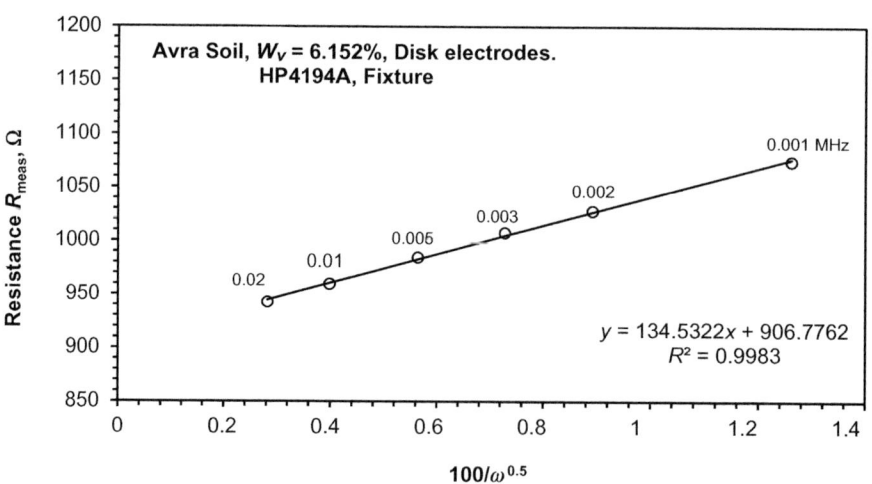

FIGURE 5.19

Resistance versus $100/\omega^{0.5}$ for a soil sample with volume wetness 6%.

R versus $\eta/\omega^{0.5}$, we considered also the reactance $X = 1/\omega C$ in the same way. Examples for the sample with volume wetness $W_v = 6.15\%$ are shown in Figs. 5.19 and 5.20. We call this procedure the Fricke–Lopatin method. The

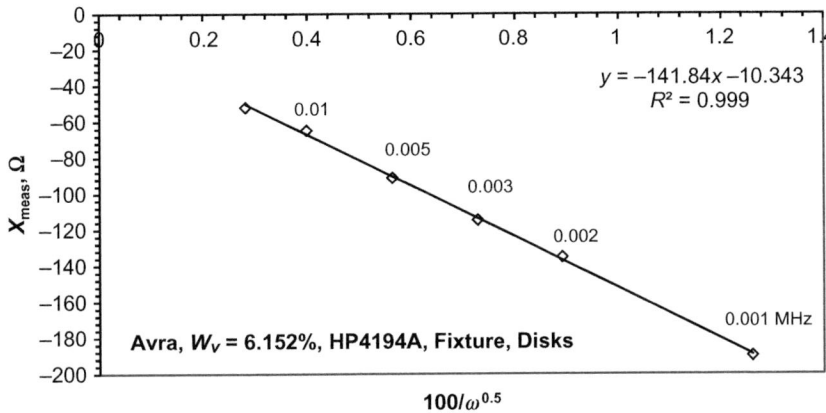

FIGURE 5.20

Reactance versus $100/\omega^{0.5}$ for a soil sample with volume wetness 6%.

values of electrode resistance and reactance at each measured frequency are calculated by using the first components of the regression equations, describing the graphs R and X versus $100/\omega^{0.5}$, analogous to Figs. 5.19 and 5.20. Thus, both the real and imaginary components of the electrode impedance were estimated and taken into account. Chapter 6, Corrections for stray parameters and error estimation, will show how the soil sample measurements are corrected for the EP impedance, along with the stray parameters of the measuring system.

References

Bartnikas, R. (Ed.), 1987. Engineering Dielectrics. Electrical Properties of Solid Insulating Materials: Measurement Techniques, vol. II B. ASTM Special Technical Publication 926, West Conshohocken, PA.

Bard, A.J., Faulkner, L.R., 2001. Chapter 10 Electrochemical Methods: Fundamentals and Applications. Wiley & Sons, New York, pp. 379−384.

Bogdanov, S.V., 1957. Methods of measuring capacitance and losses by the Q-meter at high frequencies (Russian) Pribory Tekh. Exp. 90−93.

Chang, H., Jaffe, G., 1952. Polarization in electrolytic solutions. Part I. Theory. J. Chem. Phys. 20 (7), 1071−1077.

Ferris, C.D., 1974. Chapter 2 Introduction to Bioelectrodes. Plenum Press, New York and London.

Ferris, C.D., 1978. Chapter 6 Introduction to Bioinstrumentation. The Humana Press, Clifton, NJ.

Fricke, H., 1932. The theory of electrolytic polarization. London Edinburgh Dublin Philos. Mag. J. Sci. 14, 310−318.

Hill, D.L., Hills, G.J., Young, L., Bockris, J.O.'M., 1959. A note of some measurements of Faradaic Impedances in fused salts. J. Electroanal. Chem. 1 (1), 79−83.

Jaffe, G., Rider, J.A., 1952. Polarization in electrolytic solutions. Part II. Measurements. J. Chem. Phys. 20 (7), 1077−1087.

Jones, G., Christian, S.M., 1935. The measurement of the conductance of electrolytes. VI. Galvanic polarization by alternating current. J. Am. Chem. Soc. 57, 272.

Levitskaya, T.M., Sternberg, B.K., 1996. Polarization processes in rocks, 1. Complex dielectric permittivity method. Radio Sci. 31 (4), 755−779.

Lopatin, B.A., 1971. Conductometry and Oscillometry (Translated from Russian). Israel Program for Scientific Translations, Jerusalem.

Neyman, L.R., Demirchian, K.S., 1981. (in Russian) Theoretical Foundations of Electrical Engineering. Energoizdat, St. Petersburg, Russia.

Schwan, H.P., 1963. Determination of biological impedances. In: Nastuk, W.L. (Ed.), Physical Techniques in Biological Research, vol. 6, Part B, pp. 323−407.

Schwan, H.P., 1966. Alternating current electrode polarization. Biophysik 3, 181−201.

Schwan, H.P., 1992. Linear and nonlinear electrode polarization and biological materials. Ann. Biomed. Eng. 20, 269−288.

Scott, A.H., Curtis, H.L., 1939. Edge corrections in the determination of dielectric constant. J. Res. Natl. Bur. Stand. U.S. 22, 747−775.

Smith-Rose, R.L., 1935. The electrical properties of soil at frequencies up to 100 MHz; with a note of the resistivity of ground in the United Kingdom. Proc. Phys. Soc. 47, 923−931.

Von Hippel, A.R., 1954. Dielectrics and Waves. John Wiley, New York.

Corrections for stray parameters and error estimation

6

Chapter Outline

There are two kinds of stray parameters that we take into account, when processing the experimental results of measured impedance $\hat{Z}_m = R_m + jX_m$: (1) stray parameters of the measuring system and (2) stray parameters, caused by the electrode polarization.

As described in Chapter 5, Stray parameters of the measuring system and ways of defining them, stray parameters of the measuring system include stray resistance R_{ms}, inductance L, and stray capacitance C_s. These parameters affect the experimental results in the entire frequency range, while the electrode polarization occurs mostly at low frequencies, up to $1-10$ kHz. It decreases with frequency, and above $10-100$ kHz becomes negligible.

6.1 Procedure for Removing Stray Parameters From Measurement Results

We start our data processing by first subtracting the components of stray impedance of the measuring system $\hat{Z}_L = R_{ms} + j\omega L$ and of electrode polarization impedance $\hat{Z}_{el} = R_{el} + jX_{el}$ from corresponding measured values R_m and X_m. We remove the stray capacitance C_s of the measuring system by using equations based on series or parallel equivalent circuit, shown in Figs. 5.1 and 5.2. The derivation of the equations is given in Appendix C.

For the *series equivalent circuit*, the following formulas allow us to define the sample resistance R and reactance X of the impedance $\hat{Z} = R + jX$:

$$R = \frac{R_1}{1 + 2\omega C_s X_1 + \omega^2 C_s^2 (R_1^2 + X_1^2)} \tag{6.1}$$

Electrical Spectroscopy of Earth Materials. DOI: https://doi.org/10.1016/B978-0-12-818603-9.00006-4

$$X = \frac{X_1 + \omega C_s(R_1^2 + X_1^2)}{1 + 2\omega C_s X_1 + \omega^2 C_s^2(R_1^2 + X_1^2)} \tag{6.2}$$

From these true sample values, R and X, the true values of impedance magnitude Z, $\tan \delta = R/X$, and phase angle φ can be defined, as well as the true capacitance C. Then, the sample parameters, such as ρ', ε', and σ', can be obtained, using the sample geometry (Section 4.4).

For the *parallel equivalent circuit*, we derived formulas that give us the true sample conductance G and susceptance B of the admittance $\hat{Y} = G - jB$, as follows:

$$G = \frac{R_1}{R_1^2 + X_1^2} \tag{6.3}$$

$$B = \frac{X_1}{R_1^2 + X_1^2} + \omega C_S \tag{6.4}$$

From the true sample values, G and B, the true sample capacitance C and resistance R are calculated, and the same sample parameters (σ', ε', and ρ') can be obtained. In Eqs. (6.1)–(6.4)

$$R_1 = R_m - R_{ms} - R_{el}; \quad X_1 = X_m - \omega L - X_{el} \tag{6.5}$$

For our samples, we usually used Eqs. (6.3) and (6.4), derived from the parallel equivalent circuit.

6.2 Measurement Results for Standard Materials and Error Estimation

In order to determine the accuracy of our measuring system, we measured some standard materials with known values of dielectric permittivity, such as air ($\varepsilon' = 1$), Teflon ($\varepsilon' = 2$) (Von Hippel, 1954), octanol ($\varepsilon' = 9.6-10$) (Bottreau et al., 1977; Nakamura et al., 1982), butanol ($\varepsilon' = 17.1-17.40$) (Bottreau et al., 1977; Nakamura et al., 1982; Scott and Smith, 1986), and methanol ($\varepsilon' = 32-33$) (Smyth, 1955; Kay, 1973; Burdette et al., 1980; Blackham and Pollard, 1997). The mean square error δ, in percent, was defined as a relative standard deviation from the mean for ε' (in the range where ε' is frequency independent). Table 6.1 shows the δ -values along with mean values of the relative dielectric permittivity for N data obtained at discrete frequencies with various electrodes and instruments. The difference between our mean and previously published values is less than 1%.

Fig. 6.1 shows the complex dielectric permittivity for air and Teflon versus frequency. The results that we obtained for air and Teflon are in excellent agreement for both the disk and the coaxial electrodes. We also measured some alcohols (*n*-octanol, *n*-butanol, and methanol), as standard materials.

Table 6.1 Error Estimation of Measurements With Various Instruments and Types of Electrodes

Material	HP4194A, Fixture 1 kHz to 40 MHz			HP4194A, Z-Probe 10 kHz to 100 MHz			HP4191A 1 MHz to 1 GHz			Type of Electrodes
	Mean	N	δ, %	Mean	N	δ, %	Mean	N	δ, %	
Air $\varepsilon' = 1$	0.9995	20	0.3	1.000	18	0.3–0.9	...	...	...	Disk
	1.000	18	0.5–0.8	1.000	18	0.6–1.0	1.000	37	0.23	Coaxial
Teflon $\varepsilon' = 2$	2.000	20	0.3	2.000	18	0.3–1.0	...	...	...	Disk
	2.0056	19	0.24	2.000	18	0.25	2.000	37	0.1	Coaxial
Octanol $\varepsilon' = 9.7$–10	...	...	...	...	...	...	...	...	...	Disk
	...	...	...	10.04	9	1.0	Dispersive			Coaxial
Butanol $\varepsilon' = 17.4$	...	...	...	...	...	...	...	...	...	Disk
	...	...	...	17.5	11	0.33	Dispersive			Coaxial
Methanol $\varepsilon' = 32$–33	...	...	...	31.7	12	0.95	...	...	...	Disk
	33.25	9	0.43	33.2	18	0.36	33.0	13	0.35	Coaxial
Water $\varepsilon' = 78$–80	79.4	13	1.0	79.0	11	0.61	...	...	...	Disk
	...	...	...	78.6	19	0.36	78.22	12	0.56	Coaxial

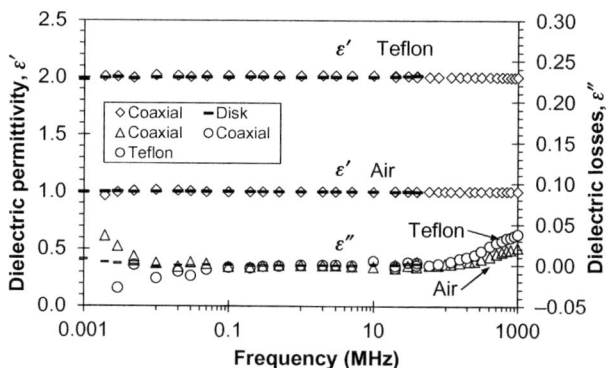

FIGURE 6.1

Dielectric permittivity versus frequency for air and Teflon.

These materials have been widely studied; however, the data from different references are not always in close agreement, particularly at high frequencies (Dalbert et al., 1949; Bottreau et al., 1977; Chahine and Bose, 1980; Nakamura et al., 1982; Scott and Smith, 1986). Nevertheless, most of the references show that at high frequencies (near 1 GHz), the data deviate from the Debye semicircle. This may indicate that a second relaxation, or at least a gradual decrease of the dielectric permittivity, takes place at higher frequencies. The optical dielectric constant is equal to the square of refraction index n. For alcohols, n^2 is of the order of two (Dalbert et al., 1949). Therefore, the value ε_∞, defined from the observed relaxation processes in some alcohols, cannot be the final value for the dielectric permittivity, which still must decrease with frequency.

Our complex dielectric permittivity data versus frequency for n-octanol are shown in Fig. 6.2A. Data from two references (Bottreau et al., 1977; Chahine and Bose, 1980) are shown for comparison. At $f = 0.1-10$ MHz, ε' is almost constant, and from 10 to 1000 MHz, a dispersion of ε' is observed, which is accompanied with a maximum in the loss factor ε''. The dielectric permittivity decreases from about $\varepsilon_s = 10$ to $\varepsilon_\infty = 3$. This behavior describes a relaxation process, which may be caused by orientation polarization of hydroxyl polar groups (−OH). As seen from Fig. 6.2B, this is a Debye relaxation, which is described with a semicircle. The critical frequency, at which the ε''_{max} occurs, is about $f_c = 130$ MHz. The relaxation time $\tau = 1.26$ ns.

Fig. 6.3 shows the real and imaginary parts of the complex dielectric permittivity versus frequency for n-butanol. There is also a relaxation process in n-butanol, but it occurs at higher frequencies (beginning at 40 MHz), in comparison with n-octanol. The frequency of ε''_{max} is about $f_c = 350$ MHz, and the values of dielectric permittivity are higher. At frequencies $f = 0.1-30$ MHz, ε' is nearly constant, and at higher frequencies, from 40 to 1000 MHz, it decreases from about $\varepsilon_s = 17$ to $\varepsilon_\infty = 5$. Our results are in good agreement with Bottreau et al. (1977).

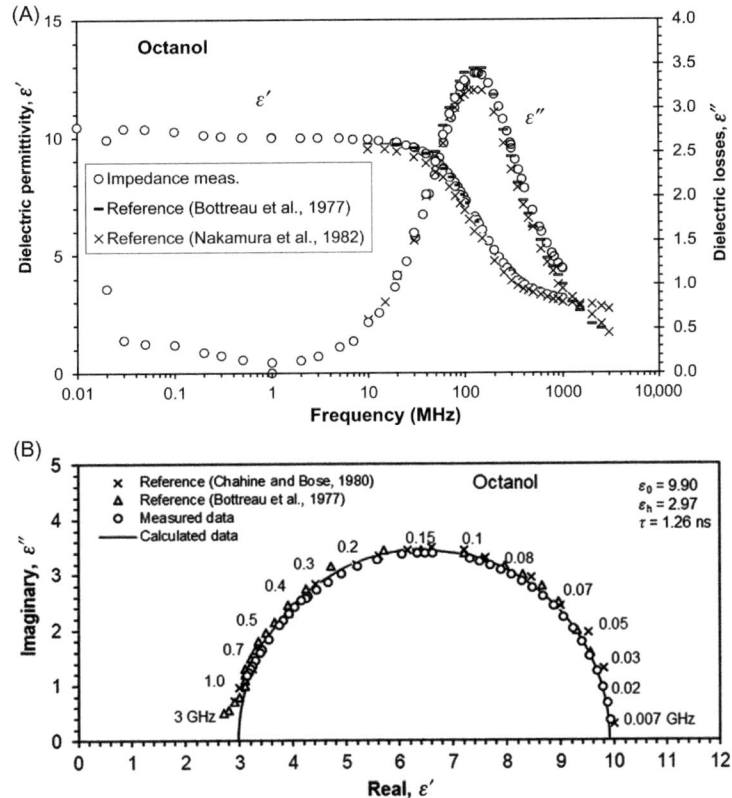

FIGURE 6.2

(A) Real dielectric permittivity ε' and loss factor ε'' versus frequency for n-octanol. (B) Semicircle diagram $\varepsilon'' = F(\varepsilon')$ for n-octanol.

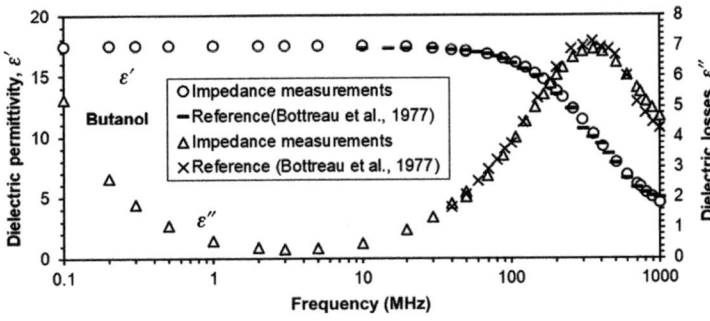

FIGURE 6.3

Real dielectric permittivity ε' and loss factor ε'' versus frequency for n-butanol.

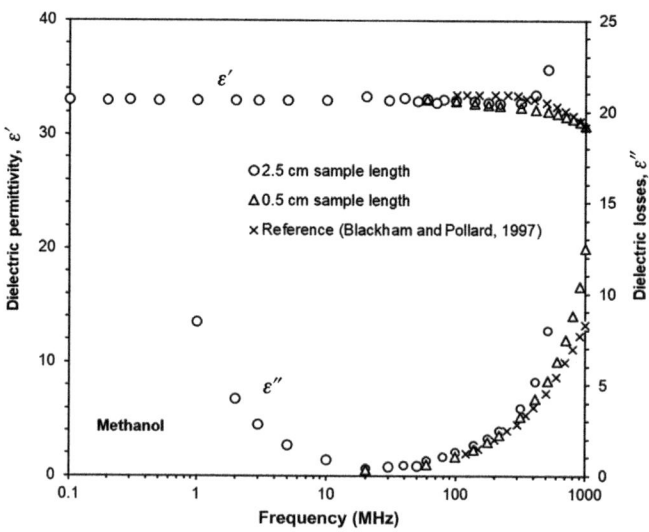

FIGURE 6.4

Real dielectric permittivity ε' and loss factor ε'' versus frequency for methanol.

For methanol, the relaxation process shows up at frequencies above 300 MHz. Up to 300 MHz, the dielectric permittivity remains almost constant, at about $\varepsilon' = 33$ (Burdette et al., 1980; Blackham and Pollard, 1997). The high-frequency range, where we could still obtain correct results, was limited to 350 MHz (Fig. 6.4). In the range from 0.1 to 300 MHz, we calculated a mean dielectric permittivity $\varepsilon' = 33.27$.

Above 350 MHz, our measurements show that ε' increases, which is unreasonable. At $f > 350$ MHz, according to Eq. (4.55), the coaxial dimensions are not small enough compared to the wavelength in this material with $\varepsilon' = 33$. The inductance therefore becomes distributed, and it is no longer appropriate to view it as a lumped parameter. For materials with high dielectric permittivity, such as methanol, using shorter samples allowed us to receive reasonable results at higher frequencies, compared with a full coaxial. The results for a 0.5-cm methanol sample are shown in Fig. 6.4.

Based on our results, materials with high dielectric permittivity can still be measured in a partially filled GR 900, 3 cm coaxial line. The frequency limit depends on the ε' value, in accordance with Eqs. (4.54) and (4 55).

We have shown that lumped-circuit methods for measuring the electrical properties of materials are appropriate over a wide frequency range. When using a coaxial sample holder, the high-frequency limit can be extended from 100 MHz up to 1 GHz for materials with low or moderate dielectric permittivity ($\varepsilon' \leq 25$ at 1 GHz). For materials with high dielectric permittivity, such as methanol, the high-frequency limit is lower (approximately 300 MHz). However, the frequency limit can be increased for high-permittivity materials by using shorter samples in a short coaxial sample holder.

The mean square error of our measuring system typically lies in the range of $\delta = 0.3\% - 1.0\%$. We applied the stray parameter correction procedures to our measured soil samples of various compositions, different locations, and with various water contents. These results will be shown in the subsequent chapters.

6.3 Revealing the Dielectric Polarization in Wet Soils

In addition to corrections for stray parameters, the data processing of our measured data includes a procedure for revealing the dielectric relaxation, which may be obscured in highly conductive materials. This procedure is based on the information that the true experimental sample parameters finally obtained from Eqs. (6.1) and (6.2) or (6.3) and (6.4) represent the total values, affected by two different mechanisms, such as (1) conduction phenomena and (2) polarization processes. In some cases, when the conduction phenomena prevail, the polarization processes are not visible. The possibility of revealing the hidden polarization processes was partly considered in Section 2.3 and Chapter 3, Methods of studying earth materials using alternating electric fields, as well as in Sternberg and Levitskaya (2001). The total loss factor ε''_{total} includes all the energy losses, as follows (Alvarez, 1973; Dukhin and Shilov, 1974; Taherian et al., 1990):

$$\varepsilon''_{total} = \varepsilon''_{diel} + \frac{\sigma'_{ohmic}}{\omega\varepsilon_0} = \frac{\sigma'_{diel} + \sigma'_{ohmic}}{\omega\varepsilon_0} \tag{6.6}$$

Similarly, the total loss tangent ($\tan\delta_{total} = \varepsilon''/\varepsilon'$) can also be expressed as consisting of two components, dielectric and conductive parts, as follows (Bartnikas, 1987):

$$\tan\delta_{total} = \tan\delta_{diel} + \frac{\sigma'_{ohmic}}{\omega\varepsilon'\varepsilon_0}. \tag{6.7}$$

Using Eq. (6.6), it is possible to extract the dielectric part ε''_{diel} by subtracting the conduction or ohmic losses $\varepsilon''_{con} = \sigma_{ohmic}/\omega\varepsilon_0$ from the measured ε''_{total}.

For each measured sample, the data-processing spreadsheet, which we used (Appendices C and D), contains columns with calculated conduction losses ε''_{con} and dielectric losses $\varepsilon''_{diel} = \varepsilon''_{total} - \varepsilon''_{con}$ at all measured frequencies. The values of static or ohmic conductivities σ'_{ohmic} were defined previously, along with the electrode polarization parameters, using the Fricke−Lopatin method described in Section 5.4. From these data, the Cole−Cole arcs $\varepsilon''_{diel} = F(\varepsilon')$ were obtained and the Cole−Cole parameters, ε_0, ε_∞, τ, and α were defined. We used these parameters in equations, shown in Fig. 3.2, for calculating the values of real ε' and imaginary ε'' in the entire frequency range. The Cole−Cole arc drawn with these calculated data approximated the experimental arc $\varepsilon''_{diel} = F(\varepsilon')$.

References

Alvarez, R., 1973. Complex dielectric permittivity in rocks: a method for its measurement and analysis. Geophysics 38 (5), 920–940.

Bartnikas, R., 1987. Engineering dielectrics, vol. 2, Part B. In: Electrical Properties of Solid Insulating Materials: Measurement Technique. ASTM Spec. Tech. Publ., vol. 926. Am. Soc. for Testing and Materials, Philadelphia, PA.

Blackham, D.V., Pollard, R.D., 1997. An improved technique for permittivity measurements using a coaxial probe. IEEE Trans. Instrum. Meas. 46 (5), 1093–1099.

Bottreau, A.M., Dutuit, Y., Moreau, J., 1977. On a multiple reflection time domain method in dielectric spectroscopy: application to the study of some normal primary alcohols. J. Chem. Phys. 66 (8), 3331–3336.

Burdette, E.C., Cain, F.L., Seals, J., 1980. In vivo probe measurement technique for determining dielectric properties at VHF through microwave frequencies. IEEE Trans. Microwave Theory Tech. 28 (4), 414–427.

Chahine, R., Bose, T.K., 1980. Comparative studies of various methods in time domain spectroscopy. J. Chem. Phys. 72 (2), 808–815.

Dalbert, M.M., Magat, M., Surdut, A., 1949. Dispersion dielectrique dans les alcools normaux. Bull. Soc. Chim. Fr. D345–D351.

Dukhin, S.S., Shilov, V.N., 1974. Dielectric Phenomena and Double Layer in Disperse Systems and Electrolytes. Halsted, New York.

Kay, R.L., 1973. Ionic transport in water and in mixed aqueous solventsChap. 4 In: Franks, F. (Ed.), Water: A ComprehensiveTreatise, vol. 3. Plenum, New York.

Nakamura, H., Mashimo, S., Wada, A., 1982. Precise and easy method of TDR to obtain dielectric relaxation spectra in GHz region. Jpn. J. Appl. Phys. 21 (7), 1022–1024.

Scott, W.R.J., Smith, G.S., 1986. Dielectric spectroscopy using monopole antennas of general electrical length. IEEE Trans. Antennas Propag. 34 (7), 919–929.

Smyth, C.P., 1955. Dielectric Behavior and Structure. McGraw-Hill, New York.

Sternberg, B.K., Levitskaya, T.M., 2001. Electrical parameters of soils in the frequency range from 1 kHz to 1 GHz, using lumped-circuit methods. Radio Sci. 36 (4), 709–719.

Taherian, M.R., Kenyon, W.E., Safinya, K.A., 1990. Measurement of dielectric response of water saturated rocks. Geophysics 55, 1530–1541.

Von Hippel, A.R., 1954. Dielectrics and Waves. John Wiley, New York.

Soils from Avra Valley, Arizona

Chapter Outline

The Avra Valley samples are from the Laboratory for Advanced Subsurface Imaging Test Site, located 32 km (20 mi.) southwest of the University of Arizona Campus at 11,415 W Ajo Way. The test site is described in detail in Sternberg et al. (1991).

The soil in Avra Valley is a relatively uniform alluvium. A sample of the soils that we used for our electrical properties measurements was sent to the Soil, Water, and Plant Analysis Laboratory in the Department of Soil, Water, and Environmental Science at the University of Arizona. Using the hydrometer method, they found that this soil contains 61% sand, 22% silt, and 17% clay (by weight). Note that this measurement is a particle-size analysis. We do not have a mineralogy analysis of the clay fraction.

We studied samples with both natural moisture and samples prepared in the laboratory with varying moisture, as described in Chapter 6, Corrections for stray parameters and error estimation. For the natural moisture tests, samples from Avra Valley were collected within a very short time after excavation (a few minutes). Using this procedure, excellent agreement was obtained between laboratory and field resistivity measurements. From this, we conclude that although the samples have been disturbed, they are representative of the in situ electrical properties (Sternberg and Levitskaya, 1998; Sternberg and Birken, 1999, and Chapter 12).

7.1 Complete Electrical Properties for Five Water Saturations

The measurement results obtained in the entire frequency range from 1 kHz to 1 GHz and the full data procedures are available in spreadsheets, for each sample

Electrical Spectroscopy of Earth Materials. DOI: https://doi.org/10.1016/B978-0-12-818603-9.00007-6

(Appendix D). Here we will present the experimental data in graphic form for samples prepared in the laboratory with five values of volume wetness (about 6%, 10%, 16%, 25%, and 45%). For each water content, we display seven figures. We show spreadsheets with the parameters that we calculated in Appendix D and another spreadsheet with the cell equations that we used for the calculations in Appendix E.

The samples were measured in two different sample holders: (1) with disk electrodes, using the Impedance Analyzer HP4194A Fixture in the frequency range 0.001–40 MHz, and (2) with coaxial electrodes, using the Impedance Analyzer HP4191A at frequencies 1–1000 MHz. The sample volume water contents (W_v) in these two sample holders may differ slightly. When analyzing the experimental results for the entire frequency range, we used the average of the two W_v values.

For each figure number, we use the following notation:

1. The chapter number (7).
2. Followed by a number (1, 2, 3, 4, 5, 6, 7) corresponding to the electrical properties being plotted (1—RMeas, 2—XMeas, 3—permittivity, 4—conductivity, 5—resistivity, 6—Argand diagram, 7—Cole–Cole diagram).
3. Followed finally by a letter (a, b, c, d, e) corresponding to the water content W_v (a—6%, b—10%, c—16%, d—25%, and e—45%).

Fig. 7.1A shows the measured resistance R_{meas} versus $100/\omega^{0.5}$ at low frequencies. The graph is linear within a good approximation ($R^2 = 0.9983$). The trend line equation, shown on this figure, allows us to define the electrode polarization resistance R_{el} versus frequency, as well as the static resistance $R_s = 907\ \Omega$. The procedure of using this "Fricke–Lopatin method" (FL method) in data processing is described in Chapter 5, Stray parameters of the measuring system and ways of defining them. From R_s and the sample geometry, we obtained the static resistivity $\rho_s = 193\ \Omega$-m and conductivity $\sigma_s = 0.0052$ S/m.

Analogous relationships of R_{meas} versus $100/\omega^{0.5}$ for samples with other values of W_v (10%, 16%, 25%, and 45%) are shown in Fig. 7.1A–E along with the regression equations. The same procedures of defining the electrode polarization resistance R_{el} versus frequency, as well as the static resistance R_s, resistivity ρ_s, and conductivity σ_s, were performed in each case. The static parameters obtained and the electrode arrangements for all the samples are shown in Table 7.1. The sample dimensions are given in spreadsheets.

Similarly, Fig. 7.2A–E shows the FL method for measured reactance X in the same frequency range for each water content. From the trend line equations, we defined the electrode polarization reactance X_{el} versus frequency for each sample.

Fig. 7.3A–E represents the real ε' and imaginary ε'' dielectric permittivity versus frequency for each Avra Valley sample. As seen, the real relative permittivity ε' and the loss factor ε'' decrease with increasing frequency, not

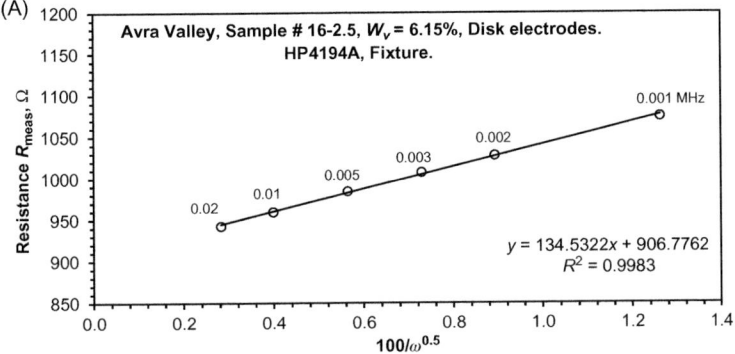

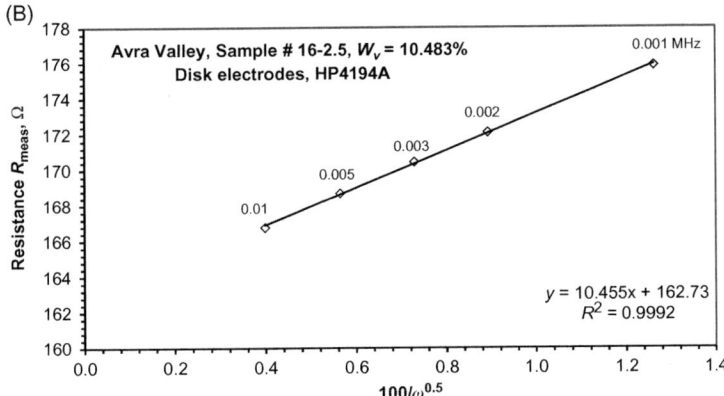

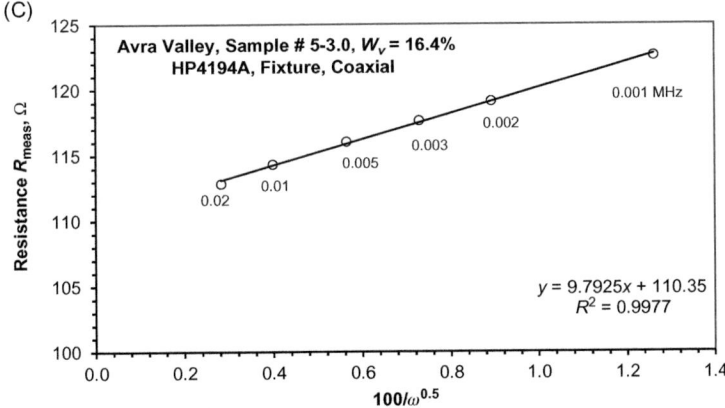

FIGURE 7.1

(A) Measured resistance versus $100/\omega^{0.5}$ for a soil sample with $W_v = 6.152\%$.
(B) Measured resistance versus $100/\omega^{0.5}$ for a soil sample with $W_v = 10.483\%$.
(C) Measured resistance versus $100/\omega^{0.5}$ for a soil sample with $W_v = 16.372\%$.

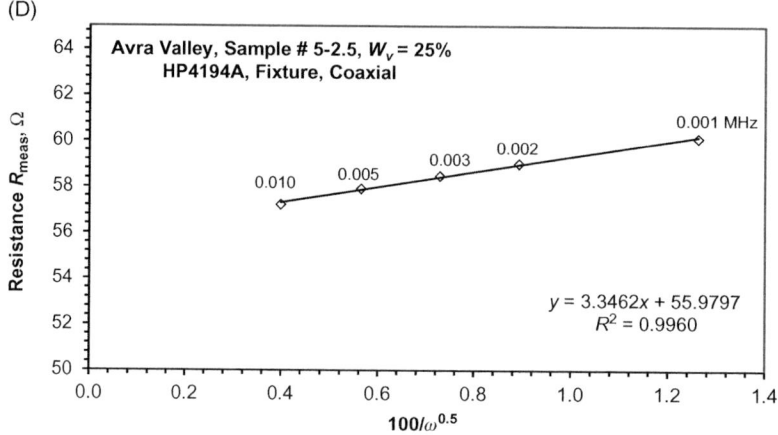

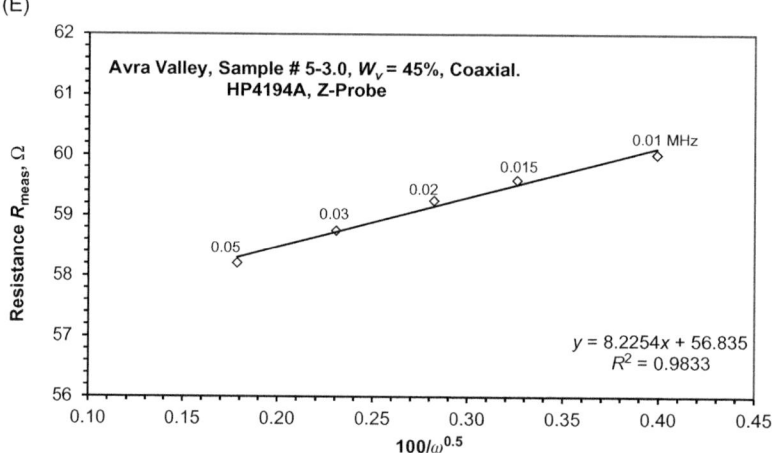

FIGURE 7.1 Continued

(D) Measured resistance versus $100/\omega^{0.5}$ for a soil sample with $W_v = 24.926\%$.
(E) Measured resistance versus $100/\omega^{0.5}$ for a soil sample with $W_v = 45.470\%$.

Table 7.1 Static Parameters Defined for Soil Samples From Avra Valley With Various Water Contents W_v by Using the Fricke–Lopatin Method

Water Content W_v, %	Instrument and Electrodes	Static Resistance R_s, Ω	Static Resistivity ρ_s, Ω m	Static Conductivity, σ_s, S/m
6.152	HP4194A, Fixture, Disks	906.7762	192.8356	0.005186
10.483	HP4194A, Fixture, Disks	162.73	34.5439	0.02895
16.372	HP4194A, Fixture, Coaxial	110.35	21.4972	0.04652
24.926	HP4194A, Fixture, Coaxial	55.98	10.9054	0.09170
45.470	HP4194A, Z-Probe, Coaxial	6.835	11.0720	0.09032

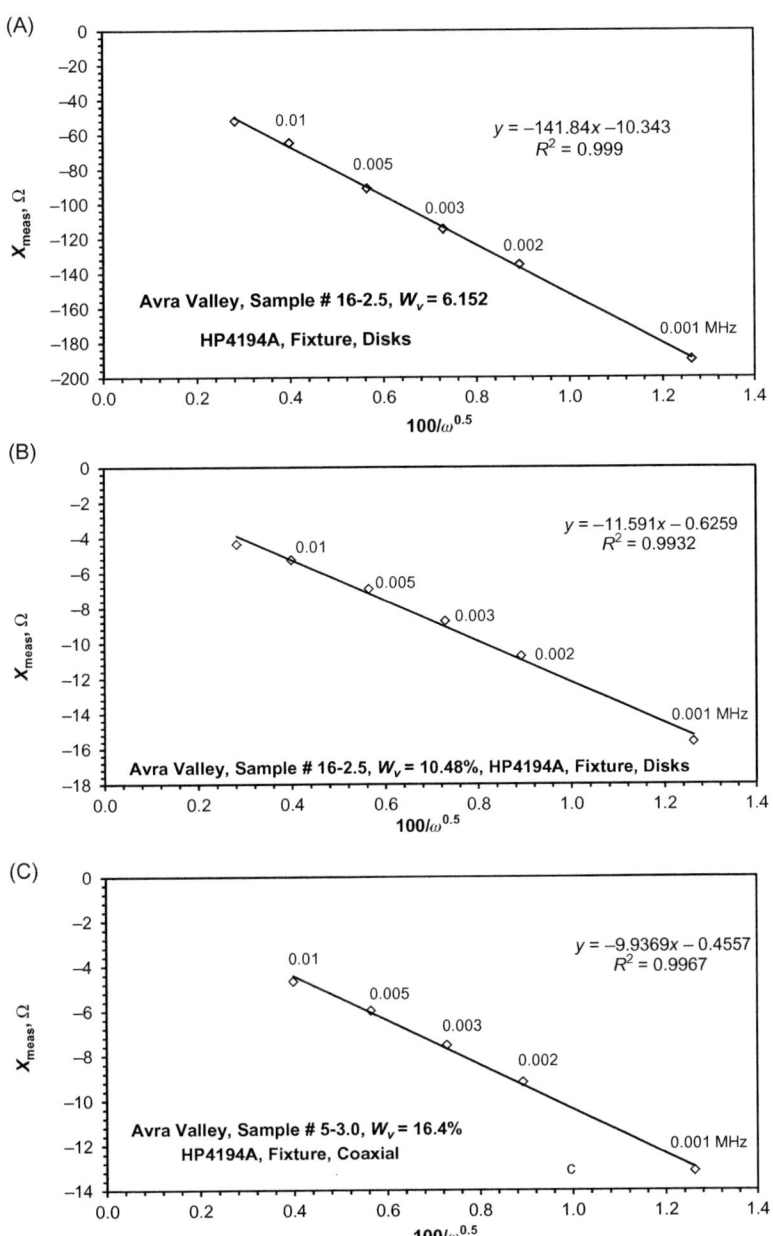

FIGURE 7.2

(A) Measured reactance versus $100/\omega^{0.5}$ for a soil sample with $W_v = 6.152\%$.
(B) Measured reactance versus $100/\omega^{0.5}$ for a soil sample with $W_v = 10.483\%$.
(C) Measured reactance versus $100/\omega^{0.5}$ for a soil sample with $W_v = 16.372\%$.

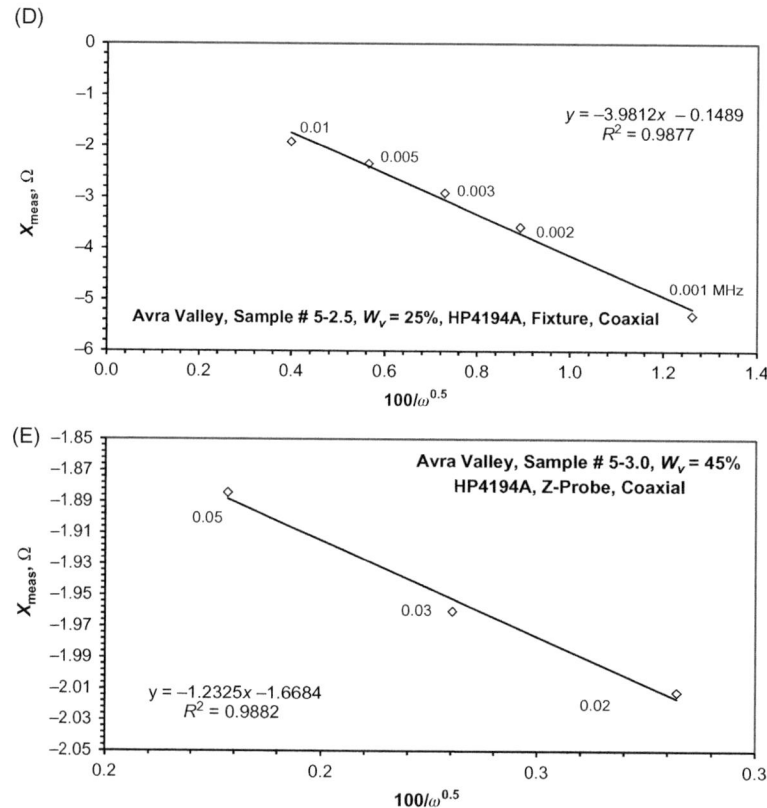

FIGURE 7.2 Continued

(D) Measured reactance versus $100/\omega^{0.5}$ for a soil sample with $W_v = 24.926\%$.
(E) Measured reactance versus $100/\omega^{0.5}$ for a soil sample with $W_v = 45.470\%$.

showing any dielectric relaxation, which is obscured by the conductivity. The loss factor ε'' versus frequency can be approximated rather well for all measured samples with a power function (Fig. 7.3A−E). Using the procedure for revealing the dielectric polarization, described in Chapter 6, Corrections for stray parameters and error estimation, we defined the dielectric component ε''_{diel} of the total loss factor ε''_{total}, obtained from the measurements. The frequency dependencies of ε''_{diel} are also shown in Fig. 7.3A−E. The relaxation maximums on these graphs indicate the existence of a dielectric polarization process.

The conductivity in the Avra Valley moist soils is rather high. The frequency dependencies of conductivity are shown in Fig. 7.4A−E. The real conductivity σ' is almost independent on frequency up to 1 MHz. Above 1 MHz, σ'

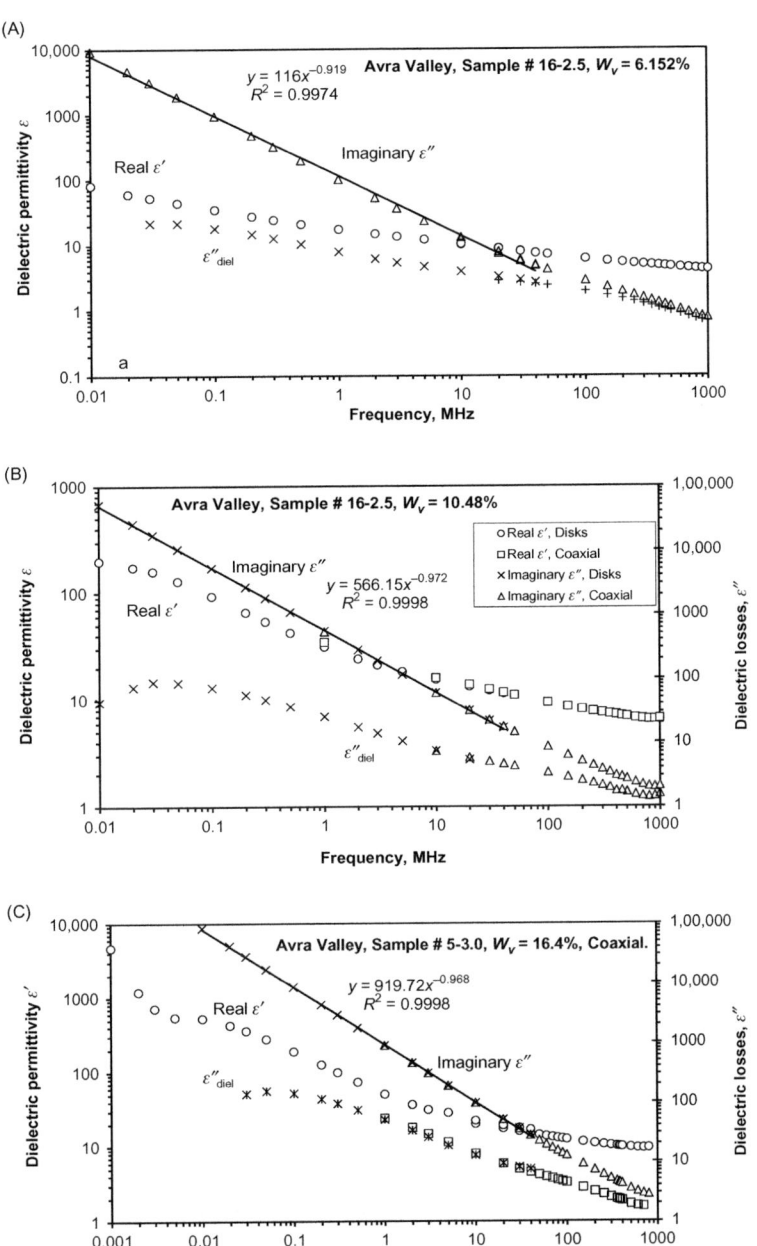

FIGURE 7.3

(A) Dielectric permittivity versus frequency for a soil sample with $W_v = 6.152\%$.

(B) Dielectric permittivity versus frequency for a soil sample with $W_v = 10.483\%$.

(C) Dielectric permittivity versus frequency for a soil sample with $W_v = 16.372\%$.

(D)

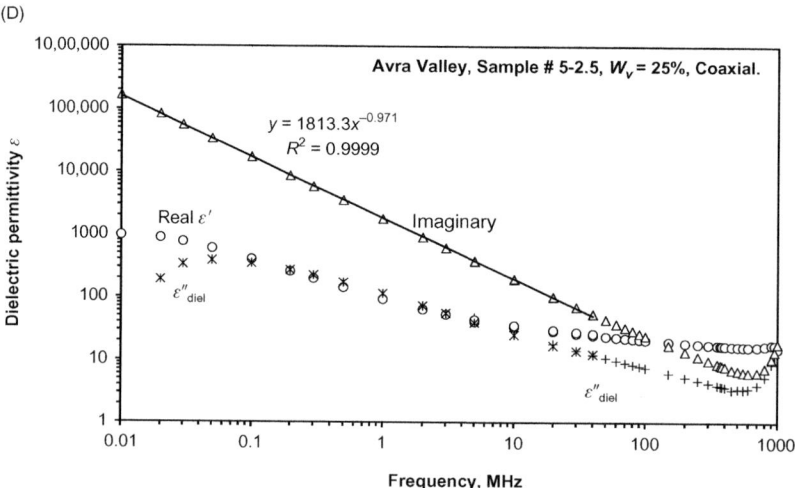

(E)

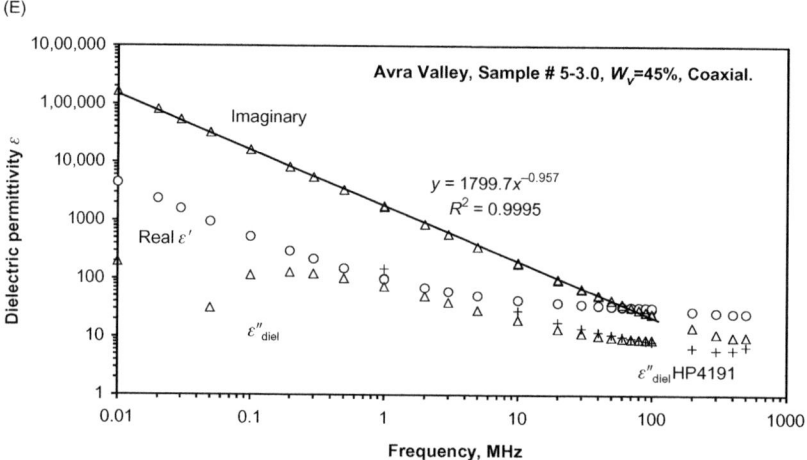

FIGURE 7.3 Continued

(D) Dielectric permittivity versus frequency for a soil sample with $W_v = 24.926\%$.
(E) Dielectric permittivity versus frequency for a soil sample with $W_v = 45.470\%$.

is increasing. The imaginary conductivity σ'' increases continuously, except very low frequencies from 0.001 to 0.01 MHz, where some samples show scattered data, others have a dip. At frequencies above 1 MHz, the imaginary conductivity σ'' versus frequency was also approximated with a power function, shown for each sample on its graph. For the sample with lowest water content (6%), the power function was applicable even below 1 MHz, down to 0.02 MHz.

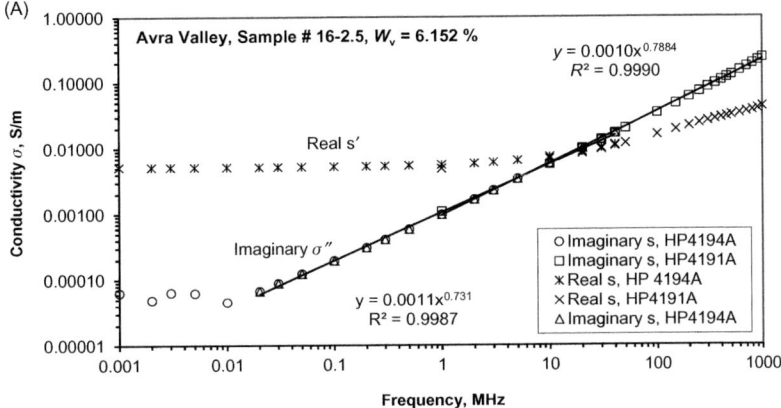

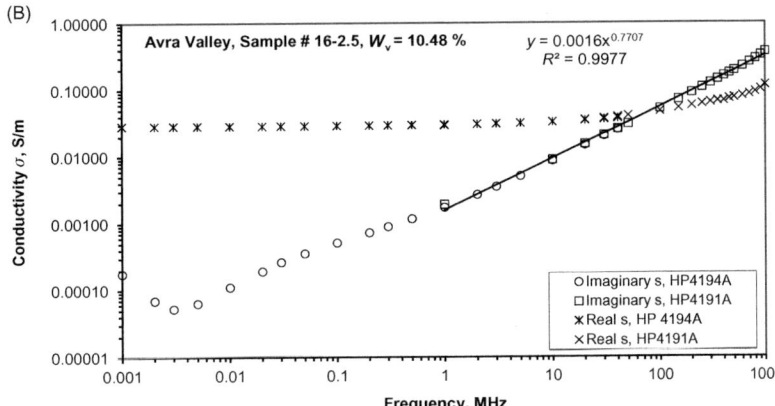

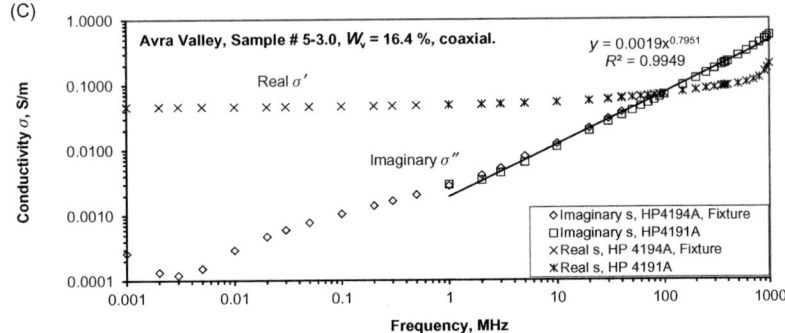

FIGURE 7.4

(A) Conductivity versus frequency for a soil sample with $W_v = 6.152\%$. (B) Conductivity versus frequency for a soil sample with $W_v = 10.483\%$. (C) Conductivity versus frequency for a soil sample with $W_v = 16.372\%$.

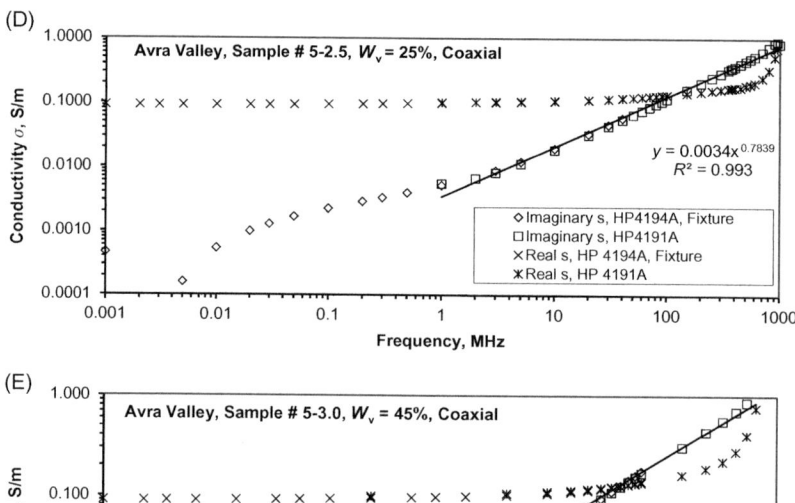

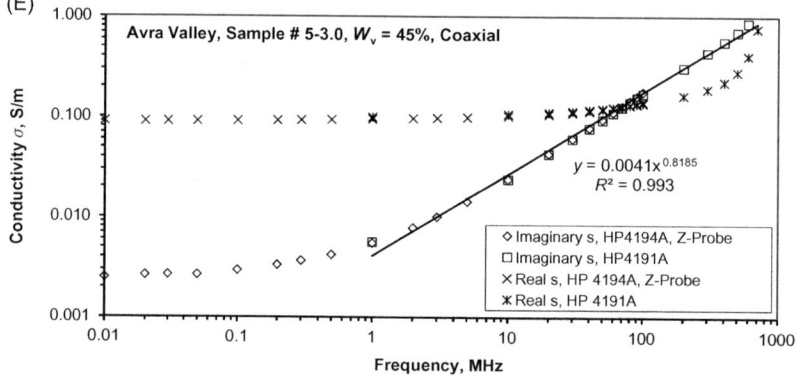

FIGURE 7.4 Continued

(D) Conductivity versus frequency for a soil sample with $W_v = 24.926\%$. (E) Conductivity versus frequency for a soil sample with $W_v = 45.470\%$.

These results are in agreement with some references, where the power law of dielectric and conduction response in an electromagnetic field of various materials is discussed (Knight, 1983; Jonscher, 1983, 1996). In particular, Jonscher (1983) showed, based on many experimental results, that the dielectric response obeys the power law in a very wide range of dielectric materials, which he called the "universal power law."

Fig. 7.5A–E shows the frequency dependencies of real ρ' and imaginary ρ'' components of complex resistivity for all Avra Valley samples. As seen, the real resistivity reveals a dispersion at some frequency above 1 MHz, characteristic for each sample, while the imaginary resistivity is passing through a maximum at the same frequency, both indicating the existence of a relaxation process.

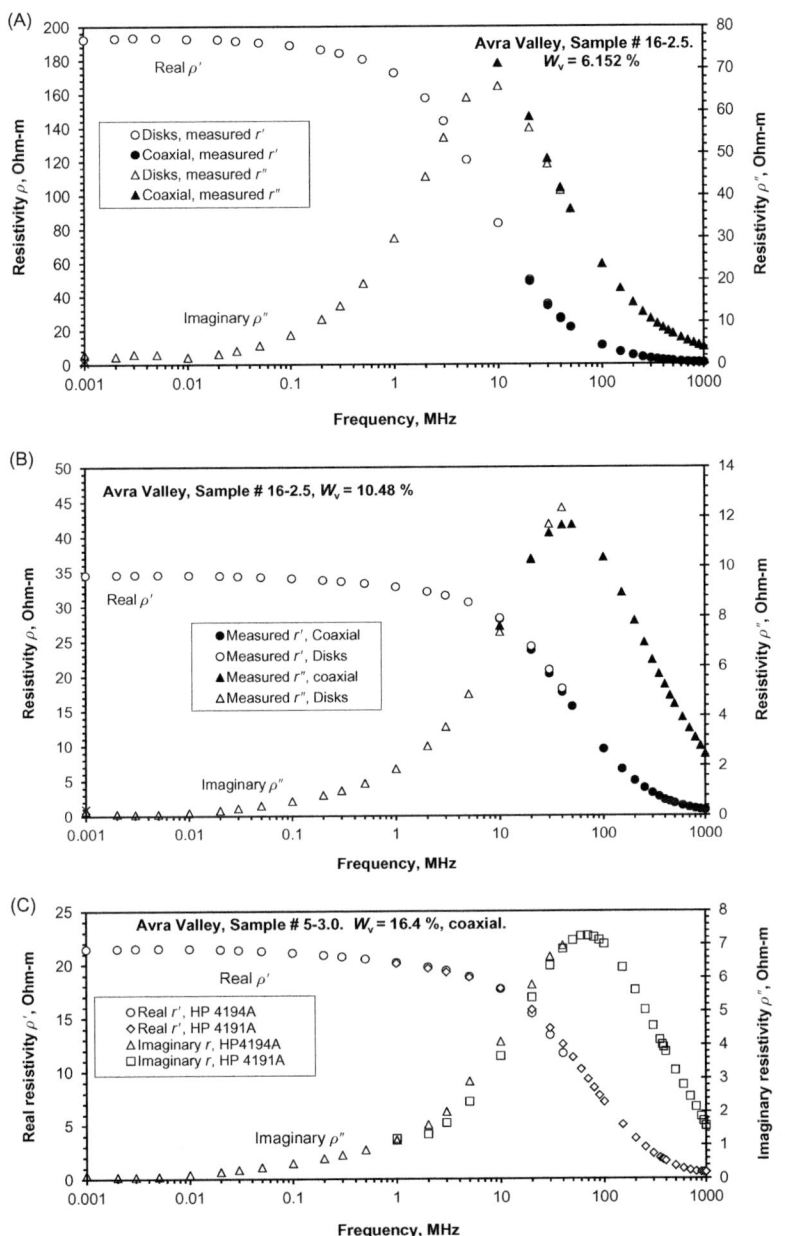

FIGURE 7.5

(A) Resistivity versus frequency for a soil sample with $W_v = 6.152\%$. (B) Resistivity versus frequency for a soil sample with $W_v = 10.483\%$. (C) Resistivity versus frequency for a soil sample with $W_v = 16.372\%$.

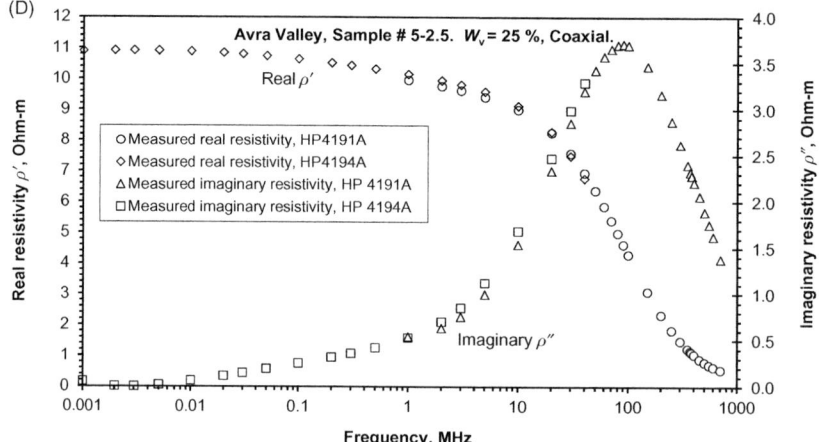

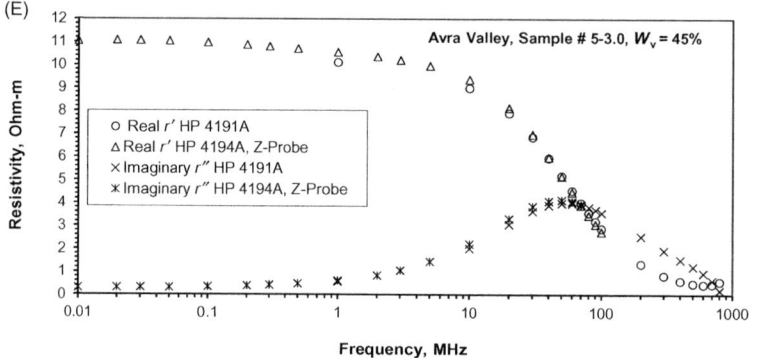

FIGURE 7.5 Continued

(D) Resistivity versus frequency for a soil sample with $W_v = 24.926\%$. (E) Resistivity versus frequency for a soil sample with $W_v = 45.470\%$.

The Argand diagrams drawn from these data are shown in Fig. 7.6A—E. They have an arc form, with the center below the abscissa. The higher the soil wetness W_v, the higher is the frequency of relaxation, at which the maximum of ρ'' occurs. The parameters of the Argand diagrams are summarized in Table 7.2.

Fig. 7.7A—E shows the Cole—Cole diagrams, revealed for the described samples, by using the procedure of removing the conduction losses σ''_{con} from the total values σ''_{total} (Chapter 6: Corrections for stray parameters and error estimation). Table 7.3 represents the Cole—Cole parameters for the Avra Valley samples measured in the full frequency range (Figs. 7.8—7.24).

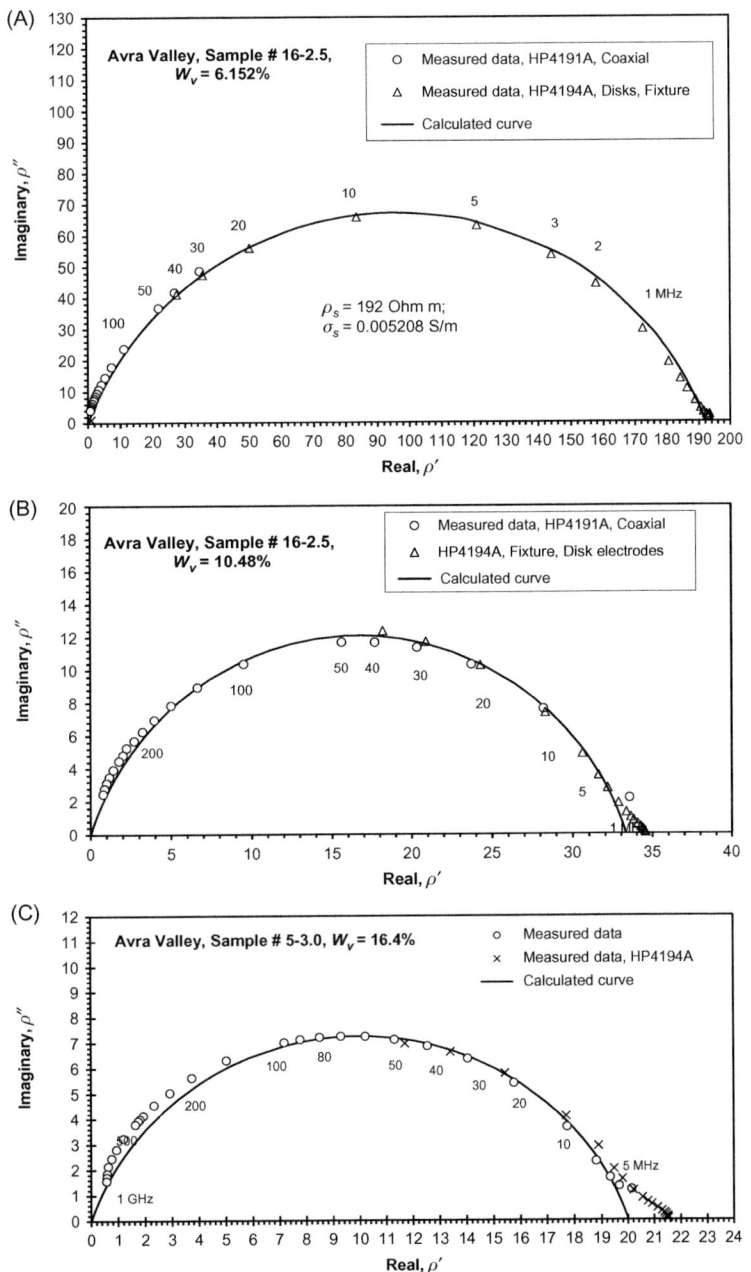

FIGURE 7.6

(A) Argand diagram for a soil sample with $W_v = 6.152\%$. (B) Argand diagram for a soil sample with $W_v = 10.483\%$. (C) Argand diagram for a soil sample with $W_v = 16.372\%$.

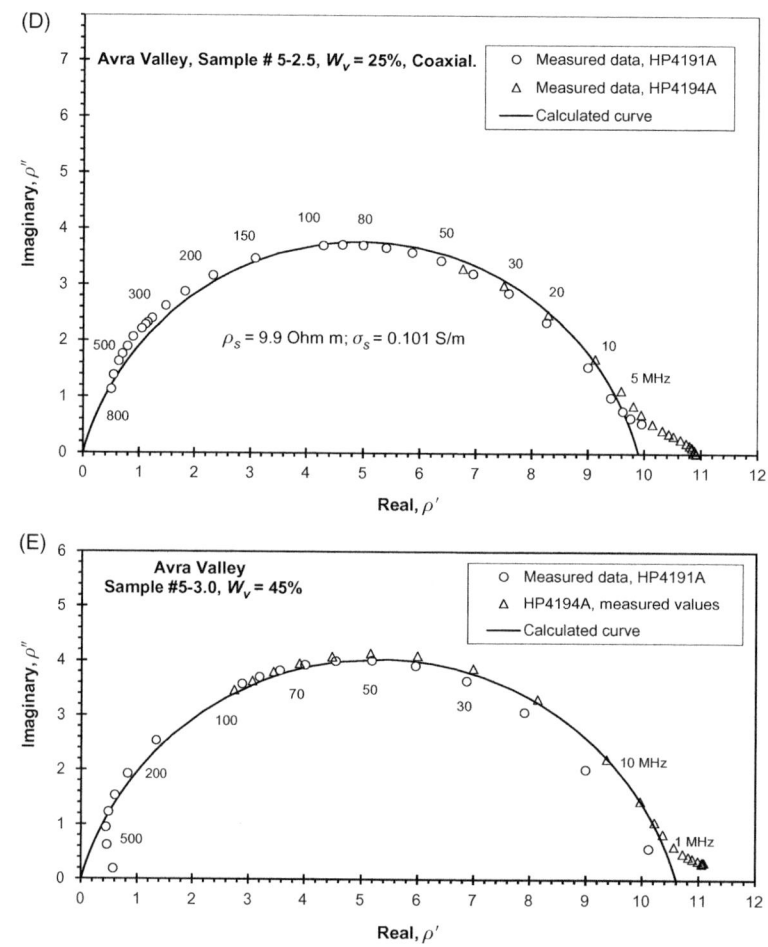

FIGURE 7.6 Continued

(D) Argand diagram for a soil sample with $W_v = 24.926\%$. (E) Argand diagram for a soil sample with $W_v = 45.470\%$.

Table 7.2 Argand Diagrams Parameters for Avra Valley Soils

Sample Wetness, W_v, %	ρ_0, Ω m	ρ_∞, Ω m	Distribution Parameter, α	Relaxation Time τ, s	Static Conductivity, σ_s, S/m
6.152	192.0	0	0.2222	$2.001 \cdot 10^{-8}$	0.0052
10.483	33.3	0	0.2000	$3.466 \cdot 10^{-9}$	0.0300
16.372	20.0	0	0.2000	$2.520 \cdot 10^{-9}$	0.0500
24.926	9.9	0	0.1722	$1.940 \cdot 10^{-9}$	0.1010
45.470	10.6	0	0.1722	$1.510 \cdot 10^{-9}$	0.0944

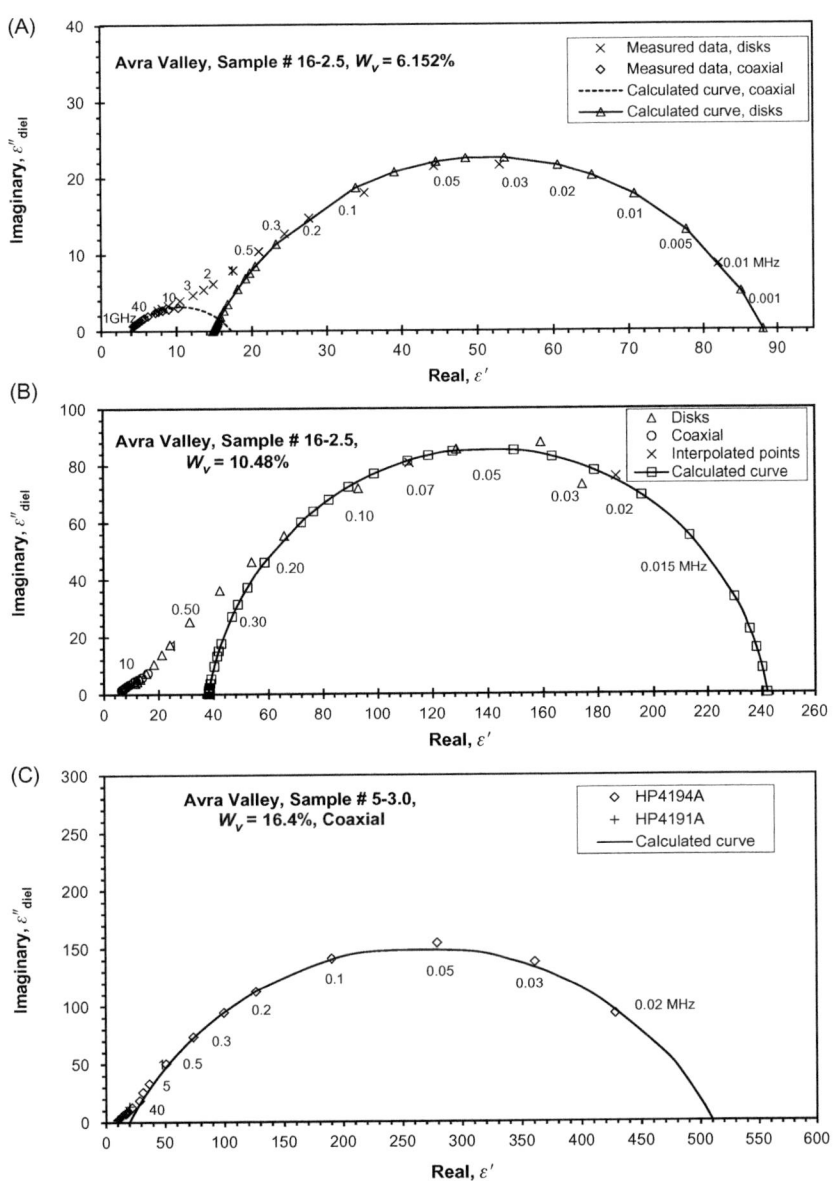

FIGURE 7.7

(A) Cole−Cole diagram revealed for a soil sample with $W_v = 6.152\%$. (B) Cole−Cole diagram revealed for a soil sample with $W_v = 10.483\%$. (C) Cole−Cole diagram revealed for a soil sample with $W_v = 16.372\%$.

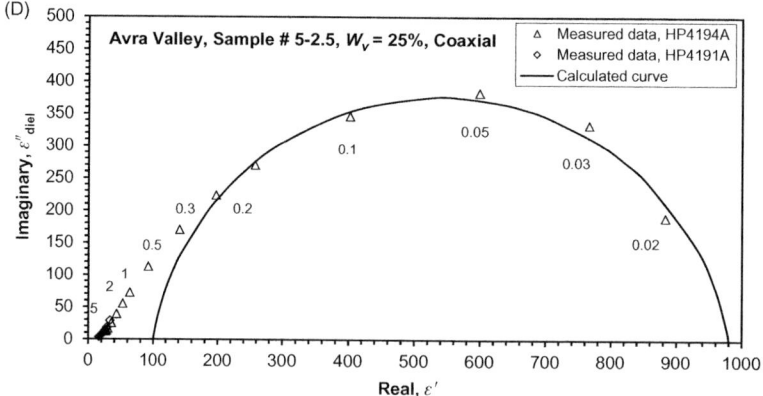

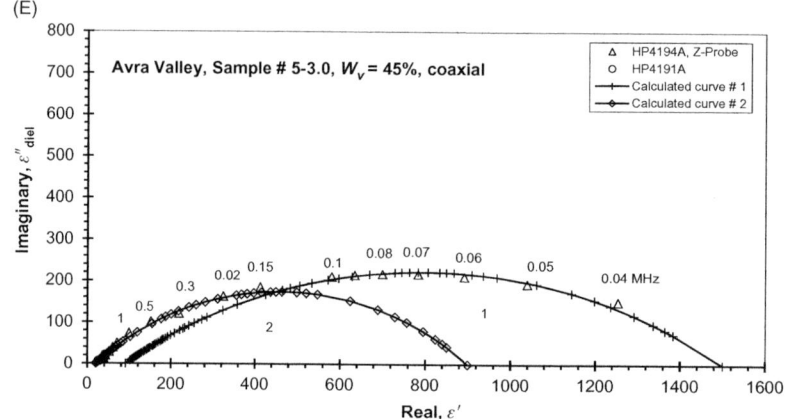

FIGURE 7.7 Continued

(D) Cole—Cole diagram revealed for a soil sample with $W_v = 24.926\%$. (E) Cole—Cole diagram revealed for a soil sample with $W_v = 45.470\%$.

Table 7.3 Cole—Cole Diagram Parameters for Avra Valley Soils

Sample Wetness W_v, %	ε_s	ε_∞	Distribution Parameter, α	Relaxation Time τ, s
6.152	88.0	3.8	0.294	$4.687 \cdot 10^{-6}$
10.483	242.0	5.0	0.111	$4.683 \cdot 10^{-6}$
16.372	510.0	7.0	0.300	$1.140 \cdot 10^{-7}$
24.926	980.0	10.0	0.100	$2.410 \cdot 10^{-6}$
45.470	1500	20.0	0.611	$2.460 \cdot 10^{-6}$

7.2 Supplemental Soil Samples From Avra Valley, Arizona

We also measured some supplemental samples over restricted frequency intervals, in order to obtain more data on static resistivity ρ_{st} and Cole–Cole parameters of dielectric permittivity. Most of the supplemental samples were measured in a parallel-plate sample holder with the HP4194A Impedance Analyzer, using the fixture, at frequencies from 0.001 to 40 MHz. Only one supplemental sample ($W_v = 11.342\%$) was measured in a coaxial sample holder at frequencies from 0.001 to 1000 MHz. Another one ($W_v = 5.791\%$) was measured in a coaxial only in a high-frequency range from 1 to 1000 MHz for defining Argand and Cole–Cole diagram parameters. Table 7.4 shows the electrode arrangements and instruments for the supplemental samples with various water contents W_v, along with their static parameters, obtained from the FL method. Tables 7.5 and 7.6 show the parameters of Argand and Cole–Cole diagrams for the same supplemental samples. In this section, we have organized the supplemental soil figures (7.8 - 7.24) as follows: chapter number, section number corresponding to the water content value (W_v), and a letter corresponding to the electrical property being plotted.

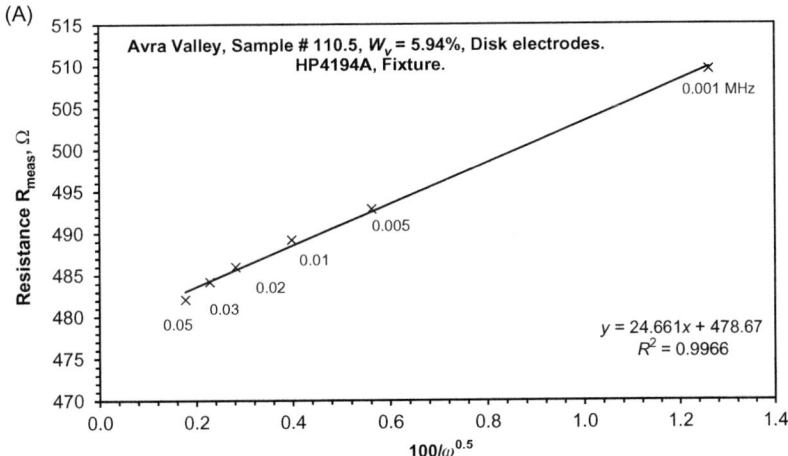

FIGURE 7.8

(A) Measured resistance versus $100/\omega^{0.5}$ for a soil sample with $W_v = 5.94\%$.

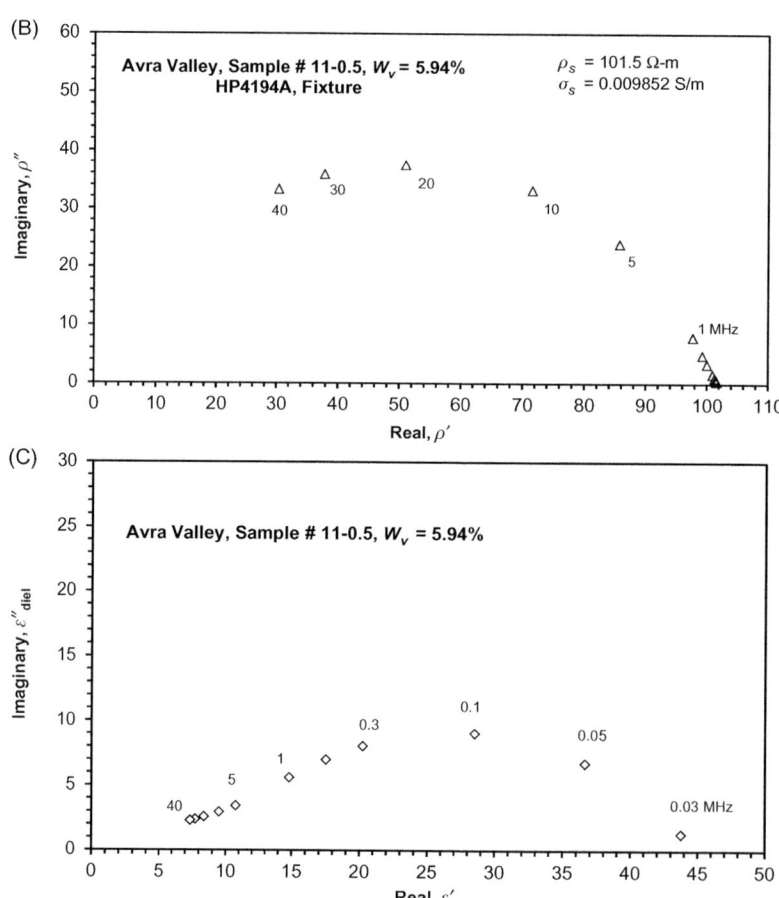

FIGURE 7.8 Continued

(B) Argand diagram for a soil sample with $W_v = 5.94\%$. (C) Cole–Cole diagram revealed for a soil sample with $W_v = 5.94\%$.

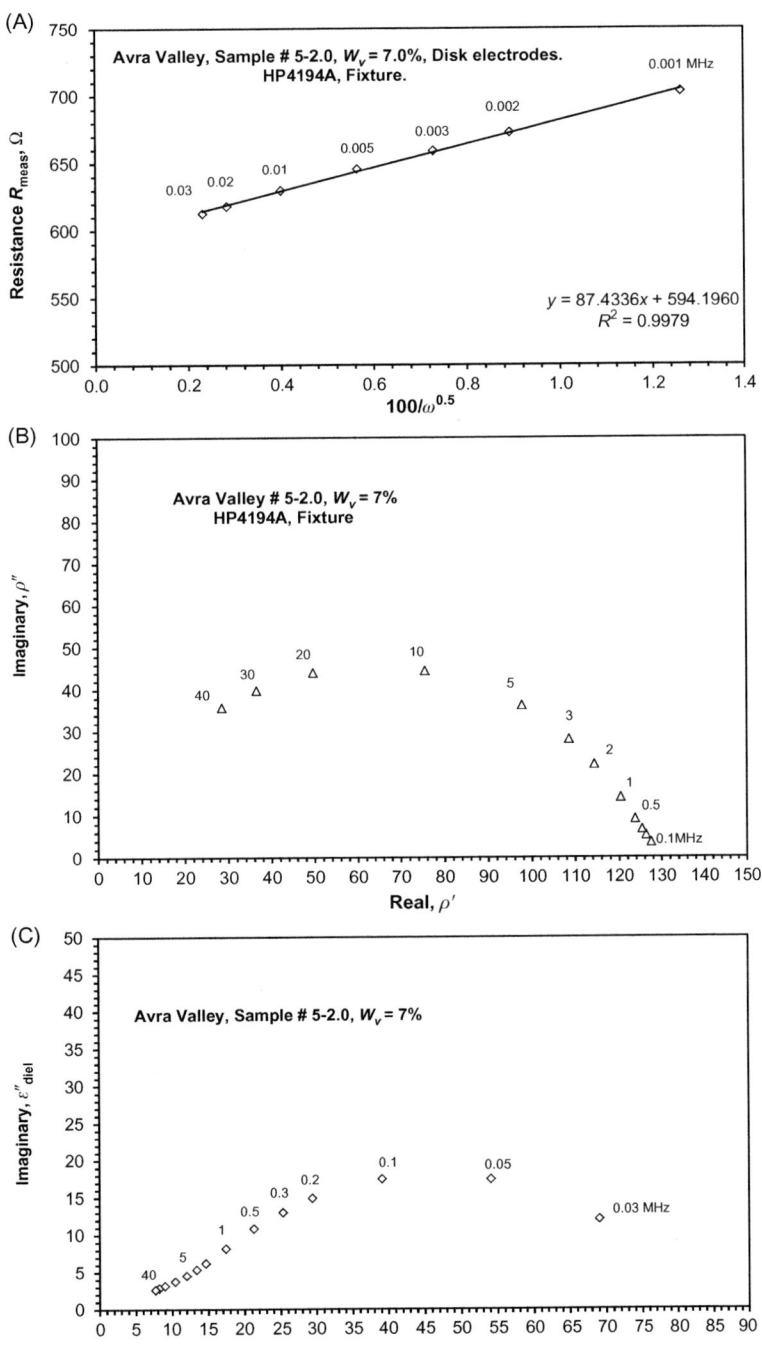

FIGURE 7.9

(A) Measured resistance versus $100/w^{0.5}$ for a soil sample with $W_v = 7\%$. (B) Argand diagram for a soil sample with $W_v = 7\%$. (C) Cole–Cole diagram revealed for a soil sample with $W_v = 7\%$.

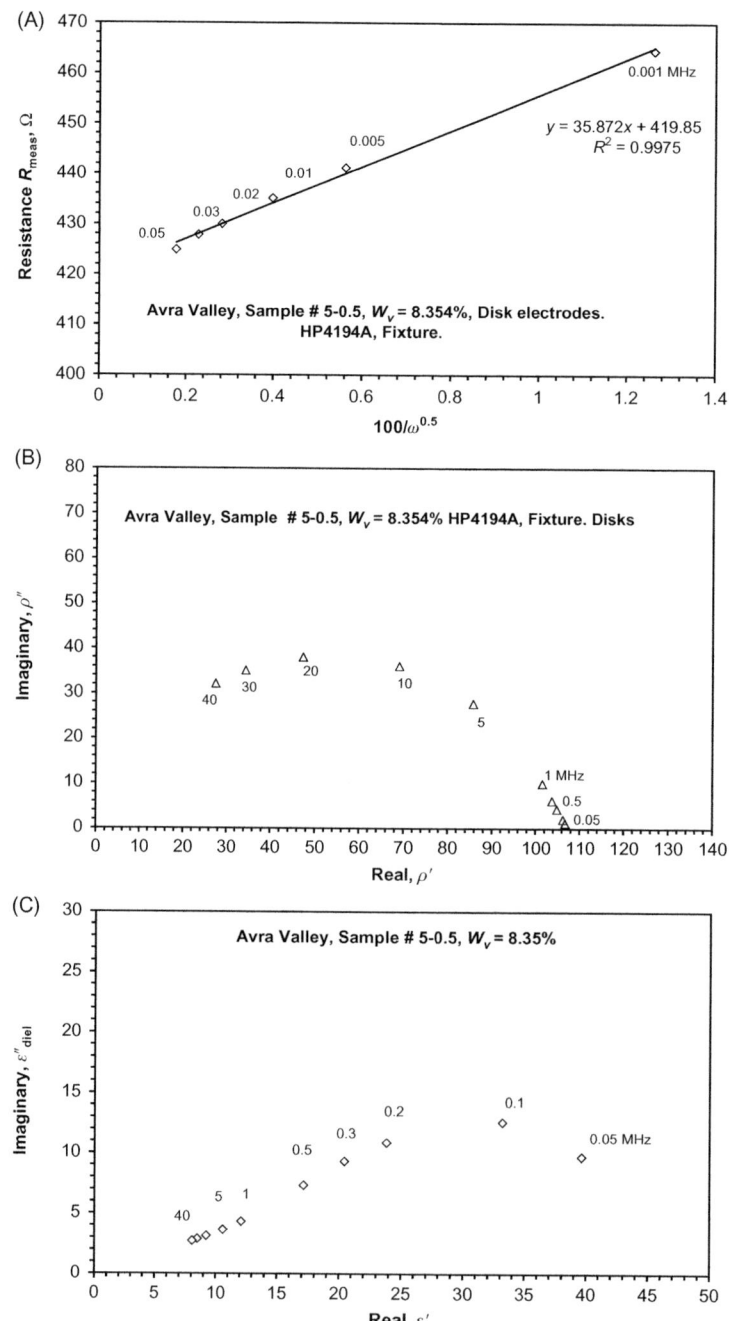

FIGURE 7.10

(A) Measured resistance versus $100/\omega^{0.5}$ for a soil sample with $W_v = 8.35\%$. (B) Argand diagram for a soil sample with $W_v = 8.35\%$. (C) Cole–Cole diagram revealed for a soil sample with $W_v = 8.35\%$.

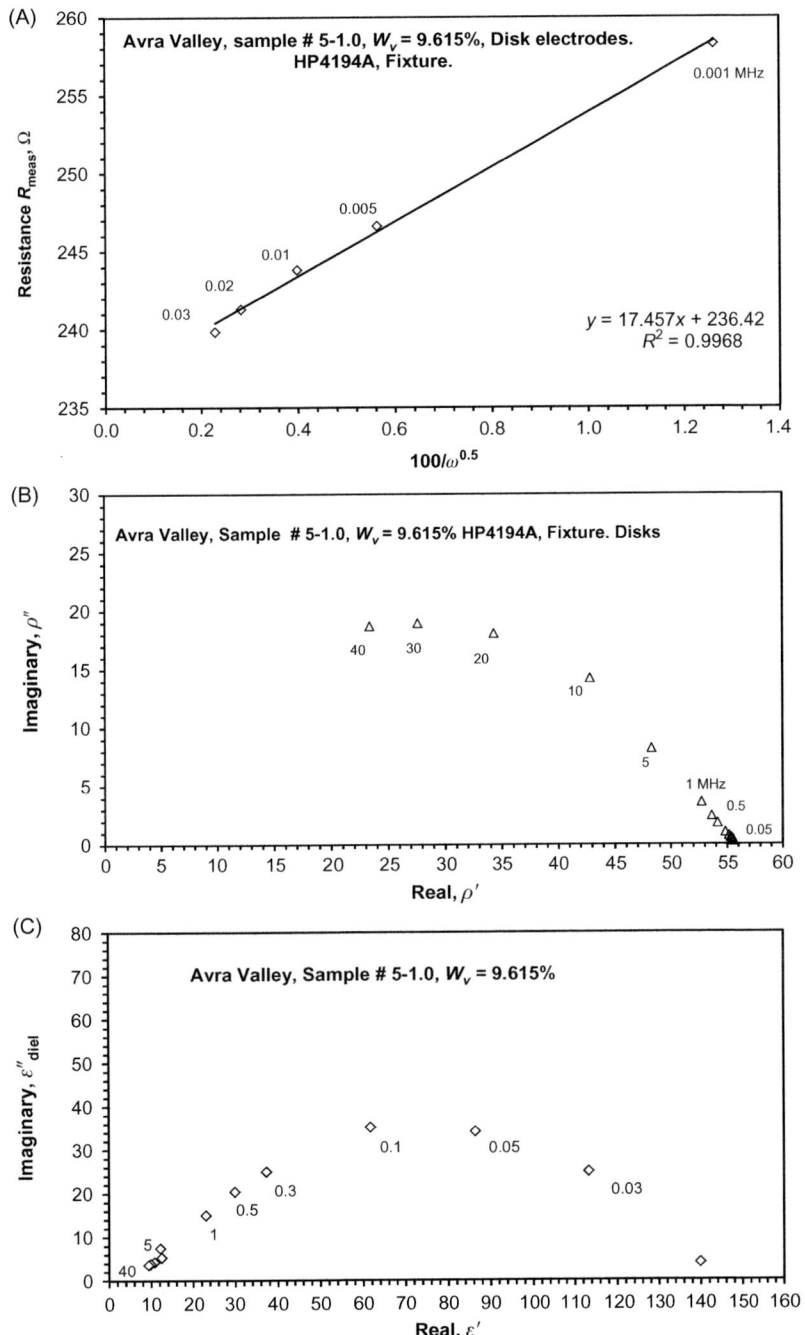

FIGURE 7.11

(A) Measured resistance versus $100/\omega^{0.5}$ for a soil sample with $W_v = 9.615\%$. (B) Argand diagram for a soil sample with $W_v = 9.615\%$. (C) Cole–Cole diagram revealed for a soil sample with $W_v = 9.615\%$.

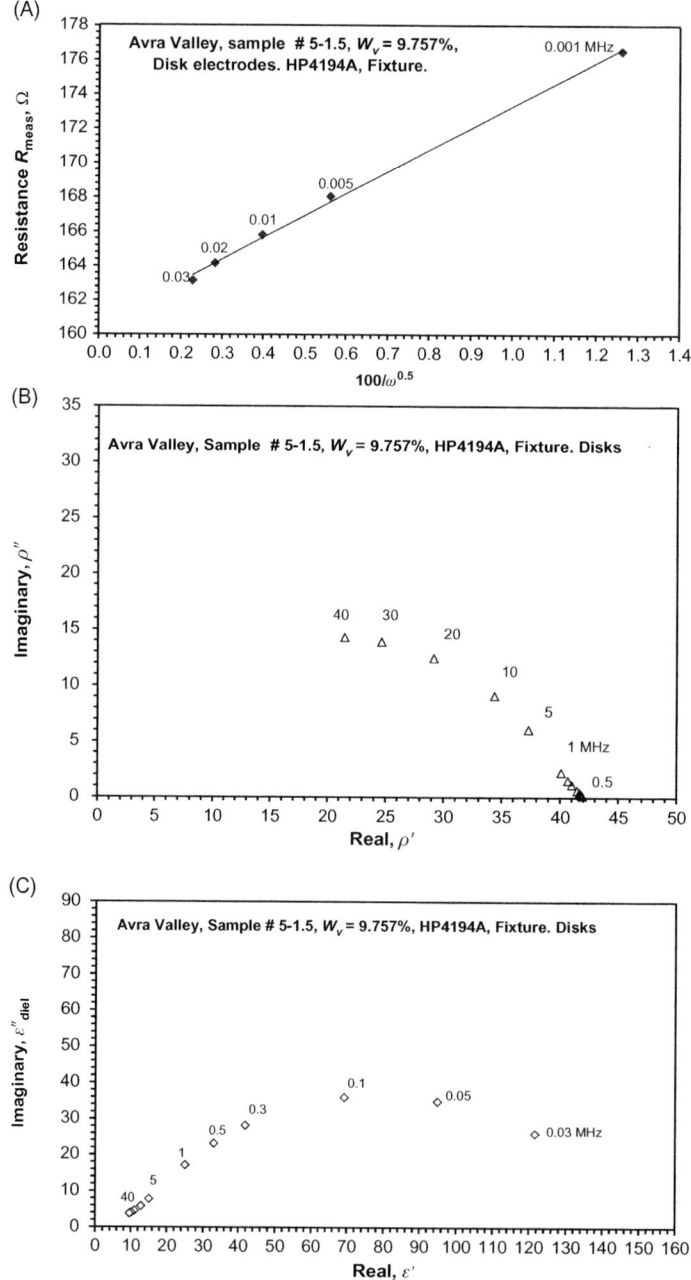

FIGURE 7.12

(A) Measured resistance versus $100/\omega^{0.5}$ for a soil sample with $W_v = 9.757\%$. (B) Argand diagram for a soil sample with $W_v = 9.757\%$. (C) Cole–Cole diagram revealed for a soil sample with $W_v = 9.757\%$.

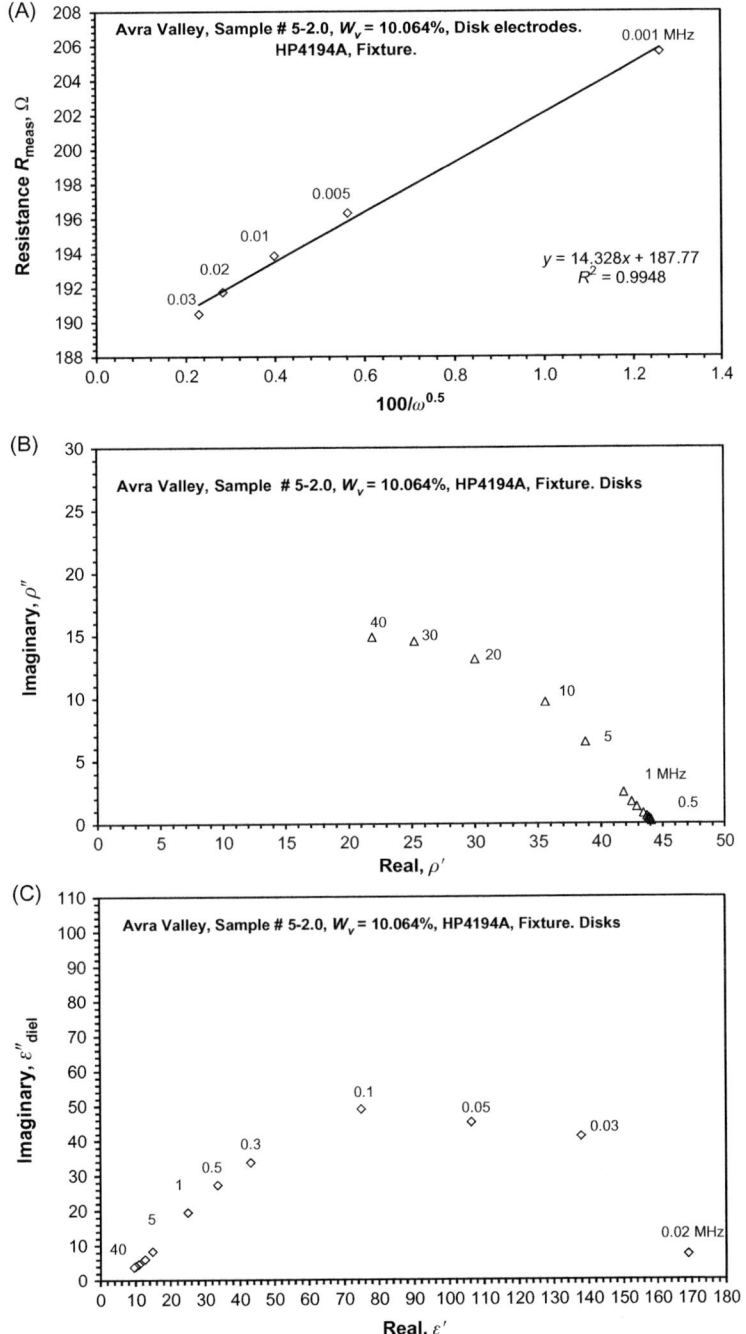

FIGURE 7.13

(A) Measured resistance versus $100/\omega^{0.5}$ for a soil sample with $W_v = 10.064\%$. (B) Argand diagram for a soil sample with $W_v = 10.064\%$. (C) Cole–Cole diagram revealed for a soil sample with $W_v = 10.064\%$.

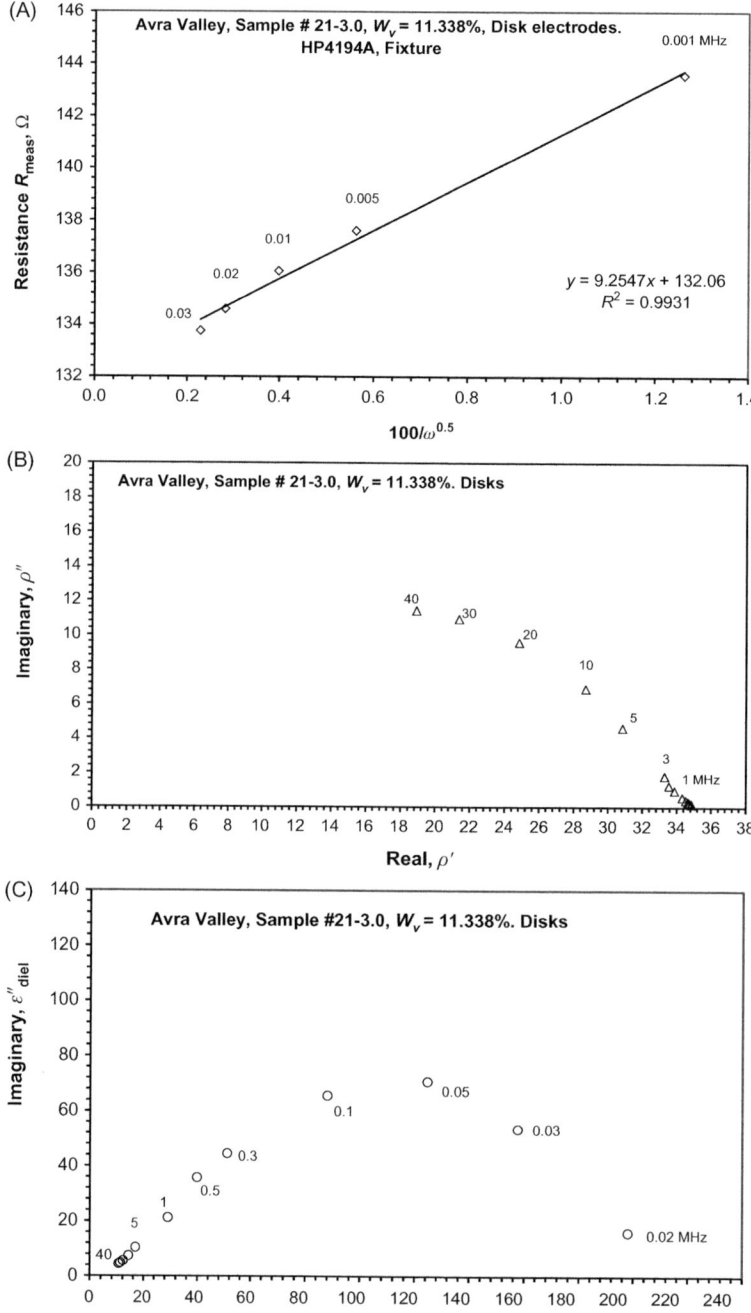

FIGURE 7.14

(A) Measured resistance versus $100/\omega^{0.5}$ for a soil sample with $W_v = 11.338\%$. (B) Argand diagram for a soil sample with $W_v = 11.338\%$. (C) Cole–Cole diagram revealed for a soil sample with $W_v = 11.338\%$.

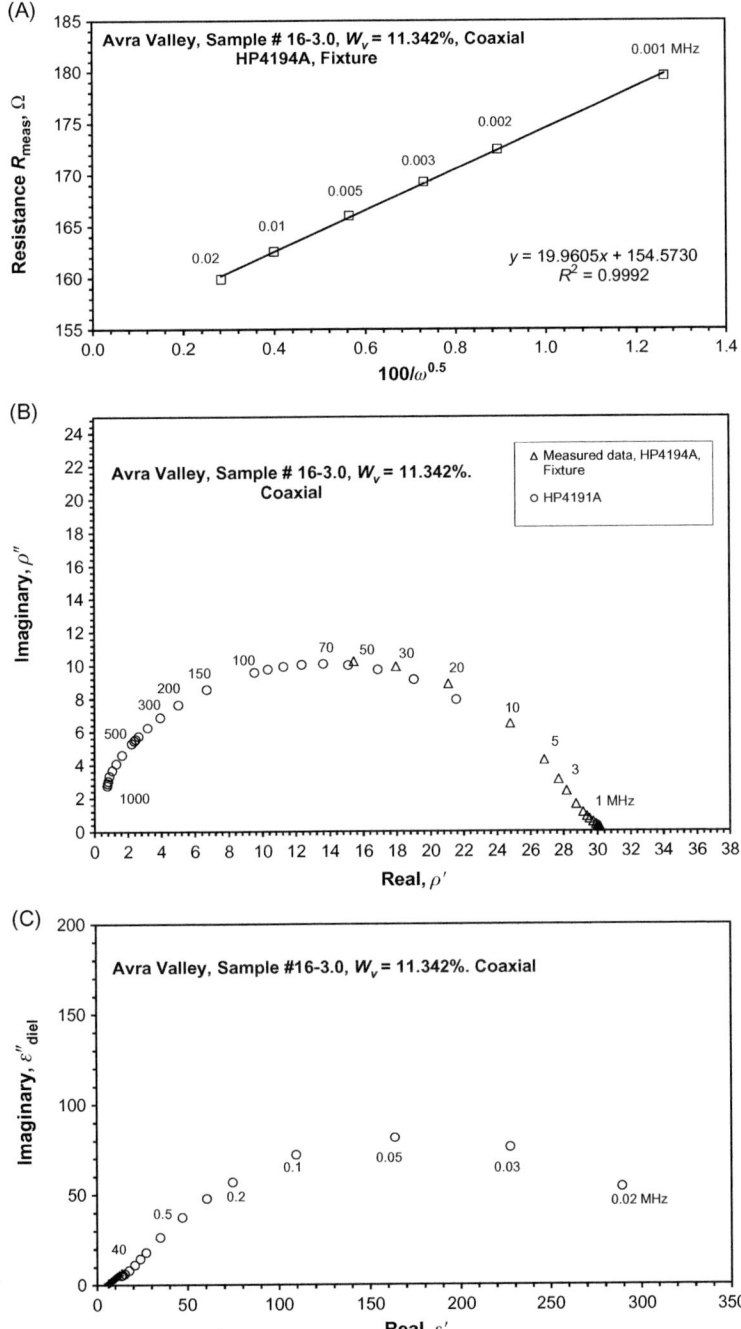

FIGURE 7.15

(A) Measured resistance versus $100/\omega^{0.5}$ for a soil sample with $W_v = 11.342\%$. (B) Argand diagram for a soil sample with $W_v = 11.342\%$. (C) Cole–Cole diagram revealed for a soil sample with $W_v = 11.342\%$.

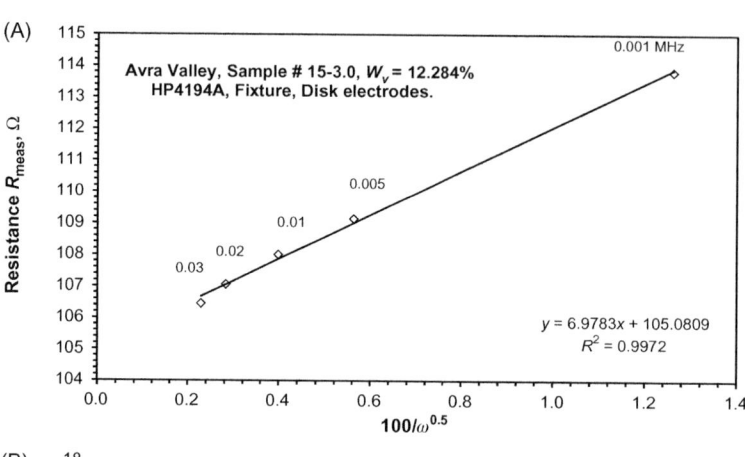

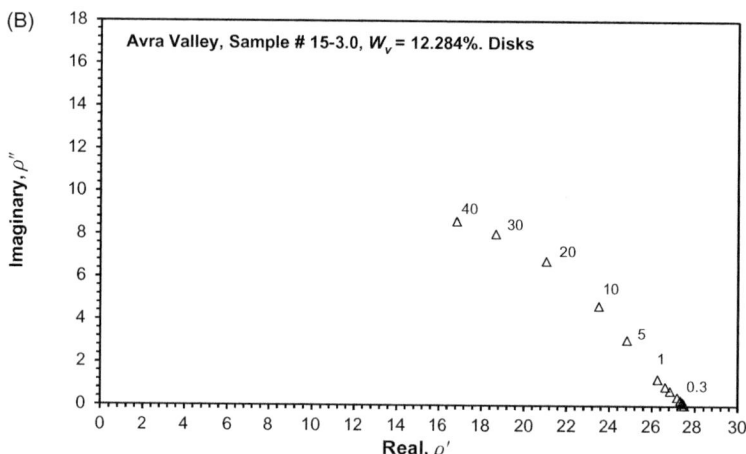

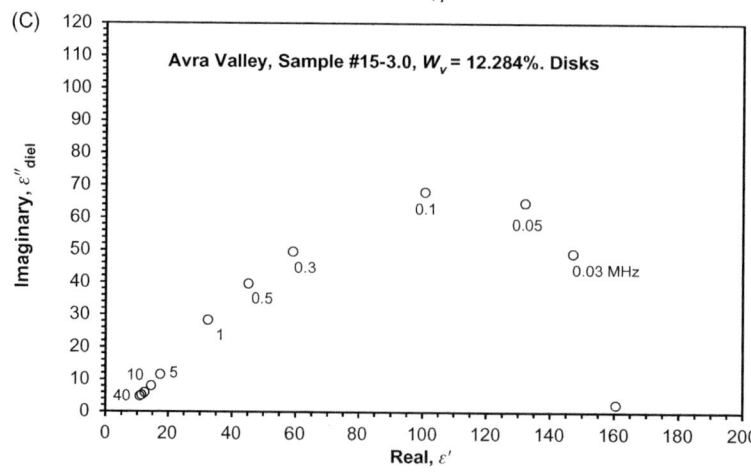

FIGURE 7.16

(A) Measured resistance versus $100/\omega^{0.5}$ for a soil sample with $W_v = 12.284\%$. (B) Argand diagram for a soil sample with $W_v = 12.284\%$. (C) Cole–Cole diagram revealed for a soil sample with $W_v = 12.284\%$.

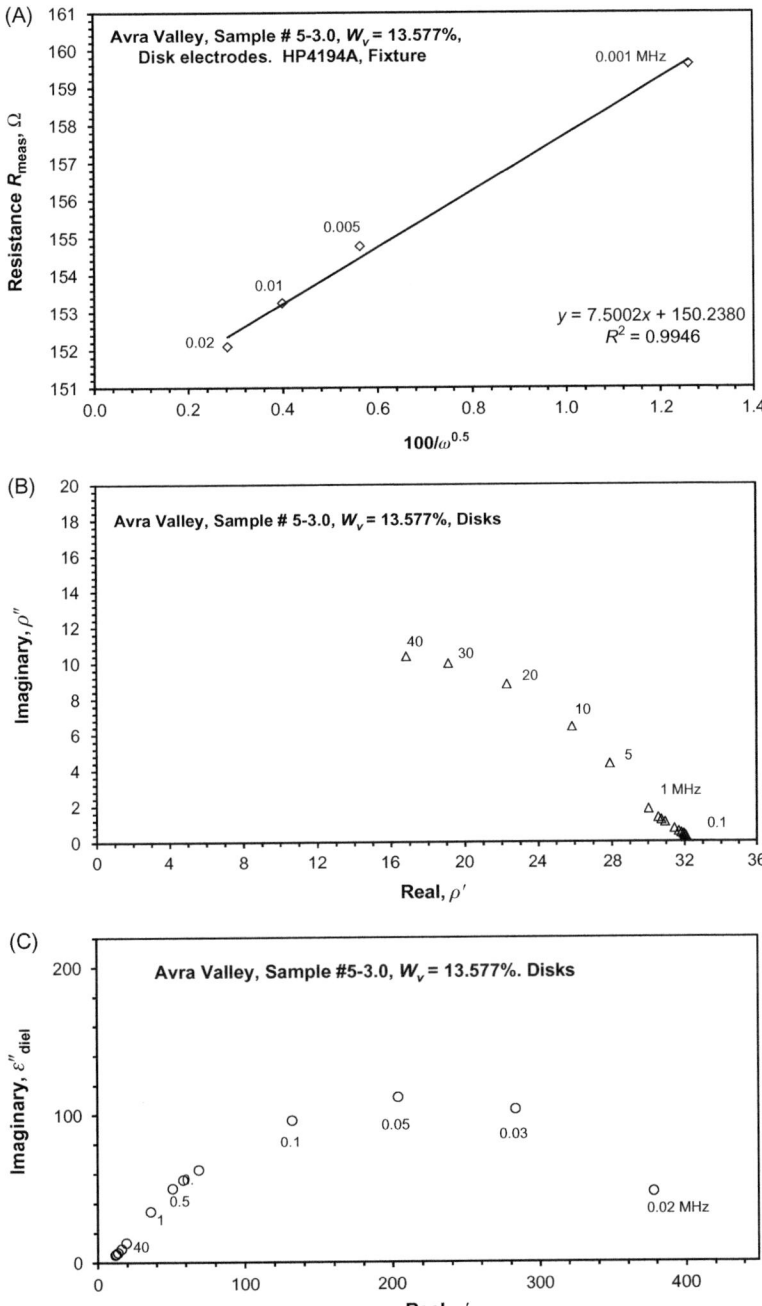

FIGURE 7.17

(A) Measured resistance versus $100/\omega^{0.5}$ for a soil sample with $W_v = 13.577\%$. (B) Argand diagram for a soil sample with $W_v = 13.577\%$. (C) Cole–Cole diagram revealed for a soil sample with $W_v = 13.577\%$.

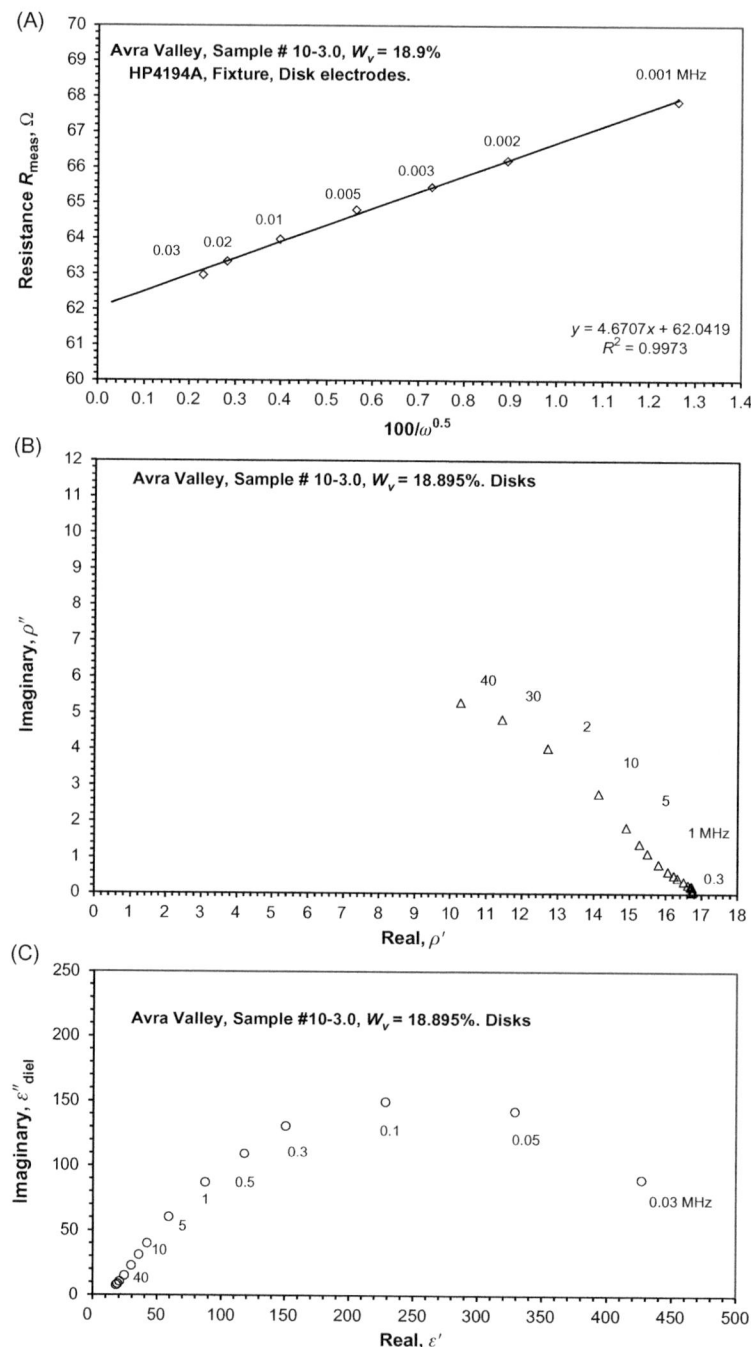

FIGURE 7.18

(A) Measured resistance versus $100/\omega^{0.5}$ for a soil sample with $W_v = 18.895\%$. (B) Argand diagram for a soil sample with $W_v = 18.9\%$. (C) Cole–Cole diagram revealed for a soil sample with $W_v = 18.895\%$.

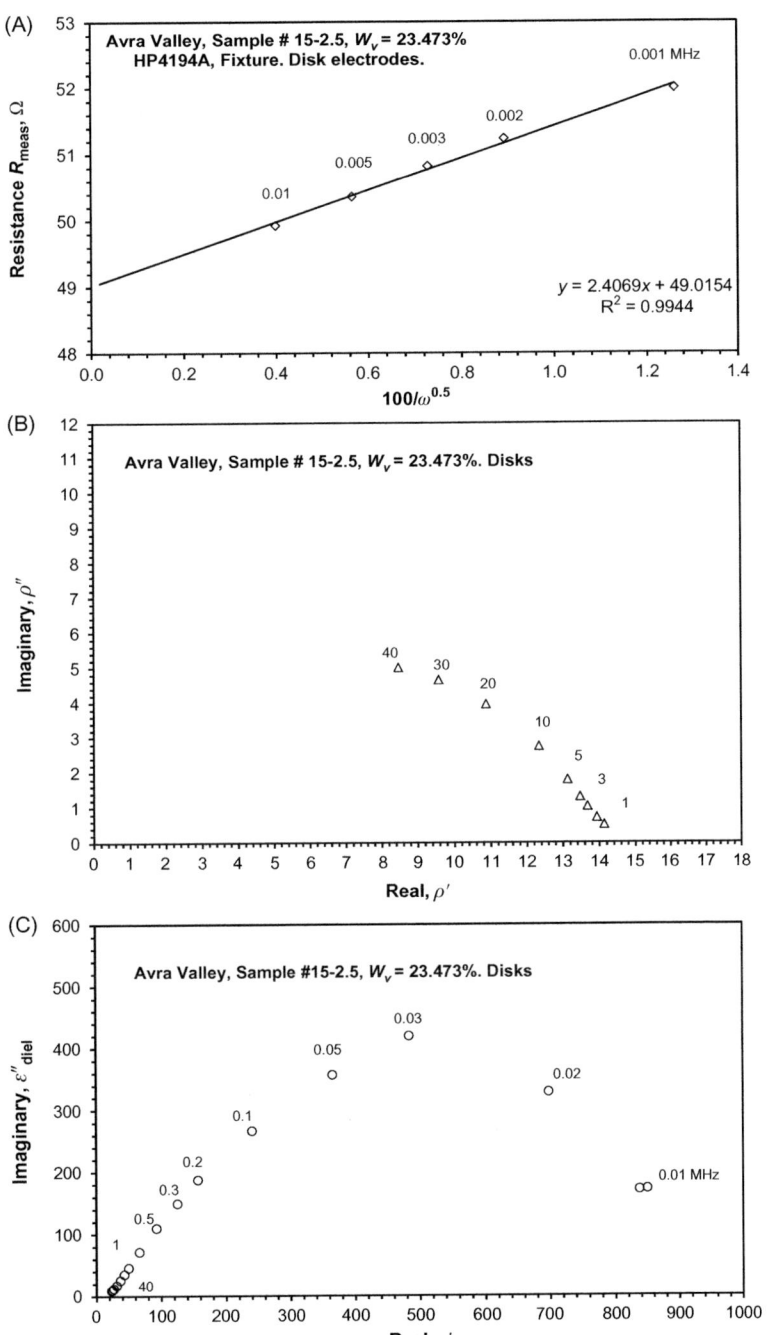

FIGURE 7.19

(A) Measured resistance versus $100/\omega^{0.5}$ for a soil sample with $W_v = 23.473\%$. (B) Argand diagram for a soil sample with $W_v = 23.473\%$. (C) Cole–Cole diagram revealed for a soil sample with $W_v = 23.473\%$.

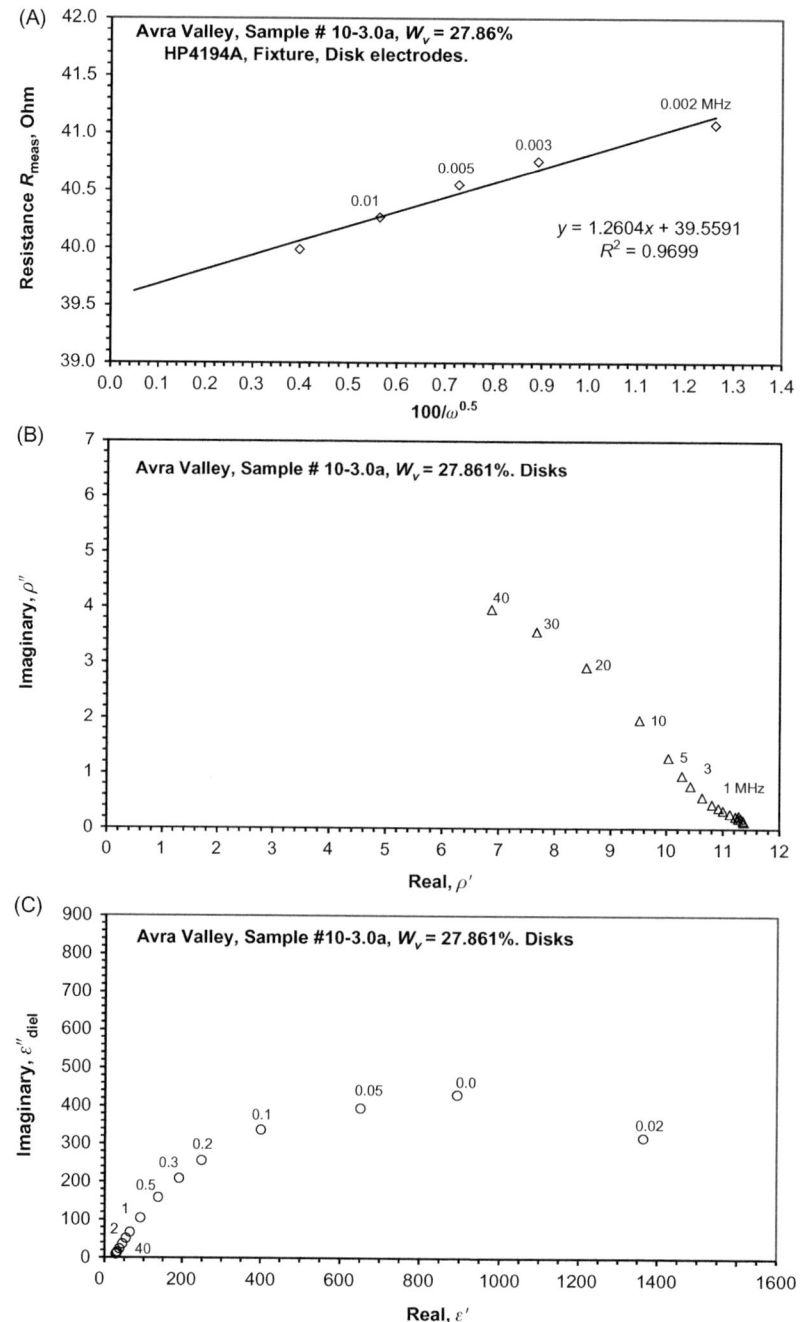

FIGURE 7.20

(A) Measured resistance versus $100/\omega^{0.5}$ for a soil sample with $W_v = 27.861\%$. (B) Argand diagram for a soil sample with $W_v = 27.861\%$. (C) Cole–Cole diagram revealed for a soil sample with $W_v = 27.861\%$.

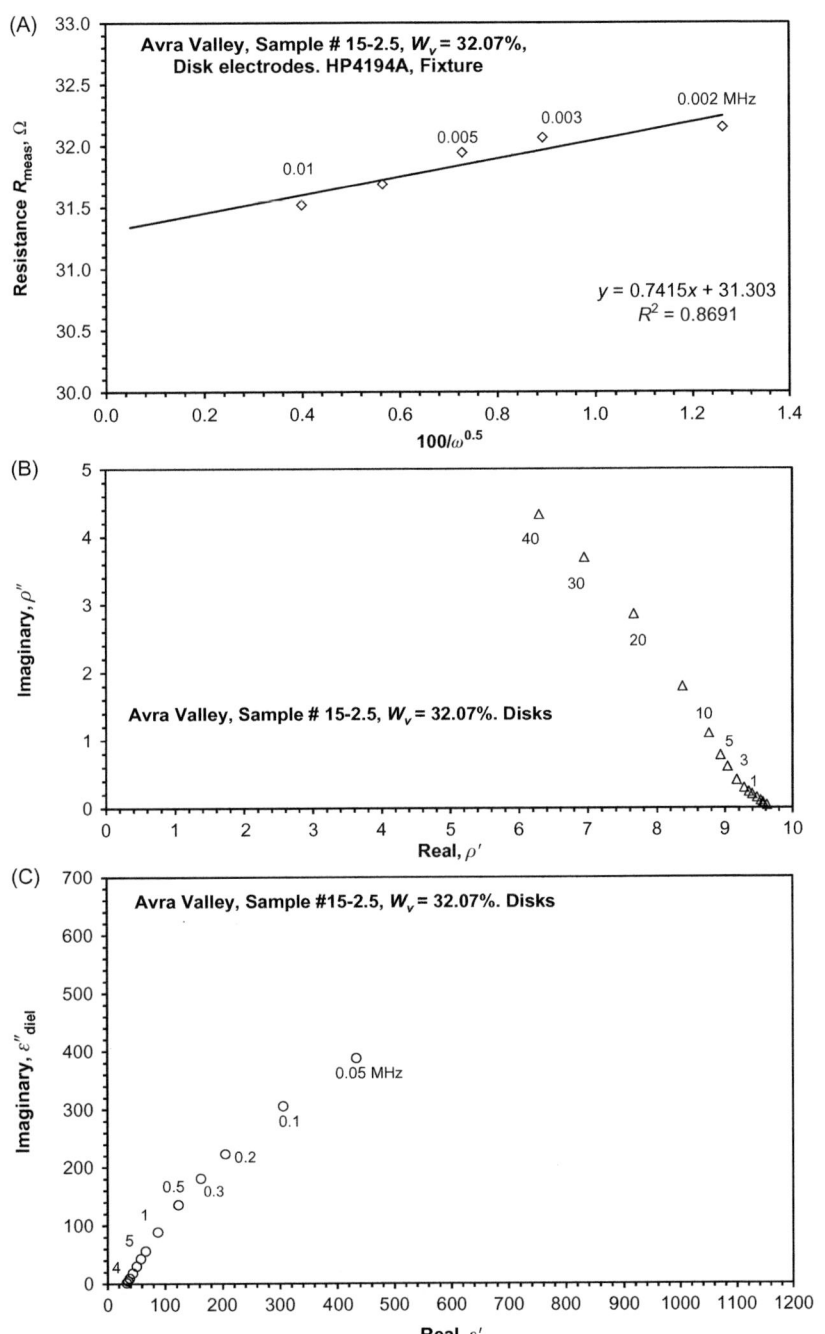

FIGURE 7.21

(A) Measured resistance versus $100/\omega^{0.5}$ for a soil sample with $W_v = 32.07\%$. (B) Argand diagram for a soil sample with $W_v = 32.07\%$. (C) Cole—Cole diagram revealed for a soil sample with $W_v = 32.07\%$.

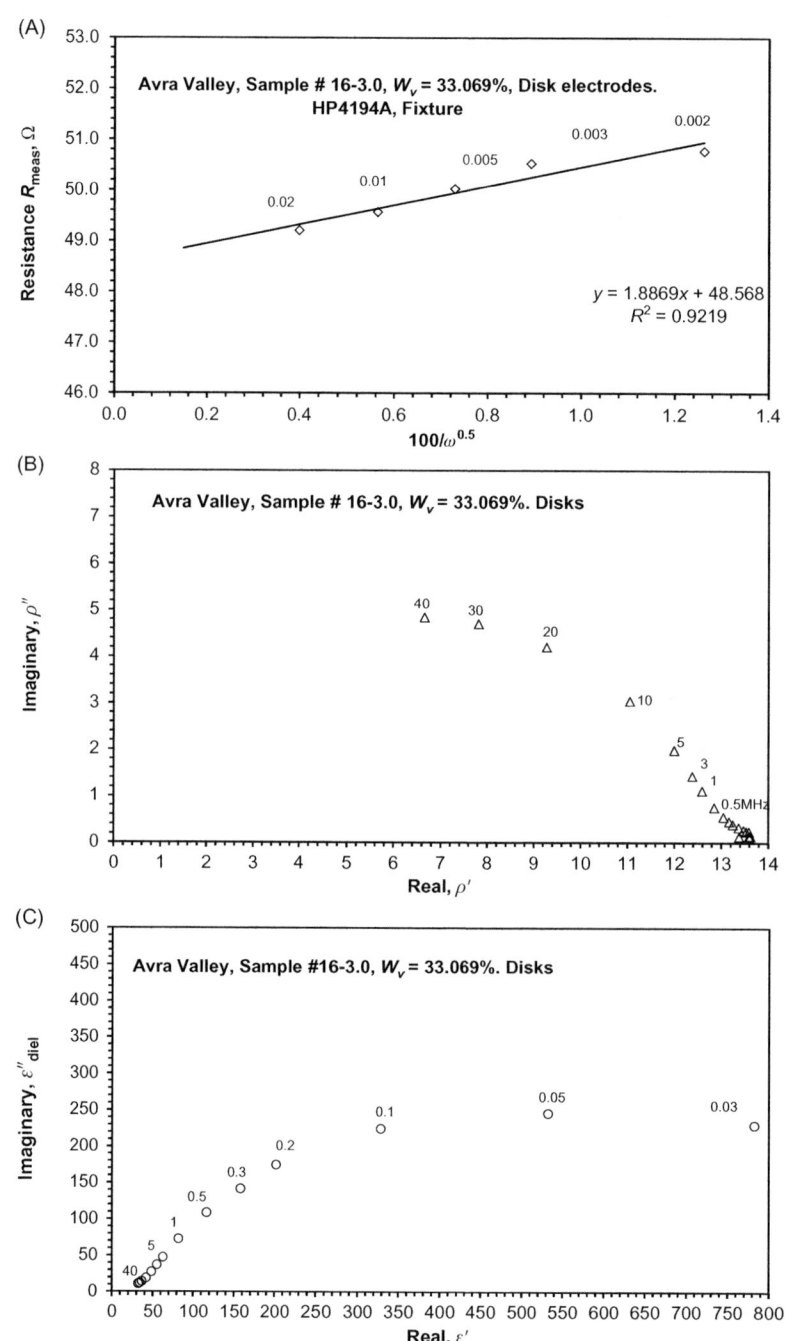

FIGURE 7.22

(A) Measured resistance versus $100/\omega^{0.5}$ for a soil sample with $W_v = 33.069\%$. (B) Argand diagram for a soil sample with $W_v = 33.069\%$. (C) Cole–Cole diagram revealed for a soil sample with $W_v = 33.069\%$.

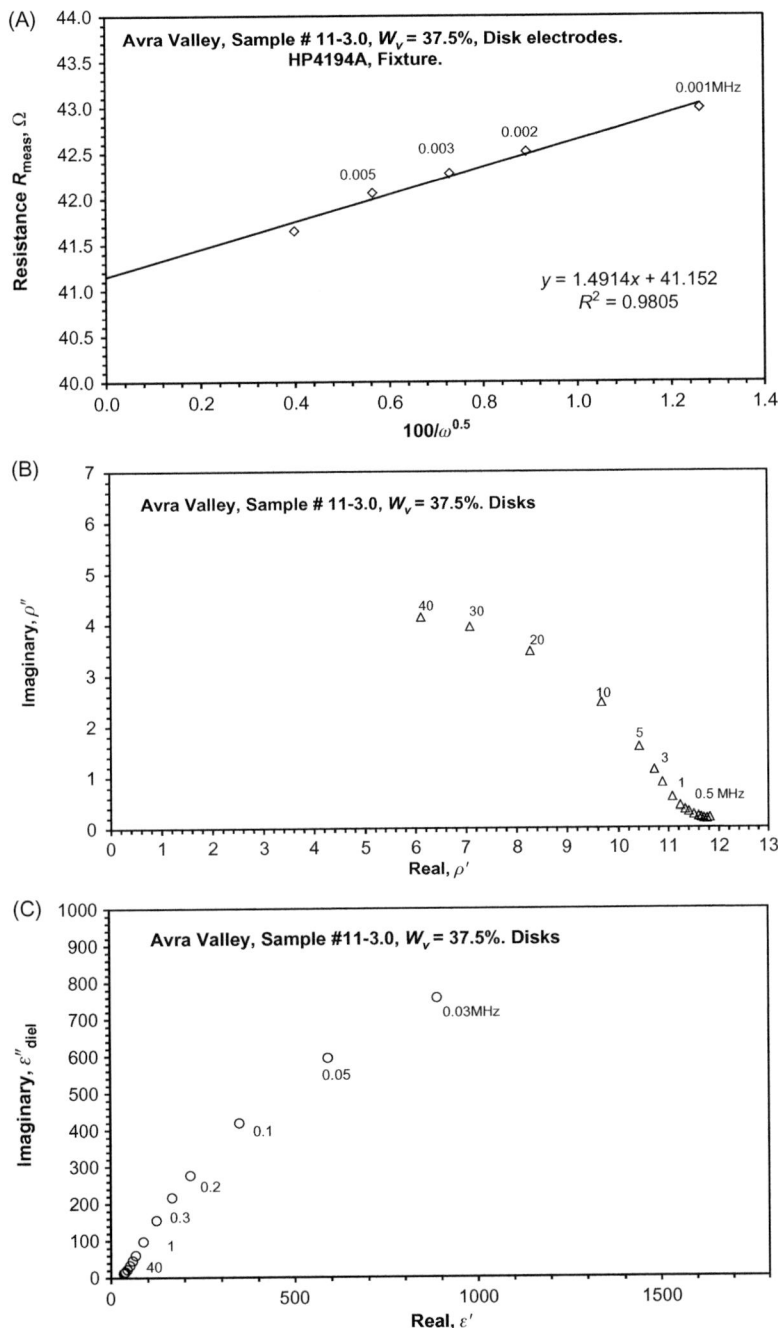

FIGURE 7.23

(A) Measured resistance versus $100/\omega^{0.5}$ for a soil sample with $W_v = 37.504$. (B) Argand diagram for a soil sample with $W_v = 37.5\%$. (C) Cole–Cole diagram revealed for a soil sample with $W_v = 37.5\%$.

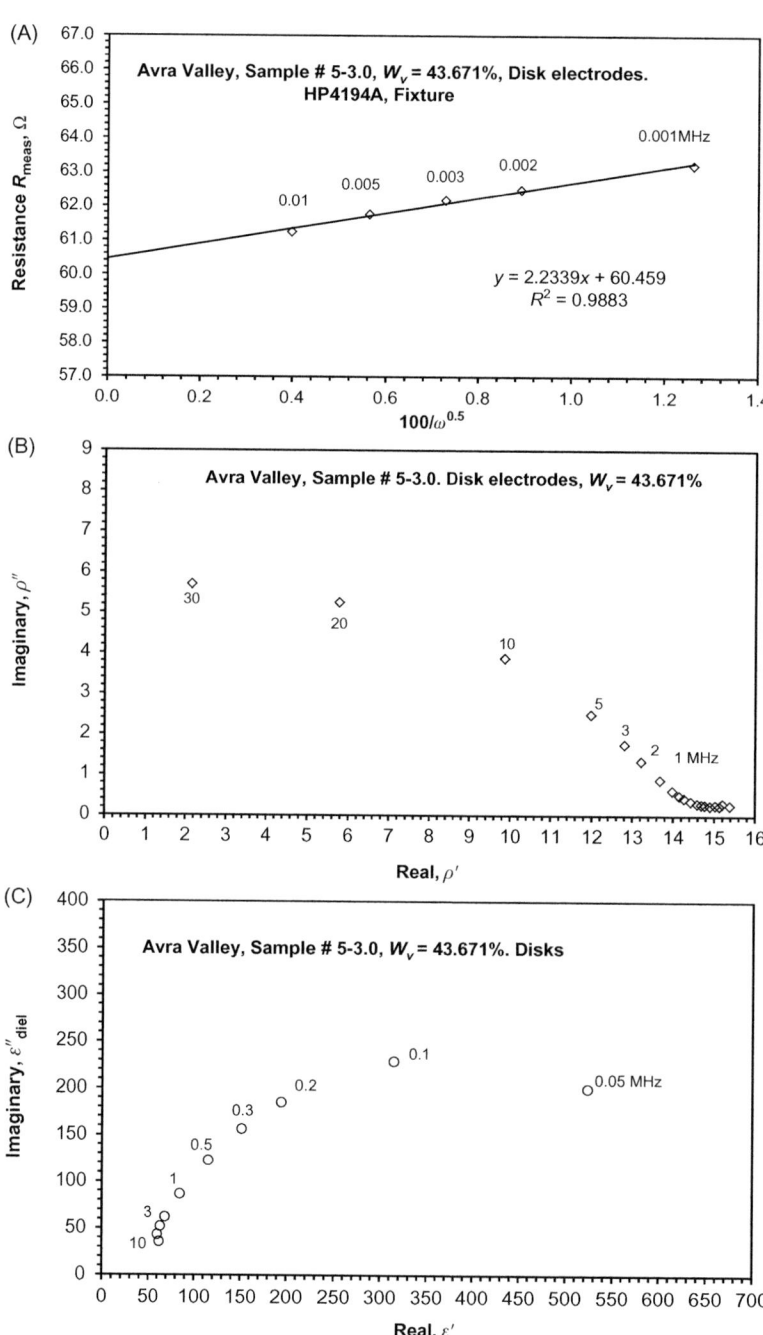

FIGURE 7.24

(A) Measured resistance versus $100/\omega^{0.5}$ for a soil sample with $W_v = 43.671\%$. (B) Argand diagram for a soil sample with $W_v = 43.67\%$. (C) Cole–Cole diagram revealed for a soil sample with $W_v = 43.671\%$.

Table 7.4 Static Parameters Defined for Supplemental Soil Samples From Avra Valley With Various Water Contents W_v by Using the Fricke–Lopatin Method

Water Content W_v, %, Sample #	Instrument and Electrodes	Static Resistance R_s, Ω	Static Resistivity ρ_s, Ω m	Static Conductivity, σ_s, S/m
5.940, # 11-0.5	HP4194A, Fixture, Disks	478.67	101.5	0.009852
7.0, # 5-2.0	HP4194A, Fixture, Disks	594.196	129.2218	0.00774
8.35, # 5-0.5	HP4194A, Fixture, Disks	419.85	21.4972	0.04652
9.615, # 5-1.0	HP4194A, Fixture, Disks	236.42	55.432	0.01804
9.757, # 5-1.5	HP4194A, Fixture, Disks	160.57	41.8519	0.02389
10.064, # 5-2.0	HP4194A, Fixture, Disks	187.77	43.956	0.02275
11.338, # 21-3.0	HP4194A, Fixture, Disks	132.06	34.7734	0.02876
11.342, # 16-3.0	HP4191A, Coaxial	154.573	30.1122	0.03321
12.284, # 15-3.0	HP4194A, Fixture, Disks	105.0809	27.4643	0.03641
13.577, # 5-3.0	HP4194A, Fixture, Disks	150.238	32.01	0.03124
18.895, # 10-3.0	HP4194A, Fixture, Disks	62.0419	16.7372	0.05975
23.473, # 15-2.5	HP4194A, Fixture, Disks	49.0154	14.7859	0.06763
27.861, # 10-3.0	HP4194A, Fixture, Disks	39.3698	11.3677	0.0877
32.07, # 15-2.5	HP4194A, Fixture, Disks	31.06	9.6246	0.1039
33.069, # 16-3.0	HP4194A, Fixture, Disks	48.016	13.6035	0.07351
37.504, # 11-3.0	HP4194A, Fixture, Disks	41.32	11.8256	0.08456
43.671, # 5-3.0	HP4194A, Fixture, Disks	60.459	14.7147	0.6796

Table 7.5 Argand Diagram Parameters for Supplemental Soil Samples From Avra Valley With Various Water Contents W_v

Water Content W_v, %, Sample #	Instrument and Electrodes	ρ_0, Ω m	ρ_∞, Ω m	Distribution Parameter, α	Relaxation Time τ, s	Static Conductivity, σ_s, S/m
5.791, # 5-2.0	HP4191A, Coaxial	163	0			0.006135
5.940, # 11-0.5	HP4194A, Fixture, Disks	101.5	0			0.009852
7.0, # 5-2.0	HP4194A, Fixture, Disks	128	0			0.007812
8.35, # 5-0.5	HP4194A, Fixture, Disks	107.5	0			0.00930
9.615, # 5-1.0	HP4194A, Fixture, Disks	55.6	0			0.01798
9.757, # 5-1.5	HP4194A, Fixture, Disks	42	0			0.02381
10.064, # 5-2.0	HP4194A, Fixture, Disks	44	0			0.02273
11.338, # 21-3.0	HP4194A, Fixture, Disks	35	0			0.02857
11.342, # 16-3.0	HP4191A, Coaxial	30	0			0.03333
12.284, # 15-3.0	HP4194A, Fixture, Disks	27.5	0			0.03636
13.577, # 5-3.0	HP4194A, Fixture, Disks	32	0			0.03125
18.895, # 10-3.0	HP4194A, Fixture, Disks	16.8	0			0.05952
23.473, # 15-2.5	HP4194A, Fixture, Disks	14.1	0			0.07092
27.861, # 10-3.0	HP4194A, Fixture, Disks	11.4	0			0.08772
32.07, # 15-2.5	HP4194A, Fixture, Disks	9.6	0			0.10417
33.069, # 16-3.0	HP4194A, Fixture, Disks	13.9	0			0.07194
37.504, # 11-3.0	HP4194A, Fixture, Disks	11.7	0			0.0855
43.671, # 5-3.0	HP4194A, Fixture, Disks	15	0			0.0667

Table 7.6 Cole–Cole Diagram Parameters for Supplemental Soil Samples From Avra Valley With Various Water Contents W_v

Water Content W_v, %, Sample #	Instrument and Electrodes	ε_s	ε_∞	Distribution Parameter, α	Relaxation Time τ, s
5.791, # 5-2.0	HP4191A, Coaxial	56	3.5		
5.940, # 11-0.5	HP4194A, Fixture, Disks	44.5	4		
7.0, # 5-2.0	HP4194A, Fixture, Disks	82	5		
8.35, # 5-0.5	HP4194A, Fixture, Disks	45	4.5		
9.615, # 5-1.0	HP4194A, Fixture, Disks	136	5		
9.757, # 5-1.5	HP4194A, Fixture, Disks	150	6		
10.064, # 5-2.0	HP4194A, Fixture, Disks	172	6		
11.338, # 21-3.0	HP4194A, Fixture, Disks	206	7		
11.342, # 16-3.0	HP4191A, Coaxial	350	5.3		
12.284, # 15-3.0	HP4194A, Fixture, Disks	168	7		
13.577, # 5-3.0	HP4194A, Fixture, Disks	420	10		
18.895, # 10-3.0	HP4194A, Fixture, Disks	495	12		
23.473, # 15-2.5	HP4194A, Fixture, Disks	905	15		
27.861, # 10-3.0	HP4194A, Fixture, Disks	1630	20		
32.07, # 15-2.5	HP4194A, Fixture, Disks		30		
33.069, # 16-3.0	HP4194A, Fixture, Disks		25		
37.504, # 11-3.0	HP4194A, Fixture, Disks		30		
43.671, # 5-3.0	HP4194A, Fixture, Disks	685	50		

References

Jonscher, A.K., 1983. Dielectric Relaxation in Solids. Chelsea Dielectrics Press, London.

Jonscher, A.K., 1996. Universal Relaxation Law. Chelsea Dielectrics Press, London.

Knight, R.J., 1983. The Use of Complex Plane Plots in Studying theElectrical Response of Rocks. J. Geomag. Geoelectr. 35, 767–776.

Sternberg, B.K., Birken, R.A., 1999. A new method of subsurface imaging – the LASI high frequency ellipticity system: Part 3. System tests and field surveys. J. Environ. Eng. Geophys. 4 (4), 227–240.

Sternberg, B.K., Levitskaya, T.M., 1998. Correlation between laboratory and in situ electrical resistivity measurements of soil. J. Environ. Eng. Geophys. 3 (2), 63–70.

Sternberg, B.K., Miletto, D.J., LaBrecque, D.J., Thomas, S.J., Poulton, M.M., 1991, The Avra Valley (Ajo Rd.) geophysical test site. In: LASI Rep. 91-2. Univ. of Arizona, Tucson.

Further Reading

Hillel, D., 1982. Introduction to Soil Physics. Academic, San Diego, CA.

Sternberg, B.K., Levitskaya, T.M., 2001. Electrical parameters of soils in the frequency range from 1 kHz to 1 GHz, using lumped-circuit methods. Radio Sci. 36 (4), 709–719.

Soil from Brookhaven, New York

The Department of Energy, Brookhaven National Laboratory is located on Long Island, New York. The soil samples from this location, which we received from Dr. Jeffrey Daniels, consist of relatively clean sand (Faust, 1963; Daniels, 1998). These samples are representative of low-loss soils, in comparison with the Avra Valley high-loss soils.

A sample of the soils that we used for our electrical properties measurements was sent to the Soil, Water, and Plant Analysis Laboratory in the Department of Soil, Water, and Environmental Science at the University of Arizona. Using the hydrometer method, they found that this soil contains 96% sand, 1% silt, and 3% clay (by weight). Note that this measurement is a particle-size analysis. We do not have a mineralogy analysis of the clay fraction.

Analogously to samples from Avra Valley, we defined the electrical parameters in the same frequency range, from 1 kHz to 1 GHz, for samples with various lab-prepared volume water contents (W_v = 8.11%, 8.69%, 9.75%, 10.85%, and 14.67%). Some of the samples were measured in a coaxial sample holder with both instruments: HP4194A, fixture and HP4191A in the entire frequency range, while some were measured with disk electrodes only over the low frequency range, using the HP4194A. The sample holder used for each sample with the low-frequency instrument HP4194A is indicated in Table 8.1, where their static parameters are shown. The static parameters, ρ_s and σ_s, were defined by using the Fricke–Lopatin method (F–L method), which is described in Chapter 5, Stray parameters of the measuring system and ways of defining them. Also, analogously to Avra Valley samples, the measured resistance R_{meas} for five samples was plotted versus $100/\omega^{0.5}$ at low frequencies in Fig. 8.1A–E. The linear regression equations are shown on the graphs. Their intercepts give the values of static resistance R_s, summarized in Table 8.1. Static parameters ρ_s and σ_s, defined from the R_s values and the samples geometry, are also shown in Table 8.1. From the regression equation, we also defined the electrode polarization resistance R_{el} versus frequency for each

Electrical Spectroscopy of Earth Materials. DOI: https://doi.org/10.1016/B978-0-12-818603-9.00008-8

Table 8.1 Static Parameters Defined for Soil Samples From Brookhaven With Various Water Contents W_v, Using the Fricke–Lopatin Method

Water Content W_v (%) Sample Number	Instrument and Electrodes	Static Resistance R_s (Ω)	Static Resistivity ρ_s (Ω m)	Static Conductivity, σ_s (S/m)
8.11 # I, 4−6′	HP4194A, fixture, coaxial	11,385	2295	0.00044
8.69 # I, 2−4′	HP4194A, fixture, coaxial	98,339	1983	0.00050
9.75 # I, 6−8′	HP4194A, fixture, coaxial	2358	456	0.00219
10.85 # II, 2−4′	HP4194A, fixture, disks	5460	1165	0.00086
14.67 # I, 6−8′	HP4194A, Z-probe, coaxial	2330	454	0.00220
Additional Samples				
8.176 # I, 4−6′	HP4194A, fixture, disks	11,917	2700	0.00037
8.552 # I, 2−4′	HP4194A, fixture, disks	10,645	2500	0.00040
8.7629 # II, 2−4′	HP4194A, fixture, disks	4374	1409	0.00071

sample, which we subtracted from our measured resistance, before further processing.

Fig. 8.2A−E shows the F−L method for measured reactance X_{meas} in the same frequency range, but only two of the curves (C and D) have a linear regression equation. For other samples, the equations are polynomial, and we could not define the intercept, using the F−L method. For these two samples, we defined the electrode polarization reactance X_{el} versus frequency.

Fig. 8.3A−E represents the frequency dependencies of real and imaginary components of dielectric permittivity ε' and ε'' for each sample studied, as well as their dielectric part ε''_{diel} of the total measured loss factor ε'', obtained by using the procedure in Chapter 6, Corrections for stray parameters and error estimation. As seen, these graphs reveal some regions of relaxation processes, while ε' and ε'' do not show any polarization, which is obscured by conductivity. The imaginary ε'' is well approximated with a power function, similar to the Avra Valley samples.

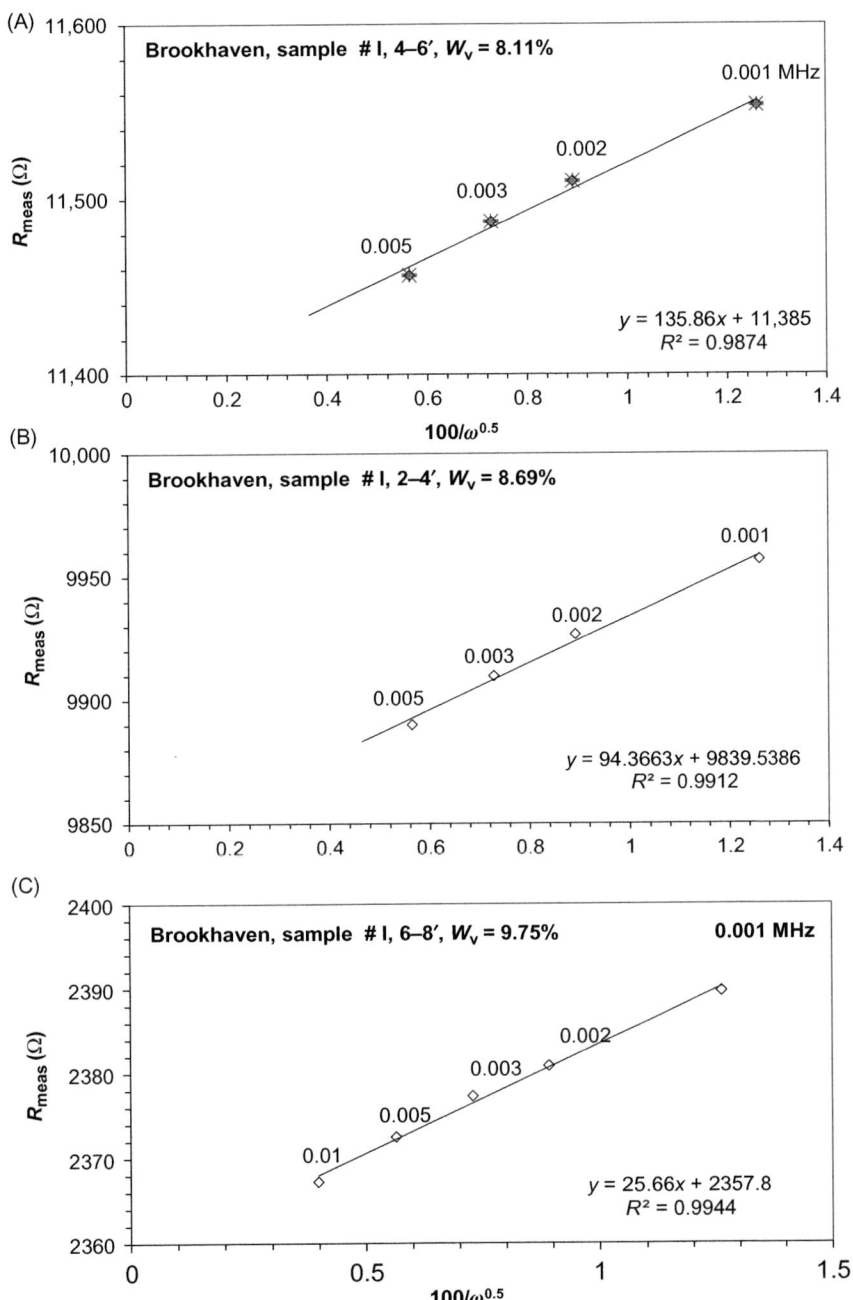

FIGURE 8.1

(A) Measured resistance versus $100/\omega^{0.5}$ for samples from Brookhaven with $W_v = 8.11\%$. (B) Measured resistance versus $100/\omega^{0.5}$ for a soil sample with $W_v = 8.69\%$. (C) Measured resistance versus $100/\omega^{0.5}$ for a soil sample with $W_v = 9.75\%$. (D) Measured resistance versus $100/\omega^{0.5}$ for samples from Brookhaven with $W_v = 10.85\%$. (E) Measured resistance versus $100/\omega^{0.5}$ for samples from Brookhaven with $W_v = 15\%$.

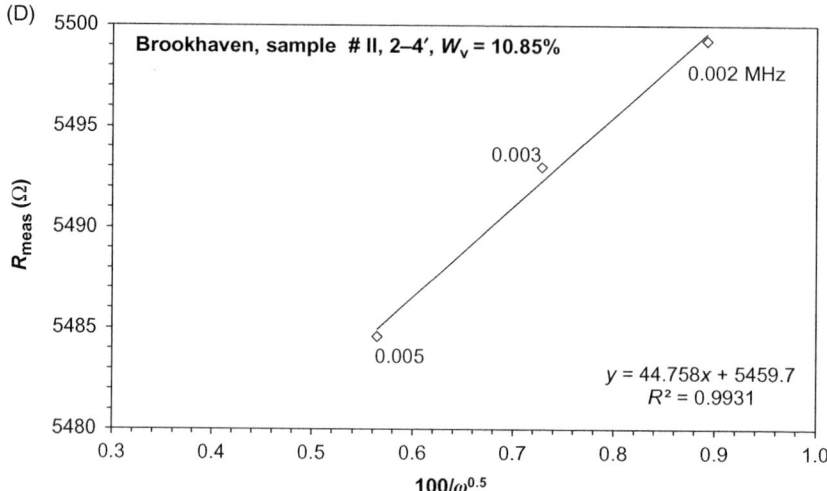

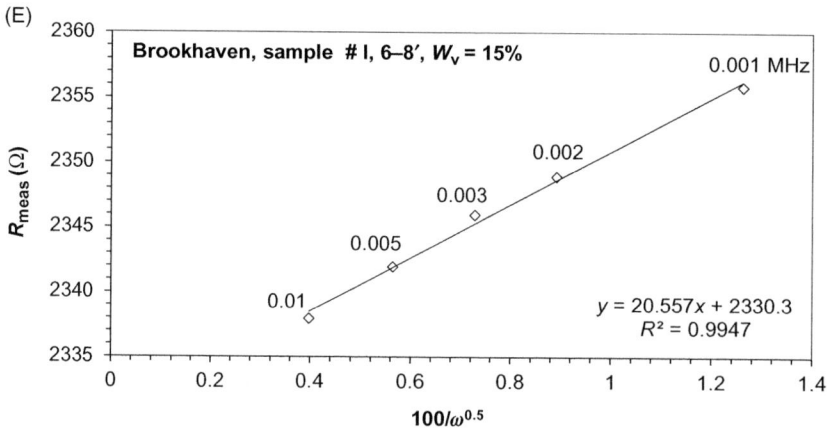

FIGURE 8.1 Continued

(A) Measured resistance versus $100/\omega^{0.5}$ for samples from Brookhaven with $W_v = 8.11\%$.
(B) Measured resistance versus $100/\omega^{0.5}$ for a soil sample with $W_v = 8.69\%$. (C) Measured resistance versus $100/\omega^{0.5}$ for a soil sample with $W_v = 9.75\%$. (D) Measured resistance versus $100/\omega^{0.5}$ for samples from Brookhaven with $W_v = 10.85\%$. (E) Measured resistance versus $100/\omega^{0.5}$ for samples from Brookhaven with $W_v = 15\%$.

The frequency dependencies of conductivity are shown in Fig. 8.4A–E. As seen, the real conductivity σ' does not change with frequency up to 10 MHz and then increases. The imaginary conductivity σ'' increases over most of the frequency range, following the power low. The regression equations are shown on the graphs.

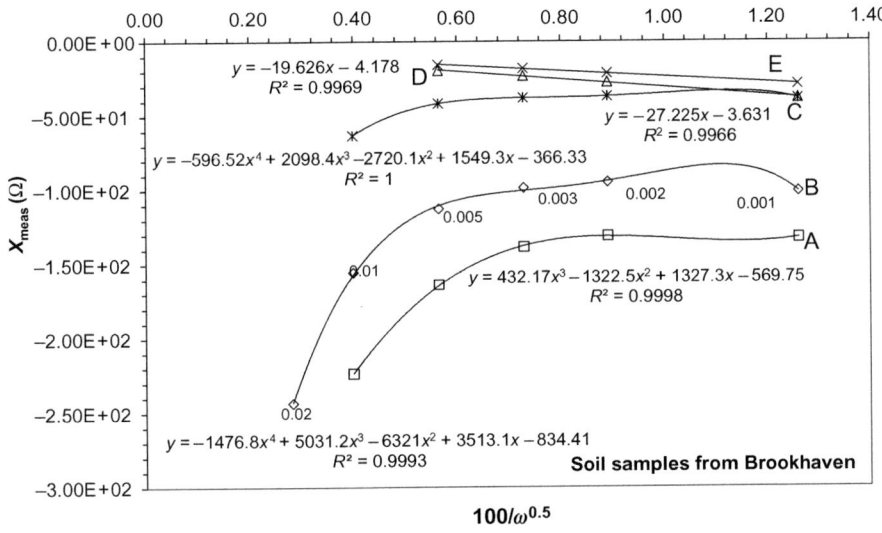

FIGURE 8.2

Measured reactance versus $100/\omega^{0.5}$ for soil samples from Brookhaven with water contents $W_v\%$: Water contents: (A) 8.11%; (B) 8.69%; (C) 9.75%; (D) 10.85%; (E) 15%.

Although we used only the high-frequency results for drawing the trend line of the σ'' curve, most of the lower frequency measurements match it as well.

Fig. 8.5A−E represents the real ρ' and imaginary ρ'' components of resistivity versus frequency. Dispersion on ρ' (f) and a maximum on ρ'' (f) at a characteristic frequency for each sample indicate a relaxation process. Argand diagrams drawn from these graphs are shown in Fig. 8.6A−E. The best fit Argand curves are also shown. From the dielectric loss factor ε''_{diel} (Fig. 8.3), we drew the Cole−Cole diagrams for our Brookhaven samples, shown in Fig. 8.7A−E, along with the calculated curves. The parameters of the Argand diagrams are summarized in Table 8.2 and the Cole−Cole parameters in Table 8.3. Figs. 8.8−8.10 illustrate the dependencies of static resistivity, static conductivity, and limit values of dielectric permittivity from soil wetness.

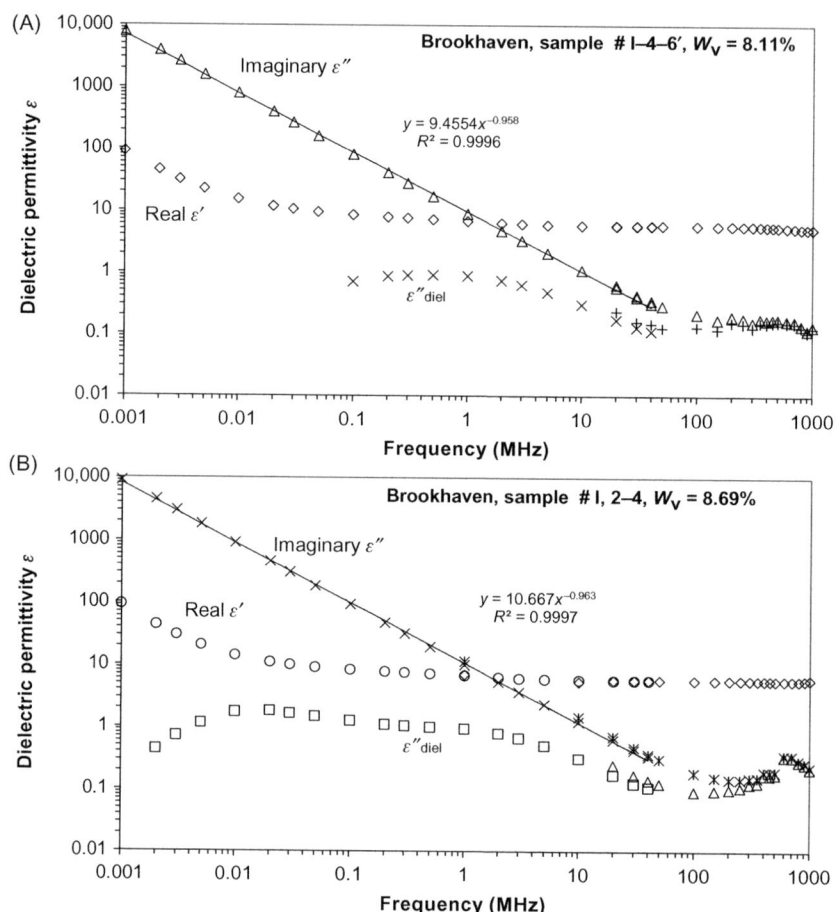

FIGURE 8.3

Dielectric permittivity versus frequency for soil samples from Brookhaven with various water contents (A) $W_v = 8.11\%$, (B) $W_v = 8.69\%$, (C) $W_v = 9.75\%$, (D) $W_v = 10.85\%$, and (E) $W_v = 14.67\%$.

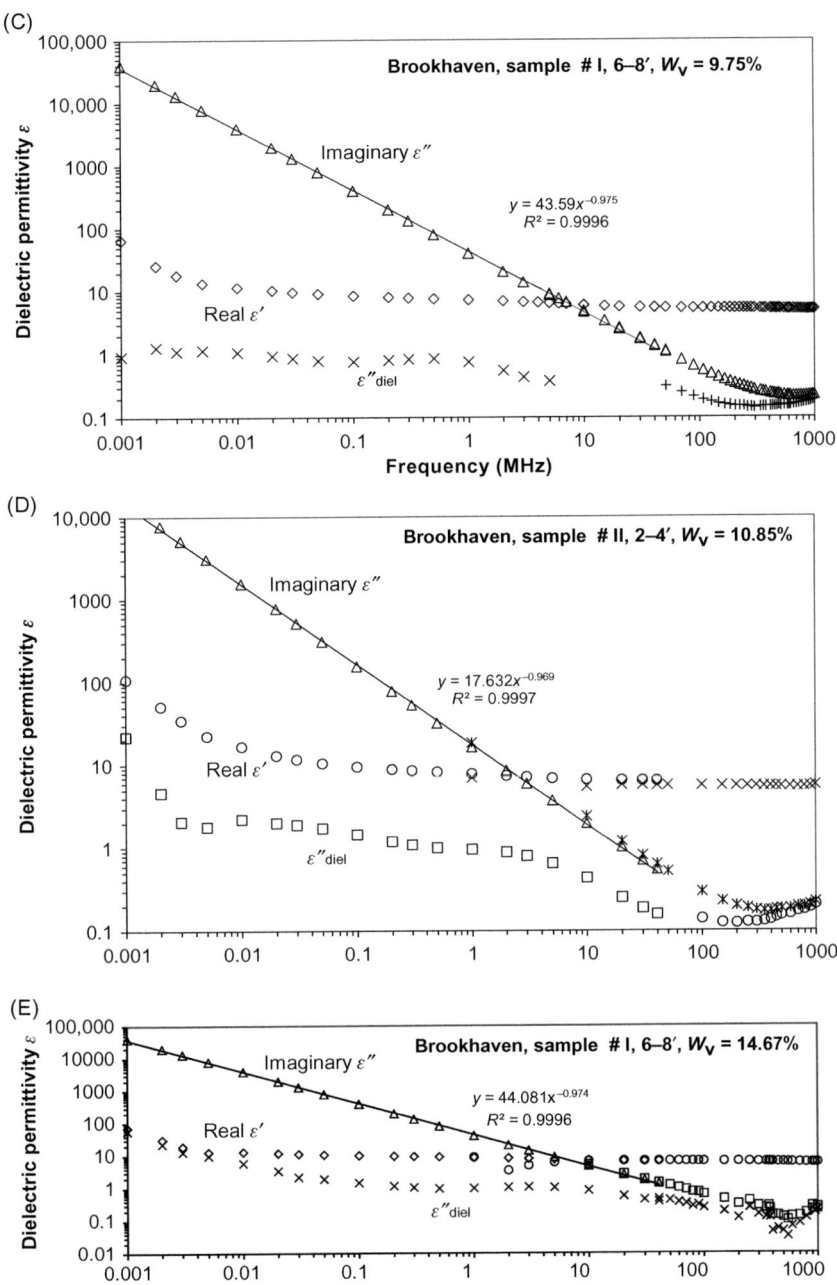

FIGURE 8.3 Continued

Dielectric permittivity versus frequency for soil samples from Brookhaven with various water contents (A) $W_v = 8.11\%$, (B) $W_v = 8.69\%$, (C) $W_v = 9.75\%$, (D) $W_v = 10.85\%$, and (E) $W_v = 14.67\%$.

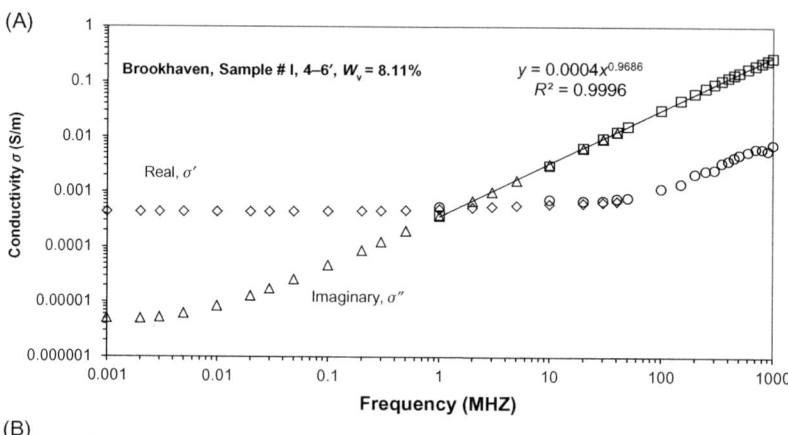

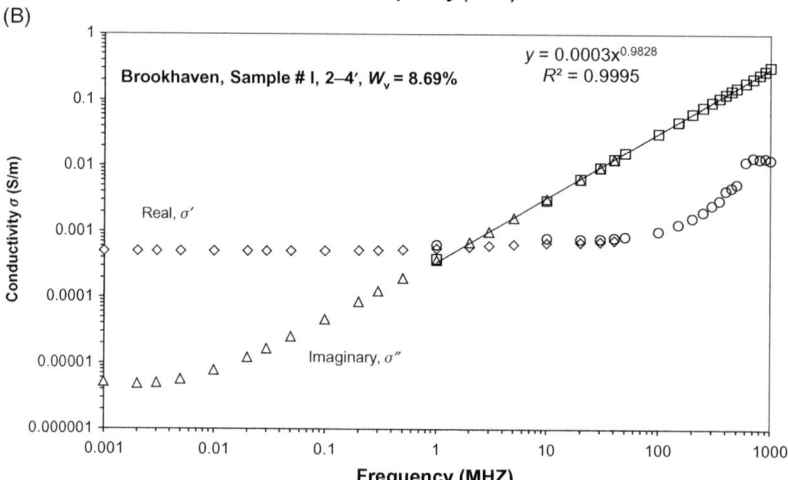

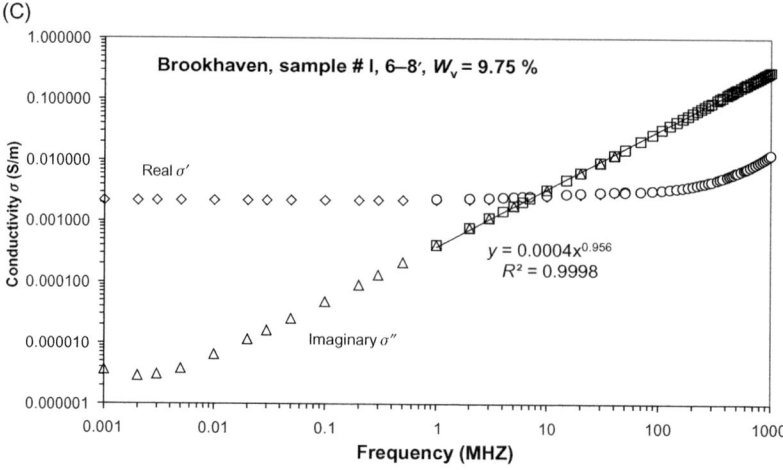

FIGURE 8.4

Conductivity versus frequency for soil samples from Brookhaven with water contents
(A) $W_v = 8.11\%$, (B) $W_v = 8.69\%$, (C) $W_v = 9.75\%$, (D) $W_v = 10.85\%$, and (E) $W_v = 14.67\%$.

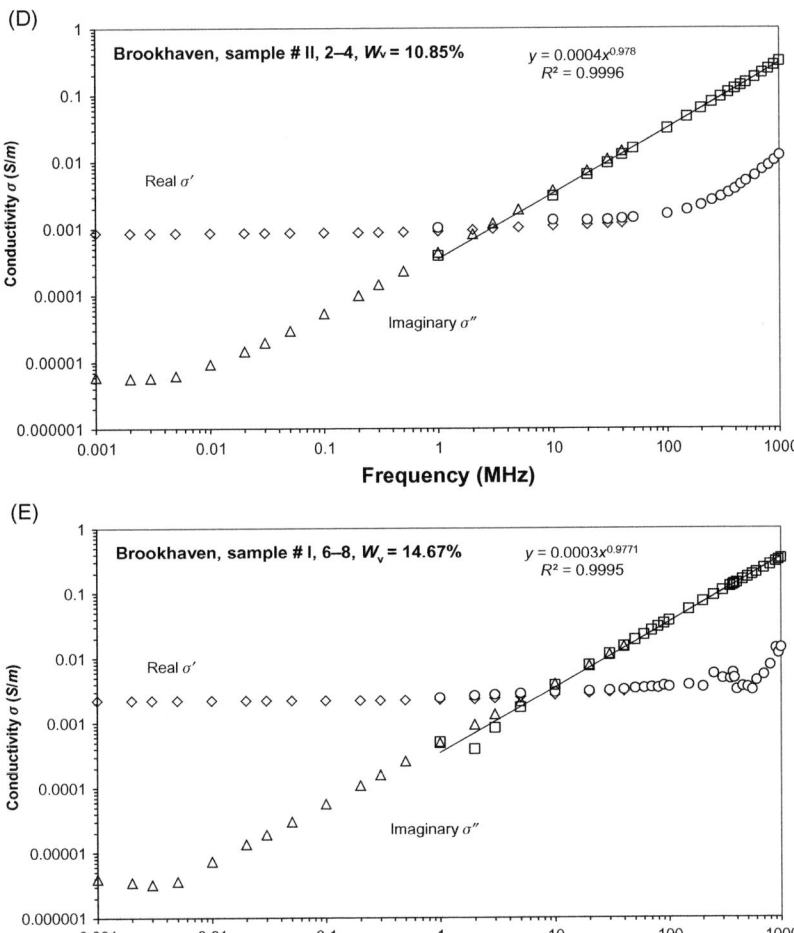

FIGURE 8.4 Continued

Conductivity versus frequency for soil samples from Brookhaven with water contents
(A) $W_v = 8.11\%$, (B) $W_v = 8.69\%$, (C) $W_v = 9.75\%$, (D) $W_v = 10.85\%$, and (E) $W_v = 14.67\%$.

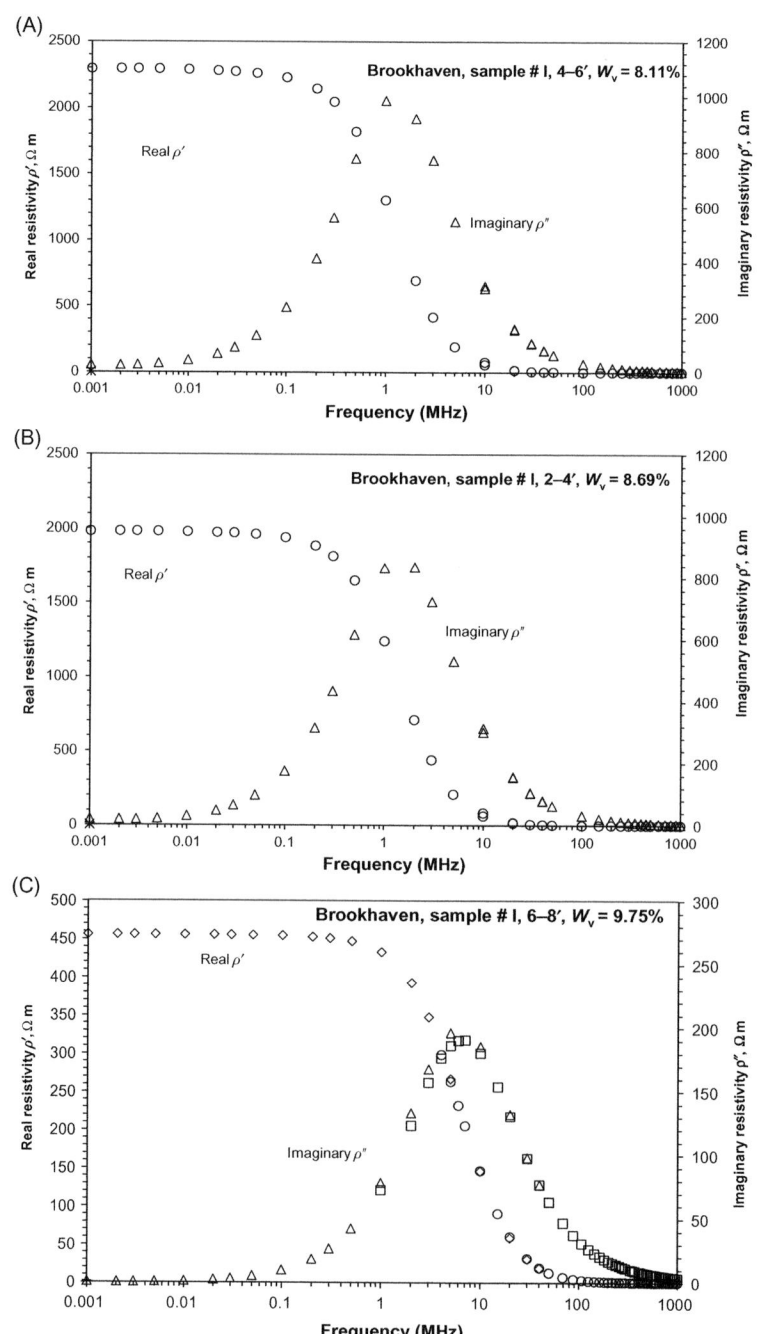

FIGURE 8.5

Resistivity versus frequency for soil samples from Brookhaven with water contents
(A) $W_v = 8.11\%$, (B) $W_v = 8.69\%$, (C) $W_v = 9.75\%$, (D) $W_v = 10.85\%$, and (E) $W_v = 15\%$.

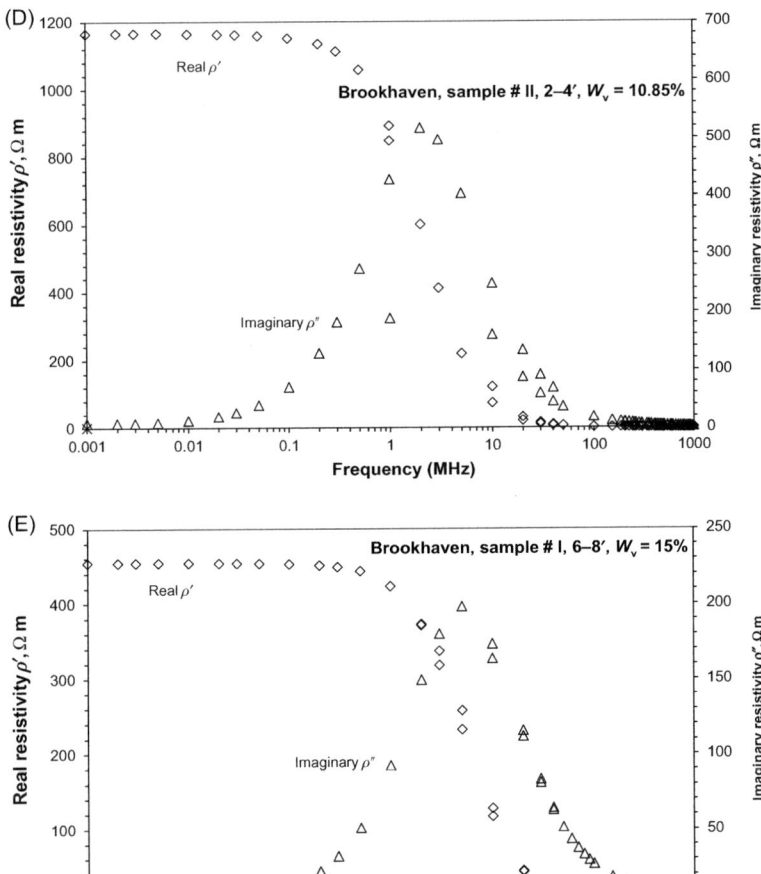

FIGURE 8.5 Continued

Resistivity versus frequency for soil samples from Brookhaven with water contents
(A) $W_v = 8.11\%$, (B) $W_v = 8.69\%$, (C) $W_v = 9.75\%$, (D) $W_v = 10.85\%$, and (E) $W_v = 15\%$.

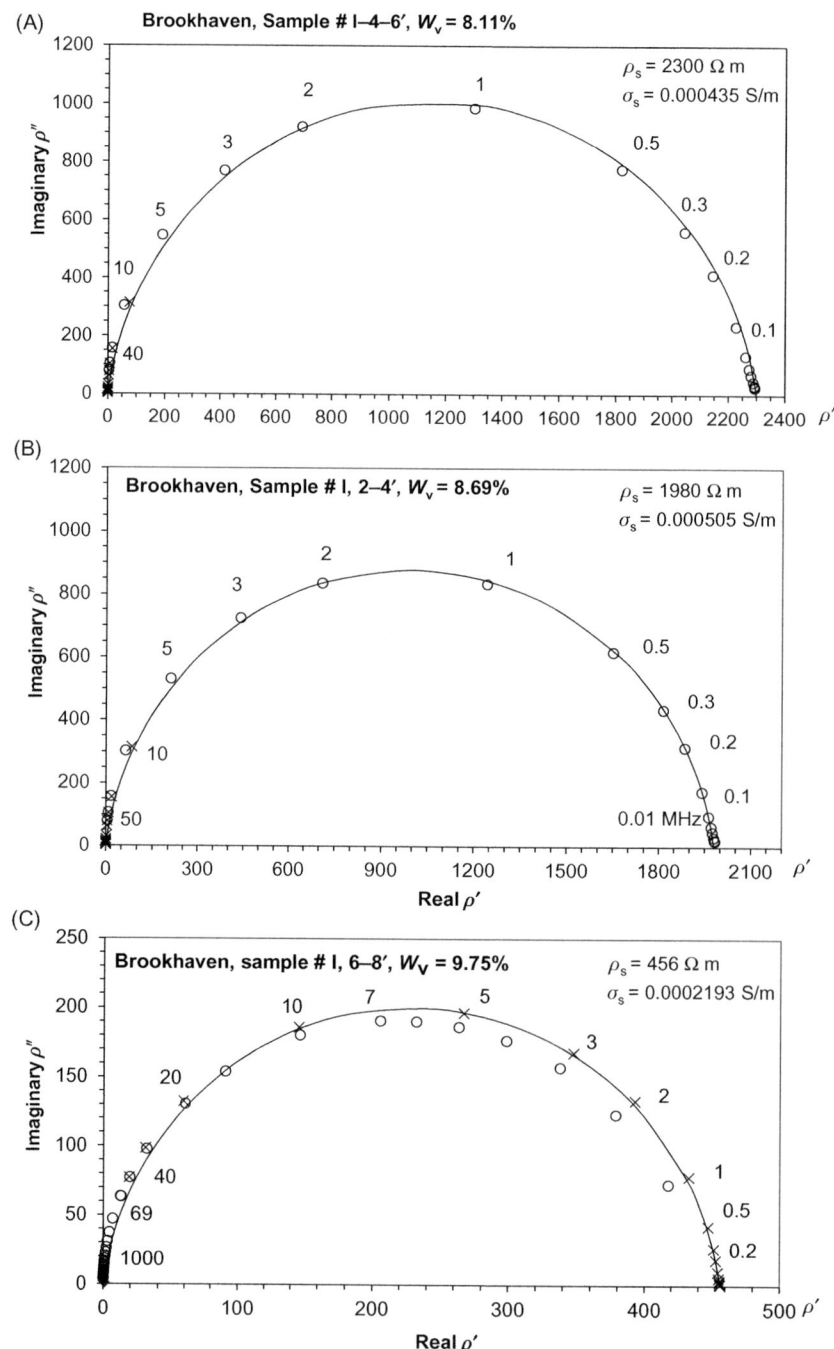

FIGURE 8.6

Argand diagrams for soil samples from Brookhaven with water contents (A) $W_v = 8.11\%$, (B) $W_v = 8.69\%$, (C) $W_v = 9.75\%$, (D) $W_v = 10.85\%$, (E) $W_v = 15\%$.

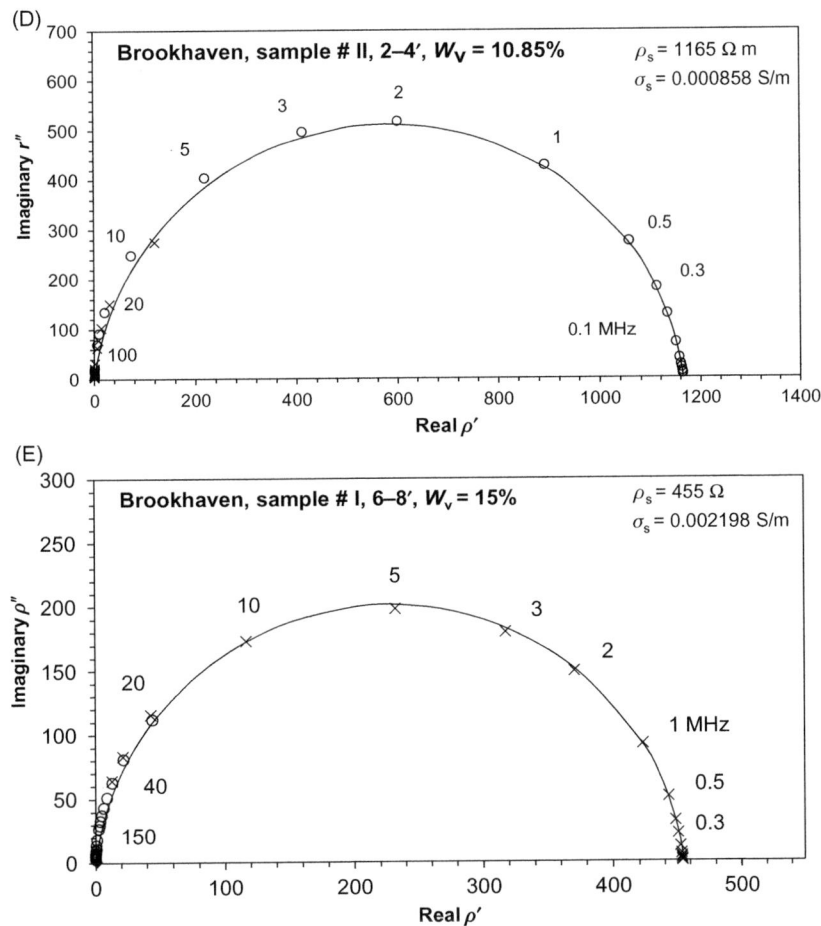

FIGURE 8.6 Continued

Argand diagrams for soil samples from Brookhaven with water contents (A) $W_v = 8.11\%$, (B) $W_v = 8.69\%$, (C) $W_v = 9.75\%$, (D) $W_v = 10.85\%$, (E) $W_v = 15\%$.

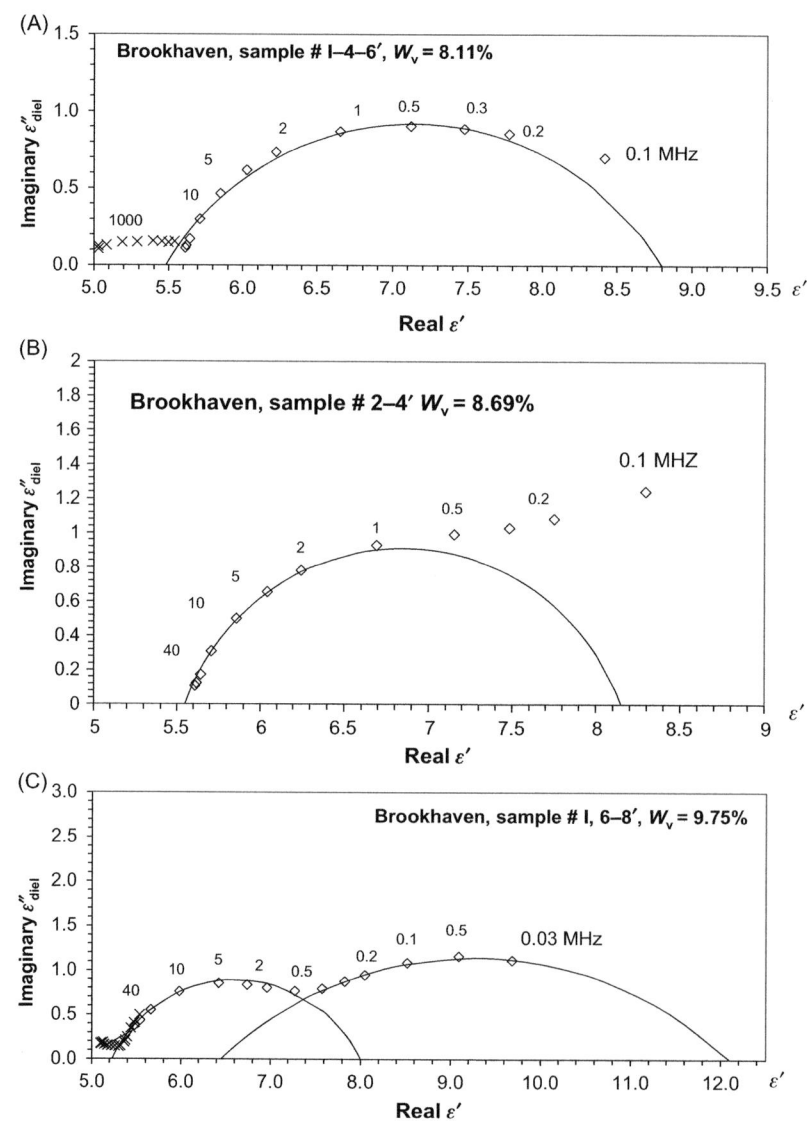

FIGURE 8.7

Cole–Cole diagrams for soil samples from Brookhaven with water contents
(A) $W_v = 8.11\%$, (B) $W_v = 8.69\%$, (C) $W_v = 9.75\%$, (D) $W_v = 10.85\%$, and (E) $W_v = 15\%$.

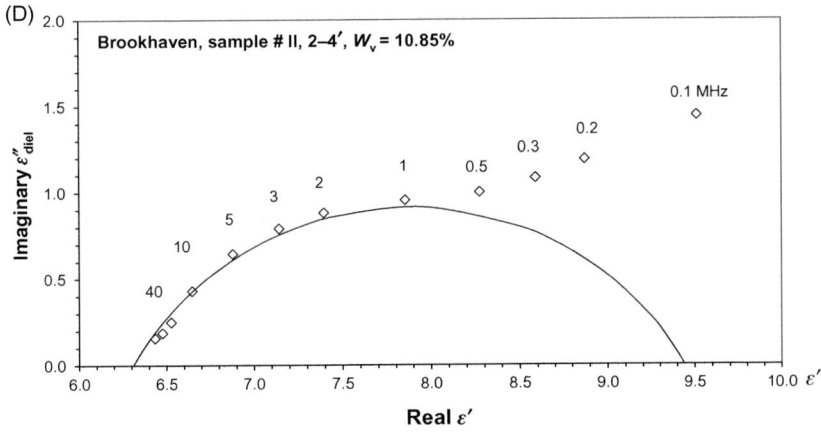

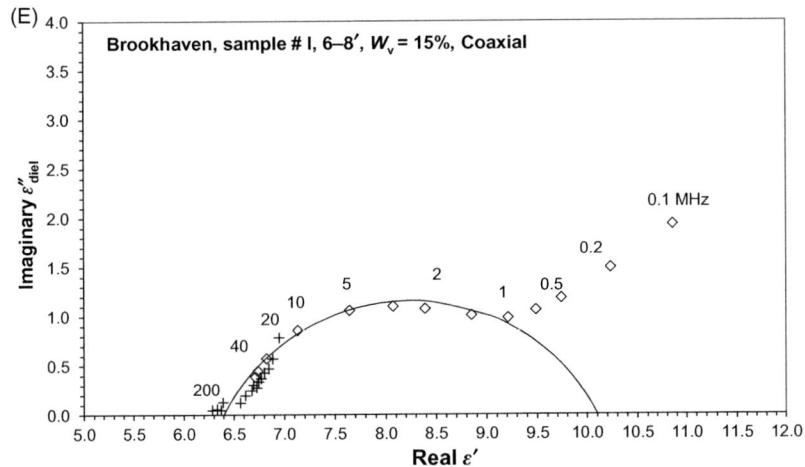

FIGURE 8.7 Continued

Cole–Cole diagrams for soil samples from Brookhaven with water contents
(A) $W_v = 8.11\%$, (B) $W_v = 8.69\%$, (C) $W_v = 9.75\%$, (D) $W_v = 10.85\%$, and (E) $W_v = 15\%$.

Table 8.2 Argand Diagrams Parameters for Brookhaven Soils

Sample Wetness W_v (%)	ρ_s (Ω m)	ρ_∞ (Ω m)	Distribution Parameter, α	Relaxation Time τ (s)	Static Conductivity σ_s (S/m)
8.11 # 1, 4−6′	2300	0	0.0833	$1.34 \cdot 10^{-7}$	0.00043
8.69 # I, 2−4′	1980	0	0.0778	$5.61 \cdot 10^{-8}$	0.00050
9.75 # I, 6−8′	456	0	0.0833	$2.52 \cdot 10^{-8}$	0.00219
10.85 # II, 2−4′	1165	0	0.0833	$1.29 \cdot 10^{-8}$	0.00086
14.67 # I, 6−8′	455	0	0.0778	$3.02 \cdot 10^{-8}$	0.00220
Additional Samples					
8.176 # 1, 4−6′	2700	0			0.00037
8.763 # II, 2−4′	1410	0			0.00071

Table 8.3 Cole−Cole Diagrams Parameters for Brookhaven Soils

Sample Wetness W_v (%)	ε_s	ε_∞	Distribution Parameter, α	Relaxation Time τ (s)
8.11 # 1, 4−6′	8.80	5.03	0.355	3.44×10^{-7}
8.69 # I, 2−4′	8.15	8.55	0.222	2.3×10^{-7}
9.75 # I, 6−8′	8.00	5.23	0.272	4.19×10^{-8}
10.85 # II, 2−4′	9.44	6.31	0.328	6.07×10^{-8}
14.67 # I, 6−8′	10.10	6.4	0.289	6.32×10^{-8}
Additional Samples				
8.176 # 1, 4−6′	2700	0		
8.763 # II, 2−4′	1410	0		

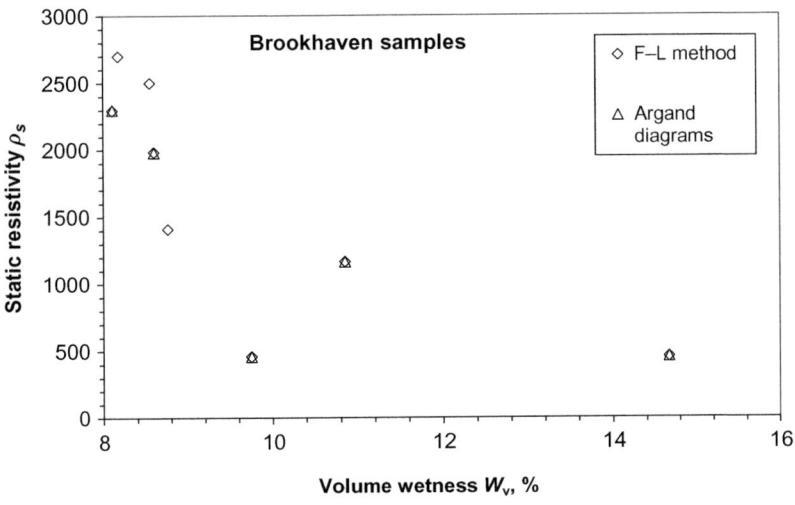

FIGURE 8.8

Static resistivity versus wetness for Brookhaven soil samples.

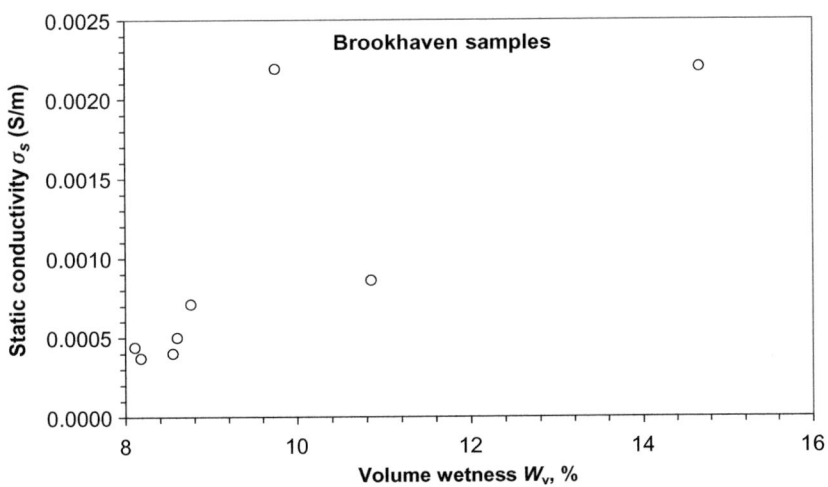

FIGURE 8.9

Static conductivity versus wetness for Brookhaven soil samples.

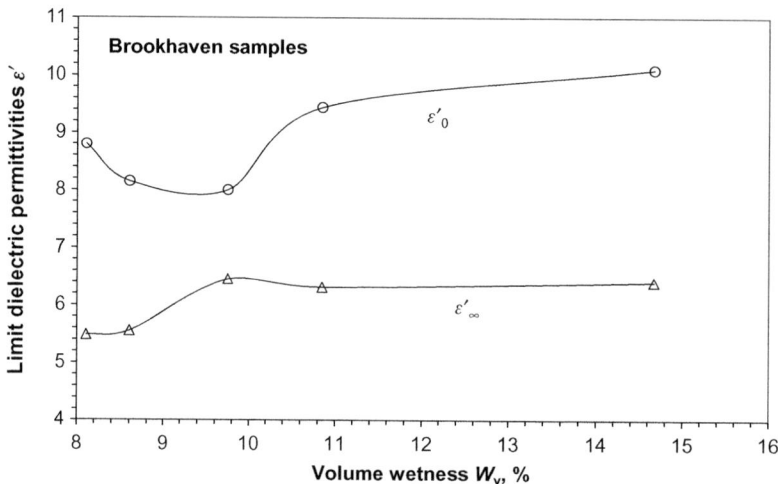

FIGURE 8.10

Static permittivity versus wetness for Brookhaven soil samples.

References

Faust, G.T., 1963. Physical Properties and Mineralogy of Selected Samples of the Sediments from the Vicinity of the Brookhaven National Laboratory, Long Island, New York.

Daniels, J.J., 1998. High resolution GPR at Brookhaven National Laboratory to delineate complex subsurface targets. JEEG 3 (1), 1–5.

Department of Soil, Water & Environmental Science. 1999. Report on Testing Results from the Soil, Water and Plant Analysis Laboratory, University of Arizona.

Soil from Columbia University, New York

The soil, which we received from Dr. Ralf Birken, Columbia University, was washed quartz sand, with no measurable clay. They used the sand in a physical modeling experiment with a 1 GHz GPR system, and our laboratory measurement results were helpful for comparison (Birken and Versteeg, 2000; Slater et al., 2002). The samples with various water contents for our electrical property measurements were prepared by adding distilled water to the dry sand. We measured three samples with volume water contents $W_v = 9.18\%$, 14.22%, and 21.28%. The measurements were performed with a coaxial sample holder, using the HP4194A Impedance Analyzer (with Z-Probe) in a frequency range from 10 kHz to 100 MHz and the HP 4191 A Impedance Analyzer at higher frequencies from 1 MHz to 1 GHz. Fig. 9.1 shows the Fricke–Lopatin graphs, that is, measured resistance R_{meas} versus $100/\omega^{0.5}$ for the three samples, which are analogous to the previous samples (from Avra Valley and Brookhaven). The static resistance, R_s, as well as static resistivity ρ_s and static conductivity σ_s defined from these graphs and their regression equations are shown in Table 9.1. The Fricke–Lopatin graphs for measured reactance X_{meas} are plotted in Fig. 9.2. From the regression equations, for the measured resistance and reactance versus $100/\omega^{0.5}$ (using the coefficients at x), we also defined the electrode polarization values R_{el} and X_{el} for these sand samples and subtracted them from the measured R_{meas} and X_{meas} values during the data processing (Chapter 6: Corrections for stray parameters and error estimation). The processed electrical parameters of the sand samples, free of electrode polarization, as well as of other stray parameters of the measuring system (Chapter 5: Stray parameters of the measuring system and ways of defining them), were plotted in the consequent figures versus frequency.

Fig. 9.3A–C represents the real and imaginary parts of dielectric permittivity for the Columbia sand with volume wetness W_v and also the dielectric components ε''_{diel} (see the procedure in Chapter 6: Corrections for stray parameters and error estimation). The dielectric ε''_{diel} graphs versus frequency reveal some relaxation areas, which are not visible in the total ε'' (f) graphs due to conduction.

Electrical Spectroscopy of Earth Materials. DOI: https://doi.org/10.1016/B978-0-12-818603-9.00009-X

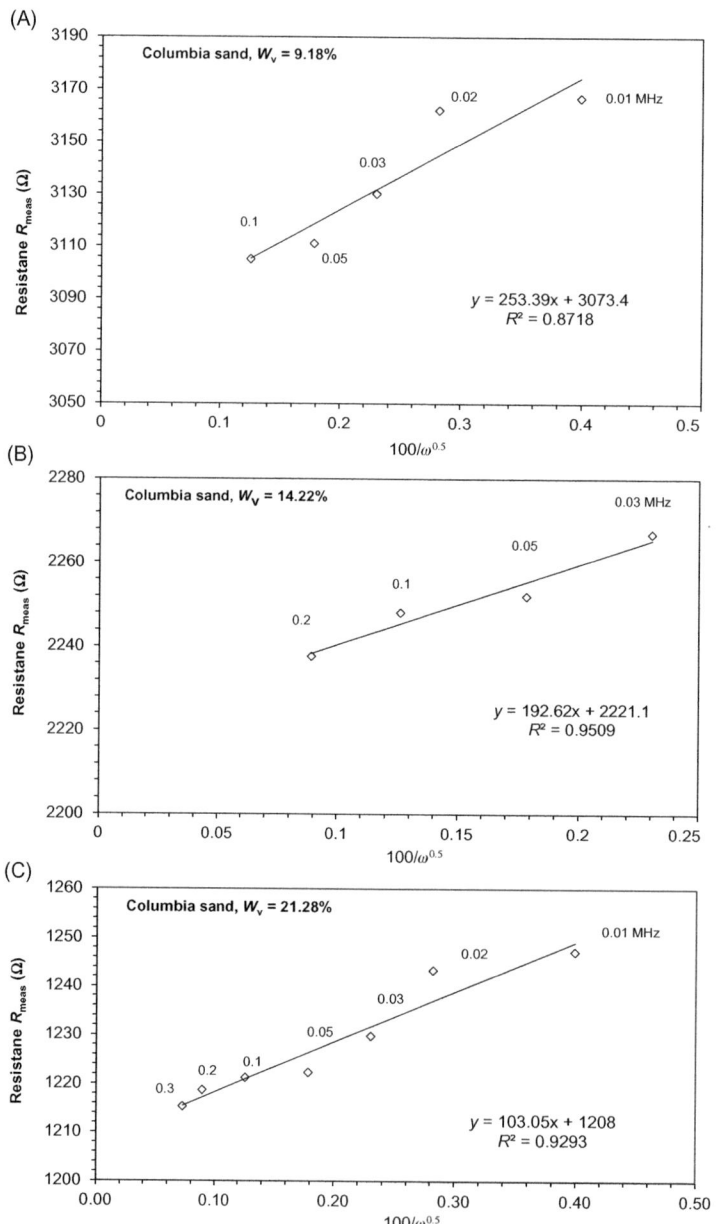

FIGURE 9.1

Measured resistance versus $100/\omega^{0.5}$ for sand samples with wetness (A) $W_v = 9.18\%$, (B) $W_v = 14.22\%$, and (C) $W_v = 21.28\%$.

Table 9.1 Static Parameters Defined for Sand Samples From Columbia With Various Water Contents W_v, Using the Fricke–Lopatin Method

Water Content W_v (%)	Instrument and Electrodes	Static Resistance R_s (Ω)	Static Resistivity ρ_s (Ω m)	Static Conductivity, σ_s (S/m)
9.18	HP4194A, Z-Probe, Coaxial	3073	596	0.00168
14.22	HP4194A, Z-Probe, Coaxial	2221	430	0.00232
21.28	HP4194A, Z-Probe, Coaxial	1208	234	0.00427

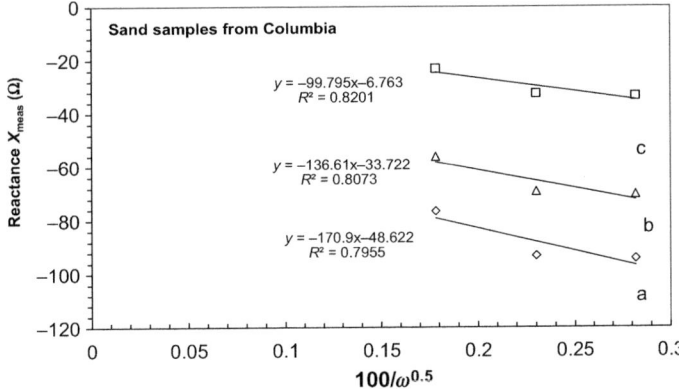

FIGURE 9.2

Measured reactance versus $100/\omega^{0.5}$ for sand samples with various wetnesses (a) $W_v = 9.18\%$, (b) $W_v = 14.22\%$, and (c) $W_v = 21.28\%$.

Similar to the previous data, the imaginary ε'' is accurately approximated with a power function in the lower frequency range, up to 100 MHz. The regression equations are given on the graphs. Analogous power functions describe the imaginary conductivity σ'' mostly in the higher frequency range, up to 1 GHz, while the lower frequency data, down to 100 kHz, also match the trend lines (Fig. 9.4). Fig. 9.5 shows the real ρ' and imaginary ρ'' components of resistivity versus frequency. Each sample reveals dispersion on the ρ' graph and a maximum on the ρ'' graph at its characteristic frequency, which is indicative of a relaxation process. These relaxation processes are represented with Argand diagrams, shown in Fig. 9.6 along with the calculated curves. Using the data of dielectric loss factor ε''_{diel} (Fig. 9.3) the Cole–Cole diagrams were plotted in Fig. 9.7. Here, the calculated curves are also shown. The parameters of Argand diagrams and the

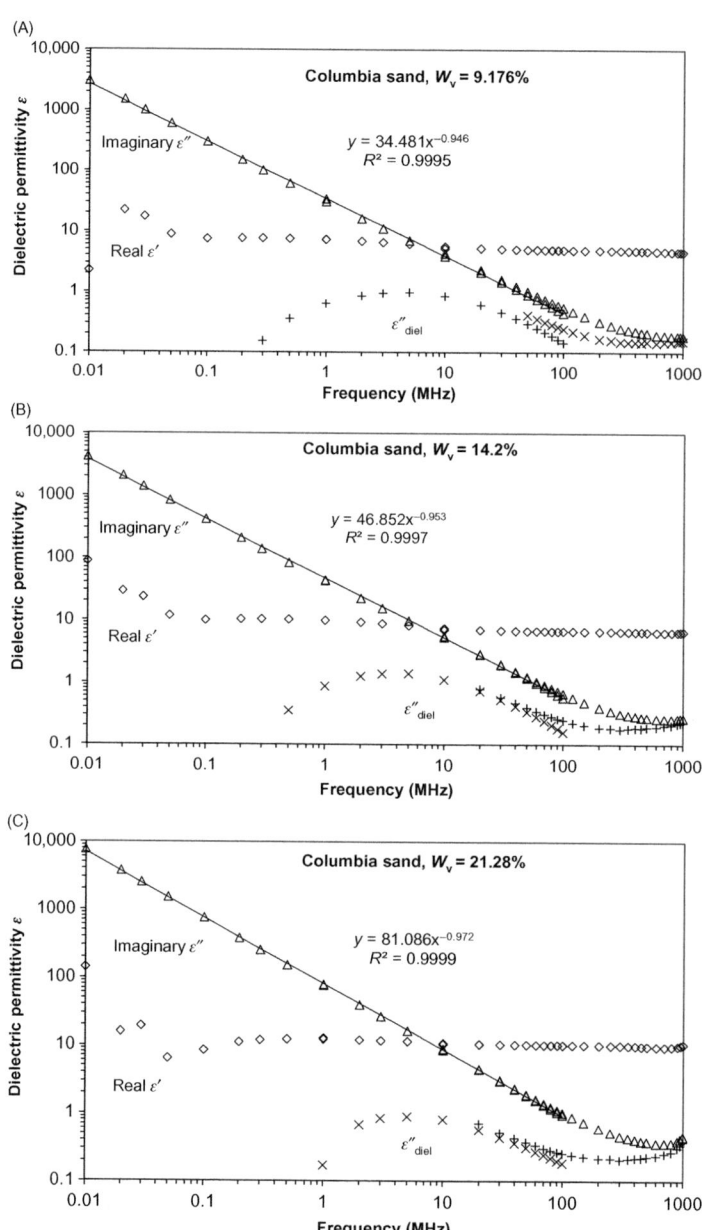

FIGURE 9.3

Dielectric permittivity versus frequency for sand samples from Columbia with volume wetness (A) $W_v = 9.176\%$, (B) $W_v = 14.2\%$, and (C) $W_v = 21.28\%$.

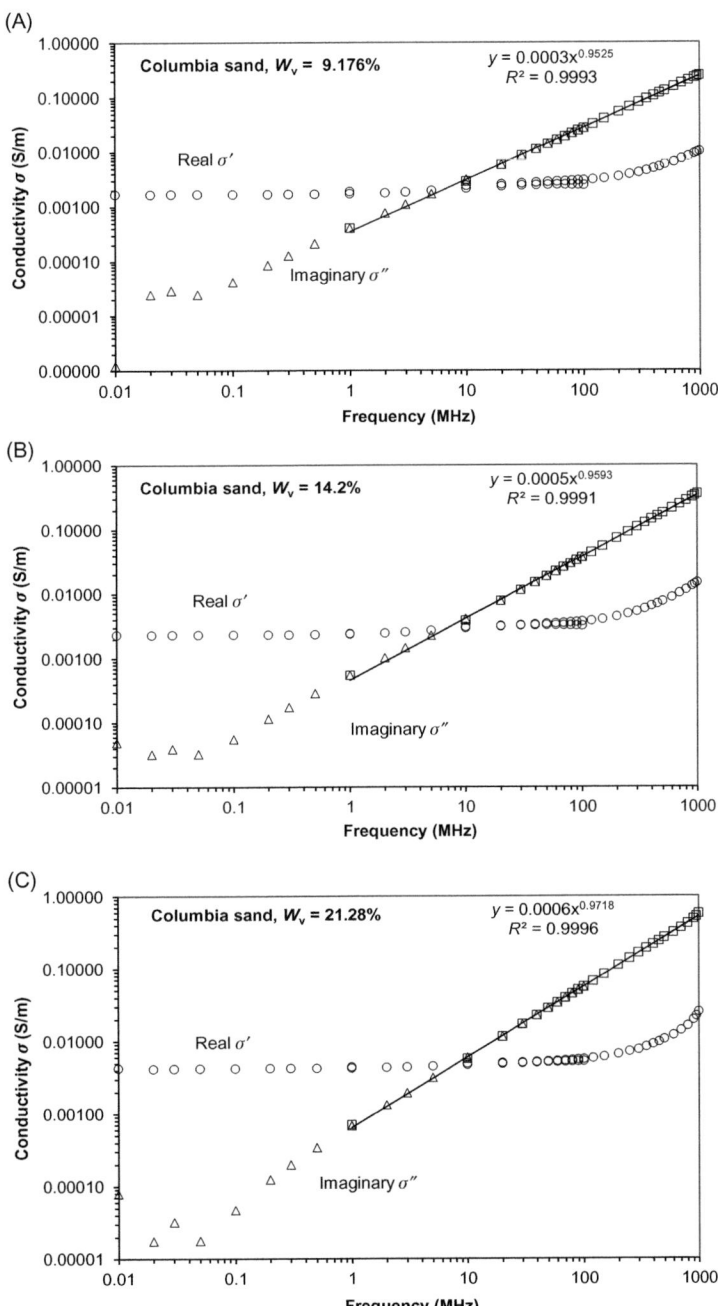

FIGURE 9.4

(A) Conductivity versus frequency for sand samples from Columbia with volume wetness $W_v = 9.176\%$, (B) $W_v = 14.2\%$, and (C) $W_v = 14.2\%$.

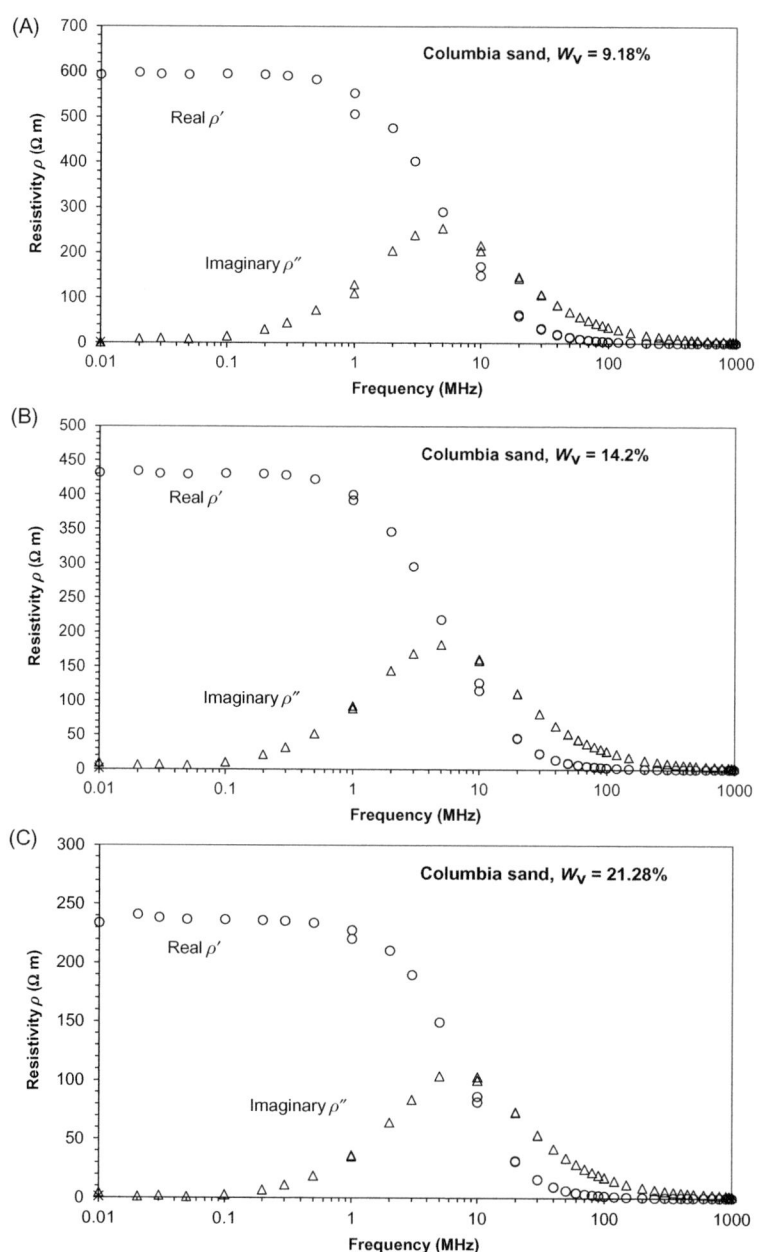

FIGURE 9.5

Resistivity versus frequency for Columbia sands with volume wetness (A) $W_v = 9.18\%$, (B) $W_v = 14.2\%$, and (C) $W_v = 21.28\%$.

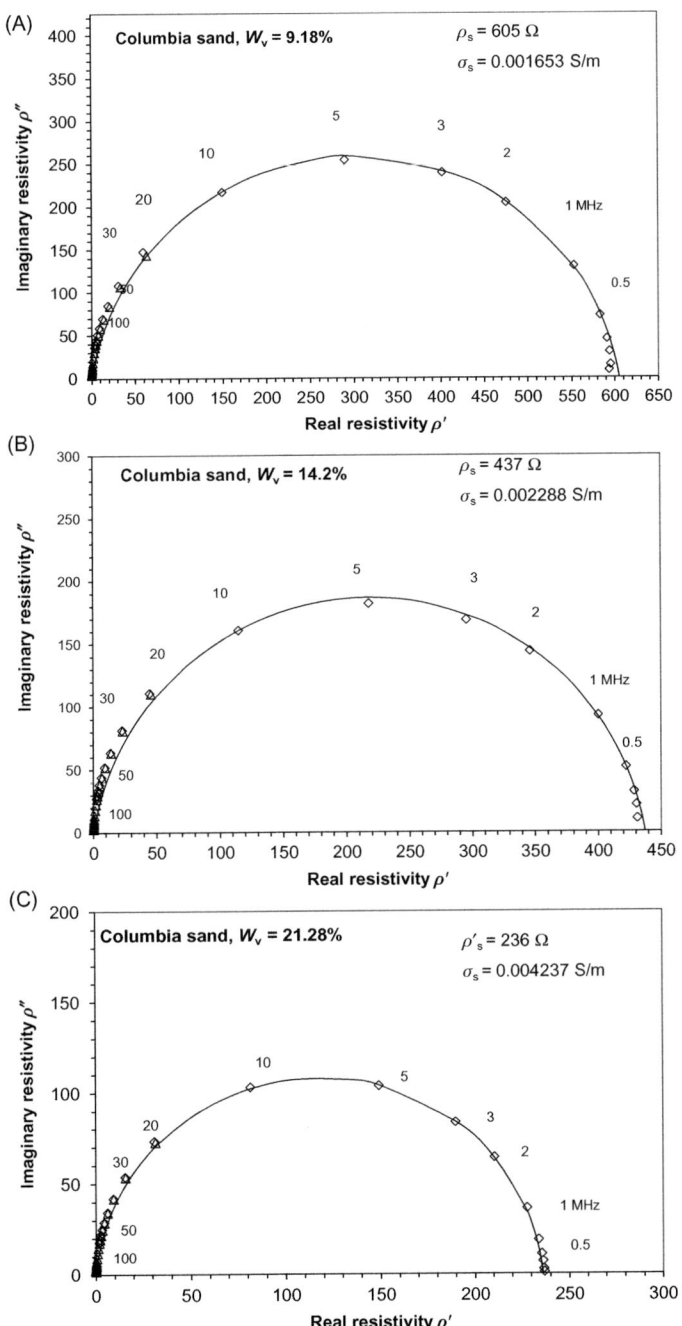

FIGURE 9.6

(A) Argand diagrams for soil samples from Columbia sands with volume wetness $W_v = 9.18\%$
(B) Argand diagrams for Columbia sands with volume wetness $= 14.2\%$. (C) Argand
diagrams for soil samples from Columbia sands with volume wetness $W_v = 21.28\%$.

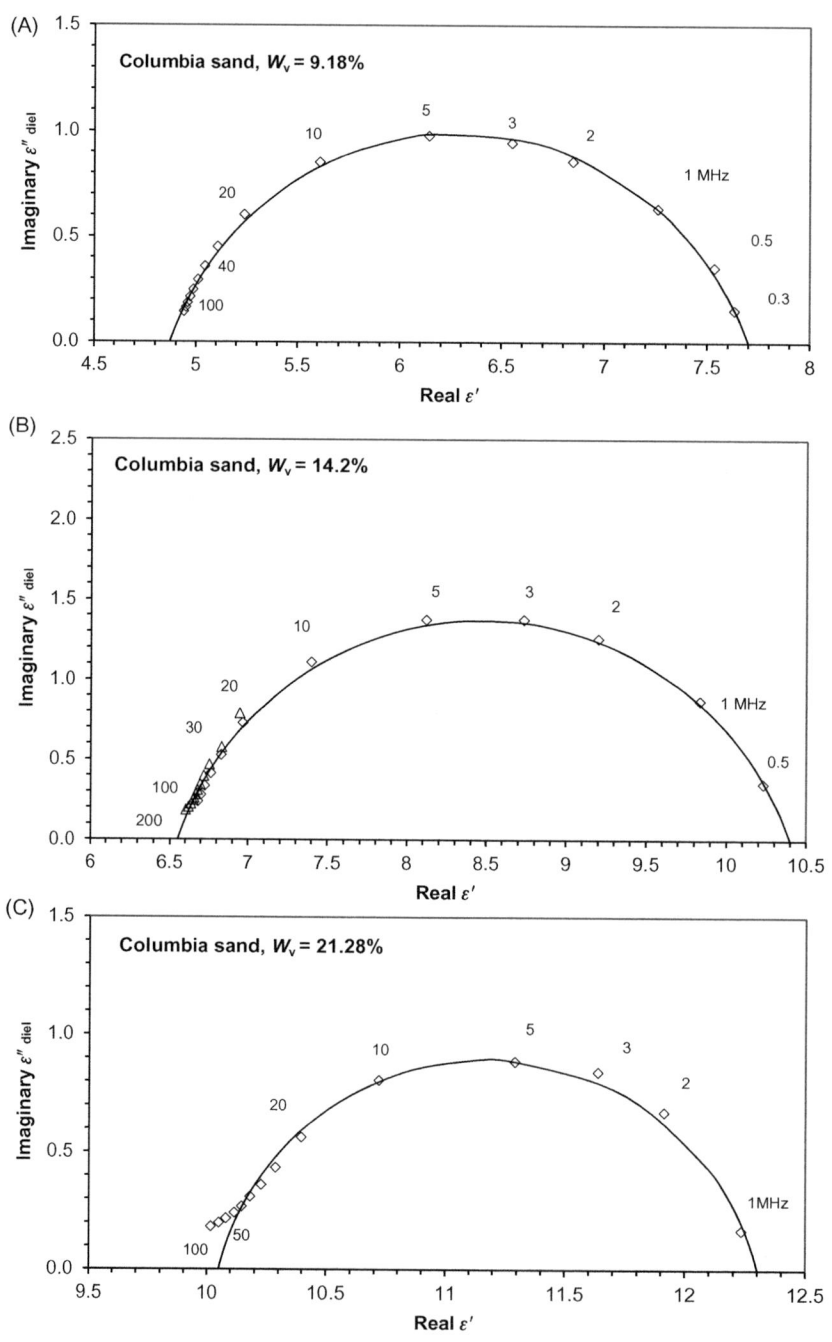

FIGURE 9.7

Cole–Cole diagrams for Columbia sands with volume wetness (A) $W_v = 9.18\%$, (B) $W_v = 14.2\%$, and (C) $W_v = 21.28\%$.

Table 9.2 Argand Diagrams Parameters for Sand Samples From Columbia With Various Water Contents W_v

Sample Wetness, W_v (%)	ρ_s (Ω m)	ρ_∞ (Ω m)	Distribution Parameter α	Relaxation Time τ (s)	Static Conductivity σ_s (S/m)
9.18	605	0	0.10	$3.214 \cdot 10^{-8}$	0.00165
14.22	437	0	0.10	$5.083 \cdot 10^{-8}$	0.00229
21.28	236	0	0.055	$2.296 \cdot 10^{-8}$	0.00424

Table 9.3 Cole–Cole Diagrams Parameters for Sand Samples From Columbia With Various Water Contents W_v

Sample Wetness, W_v (%)	ε_s (Ω m)	ε_∞ (Ω m)	Distribution Parameter α	Relaxation Time τ (s)
9.18	7.070	4.87	0.222	$3.645 \cdot 10^{-8}$
14.22	10.4	6.55	0.211	$4.384 \cdot 10^{-8}$
21.28	12.3	10.05	0.144	$2.832 \cdot 10^{-8}$

Cole–Cole parameters are summarized in Tables 9.2 and 9.3, respectively. It is seen from the tables how the parameters defined for the sand samples depend on their wetness.

References

Birken, R.A., Versteeg, R., 2000. Use of four dimensional ground penetrating radar and advanced visualization methods to determine subsurface fluid flow. J. Appl. Geophys. 43 (2), 148–155.

Slater, L., Versteeg, R., Binley, A., Cassiani, G., Birken, R., Sandberg, S., 2002. A 3D ERT study of solute transport in a large experimental tank. J. Appl. Geophys. 49 (4), 211–229.

Soils from Fort Huachuca, Arizona, Antenna Test Facility

Soil samples from the Fort Huachuca, Antenna Test Facility (ATF), were sent to us by Allison Kipple. This work was related to their modeling of the antennas at this facility.

A sample of the soils that we used for our electrical properties measurements was sent to the Soil, Water, and Plant Analysis Laboratory in the Department of Soil, Water, and Environmental Science at the University of Arizona. Using the hydrometer method, they found that this soil contains 81% sand, 8% silt, and 11% clay (by weight). Note that this measurement is a particle-size analysis. We do not have a mineralogy analysis of the clay fraction.

We show three samples with volume water contents $W_v = 7.2\%$, 14.6%, and 25%. The electrical properties of wet soil samples were measured in a frequency range $f = 10 \text{ kHz} - 100 \text{ MHz}$ using a parallel-plate sample holder with disk electrodes and the HP4194A Impedance Analyzer with Z-Probe. This is indicated in Table 10.1, along with the static parameters, defined for these samples from their Fricke−Lopatin graphs R_{meas} versus $100/\omega^{0.5}$ (Fig. 10.1), analogous to the soil samples from other locations (Chapter 7, Soils from Avra Valley, Arizona, Chapter 8, Soil from Brookhaven, New York, Chapter 9, Soil from Columbia University, New York). The Fricke−Lopatin graphs for the measured reactance X_{meas} are shown in Fig. 10.2. From regression equations of the graphs, displayed

Table 10.1 Static Parameters Defined for Antenna Test Facility Samples From Arizona With Various Water Contents W_v Using the Fricke−Lopatin Method

Water Content W_v, % Sample Number	Instrument and Electrodes	Static Resistance R_s, Ω	Static Resistivity ρ_s, Ω m	Static Conductivity, σ_s, S/m
7.2 # 11	HP4194A, Z-Probe, Disks	600.96	166.96	0.0060
14.6 # 11	HP4194A, Z-Probe, Disks	336.6	90.7	0.0110
25.0 # 11	HP4194A, Z-Probe, Disks	271.75	75.15	0.0133

Electrical Spectroscopy of Earth Materials. DOI: https://doi.org/10.1016/B978-0-12-818603-9.00010-6

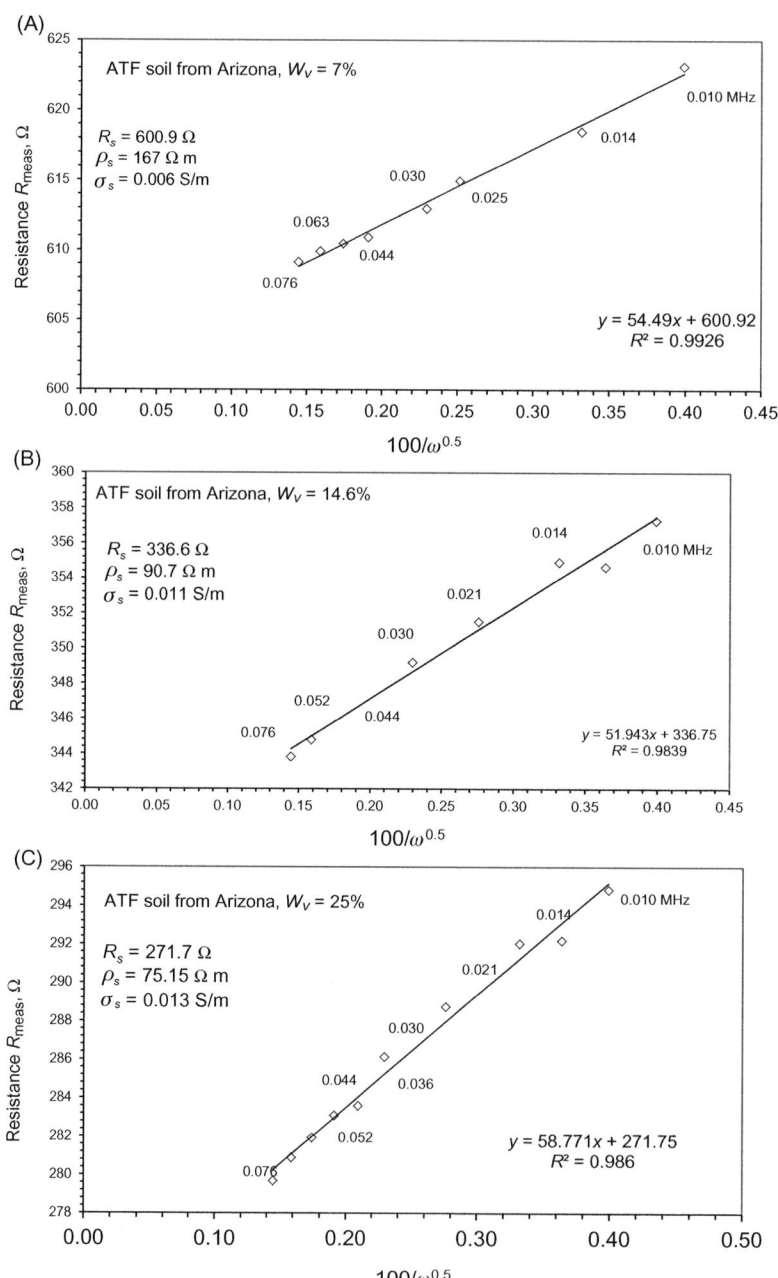

FIGURE 10.1

(A) Measured resistance versus $100/\omega^{0.5}$ for ATF soil from Arizona with wetness $W_v = 7\%$. (B) Measured resistance versus $100/\omega^{0.5}$ for ATF soil from Arizona with wetness $W_v = 14.6\%$. (C) Measured resistance versus $100/\omega^{0.5}$ for ATF soil from Arizona with wetness $W_v = 25\%$. *ATF*, Antenna Test Facility.

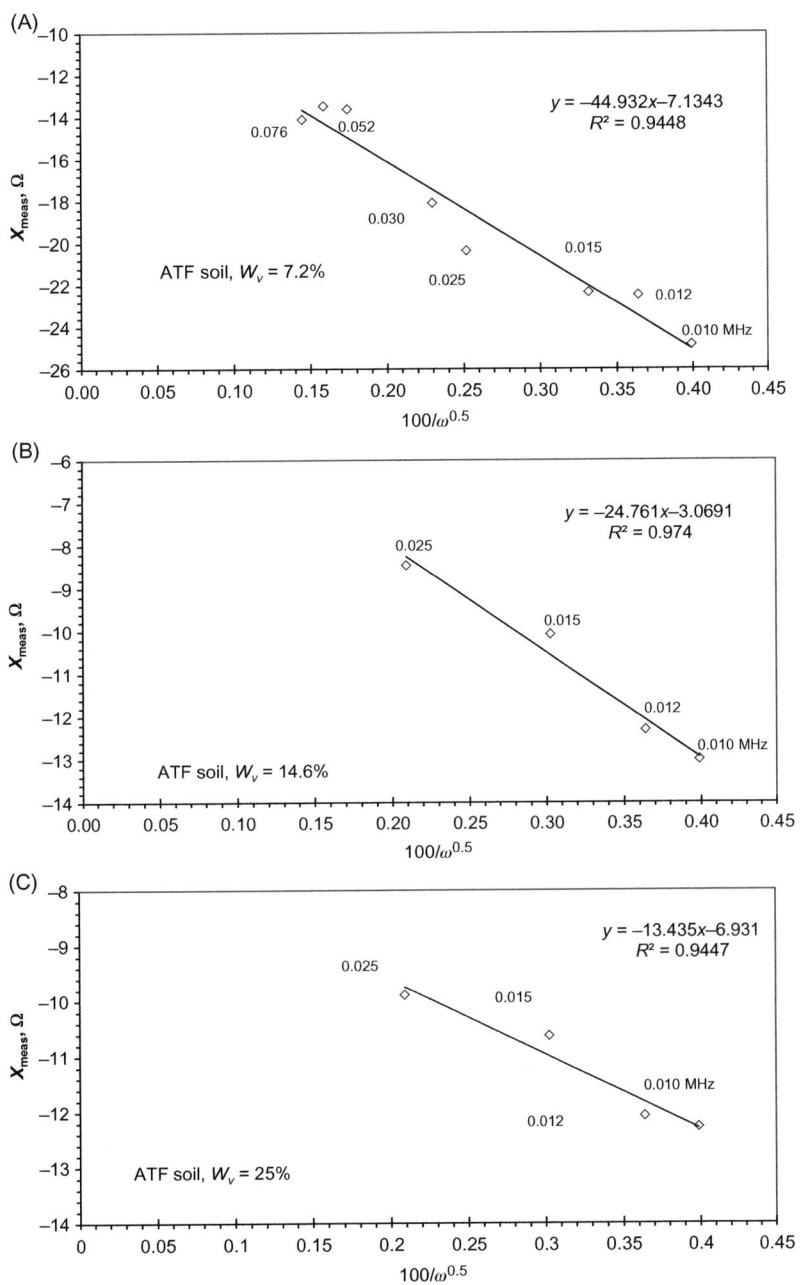

FIGURE 10.2

(A) Measured reactance versus $100/\omega^{0.5}$ for ATF soil from Arizona with wetness $W_v = 14.6\%$. (B) Measured reactance versus $100/\omega^{0.5}$ for ATF soil from Arizona with wetness $W_v = 14.6\%$. (C) Measured reactance versus $100/\omega^{0.5}$ for ATF soil from Arizona with wetness $W_v = 25\%$. *ATF*, Antenna Test Facility.

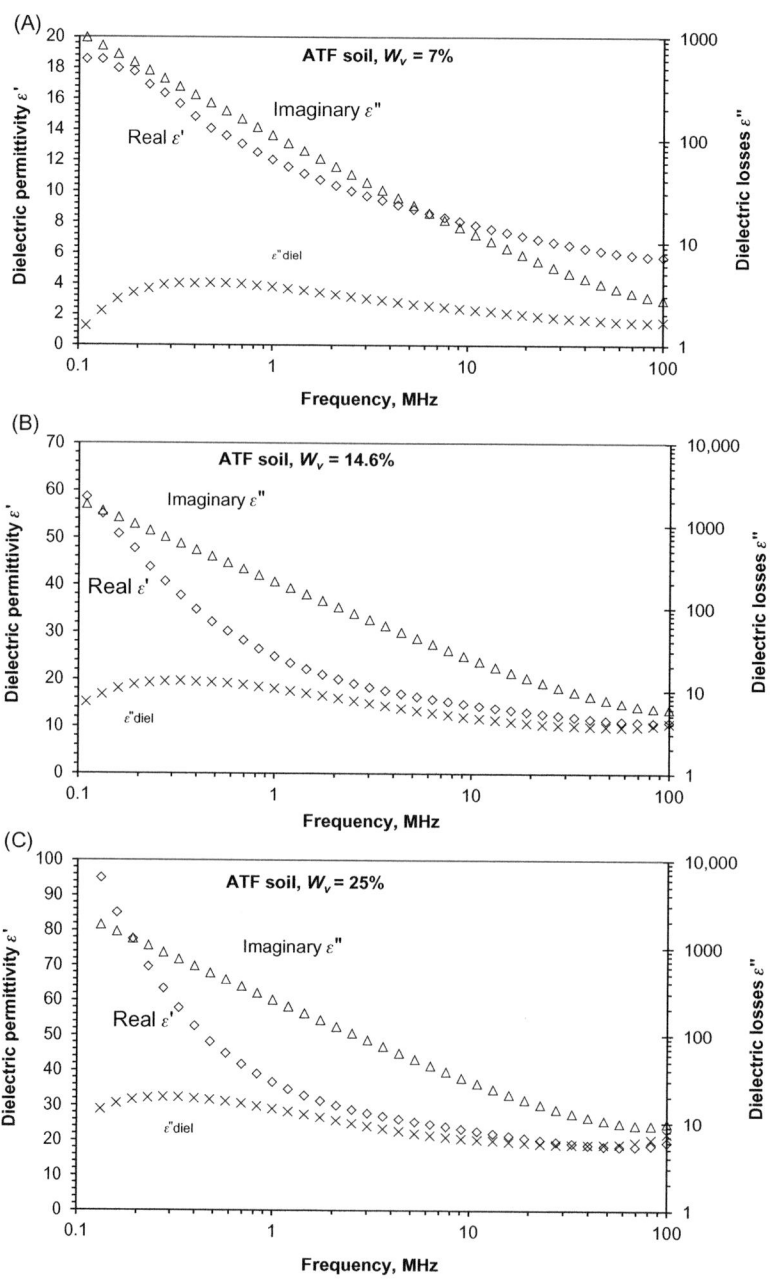

FIGURE 10.3

(A) Dielectric permittivity versus frequency for ATF soil from Arizona with wetness $W_v = 7\%$. (B) Dielectric permittivity versus frequency for ATF soil from Arizona with wetness $W_v = 14.6\%$. (C) Dielectric permittivity versus frequency for ATF soil from Arizona with wetness $W_v = 25\%$. *ATF*, Antenna Test Facility.

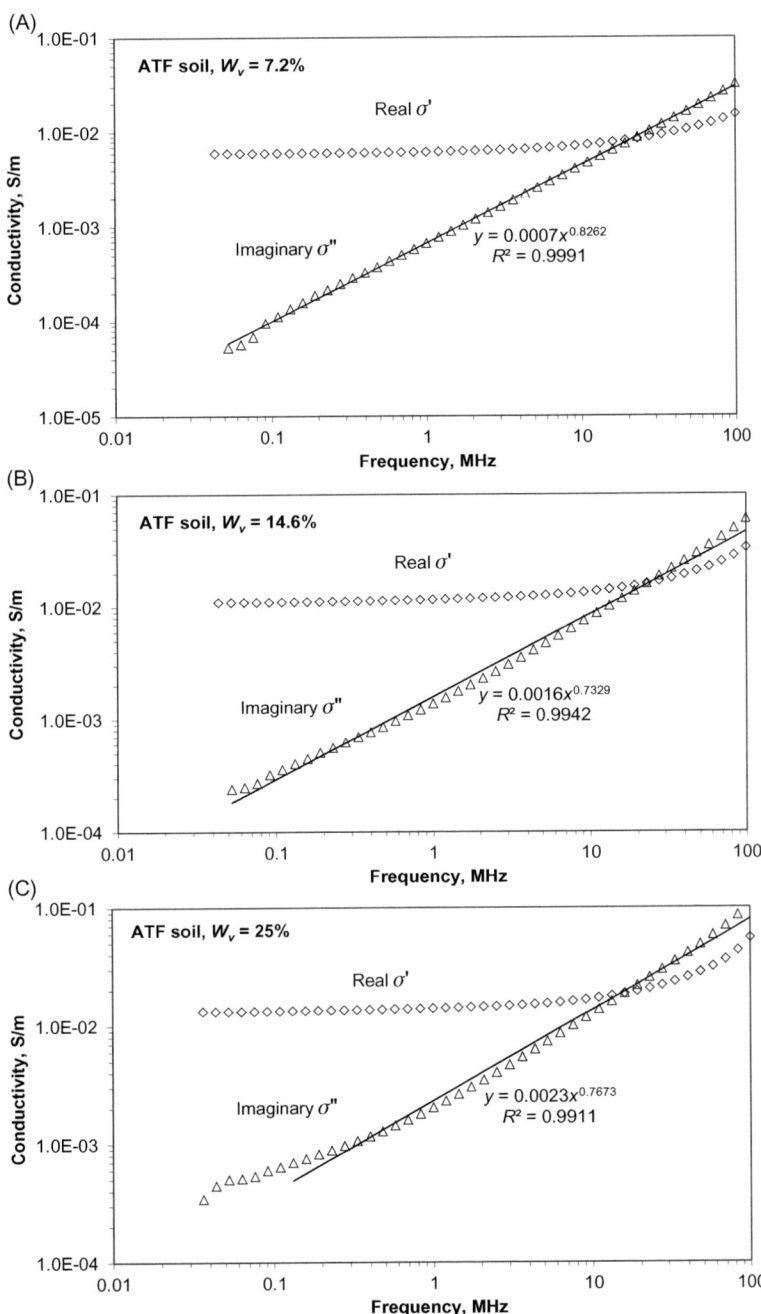

FIGURE 10.4

(A) Conductivity versus frequency for ATF soil sample with water content $W_v = 7.2\%$.
(B) Conductivity versus frequency for ATF soil samples with water contents $W_v = 14.6\%$.
(C) Conductivity versus frequency for ATF soil samples with water contents $W_v = 25\%$.
ATF, Antenna Test Facility.

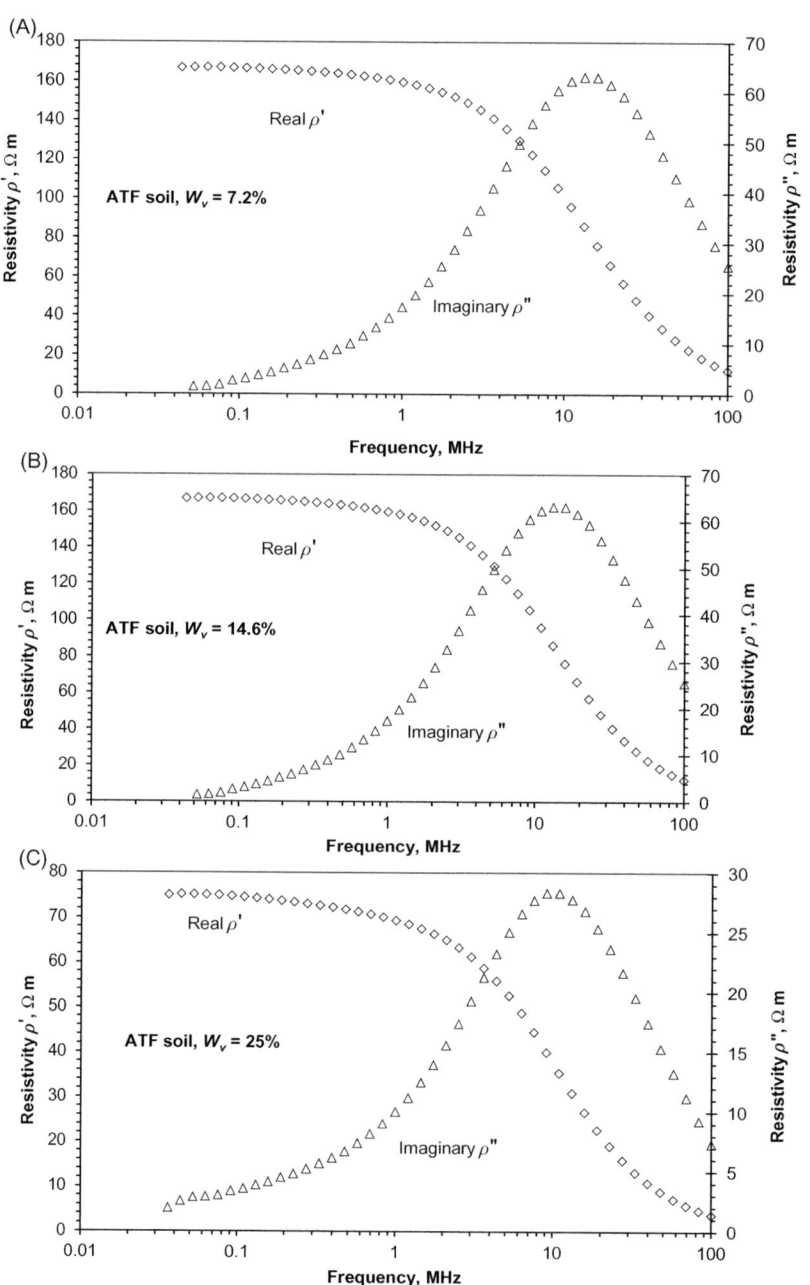

FIGURE 10.5

(A) Resistivity versus frequency for ATF soil from Arizona with wetness $W_v = 7.2\%$.
(B) Resistivity versus frequency for ATF soil from Arizona with wetness $W_v = 14.6\%$.
(C) Resistivity versus frequency for ATF soil from Arizona with wetness $W_v = 25\%$. *ATF,*
Antenna Test Facility.

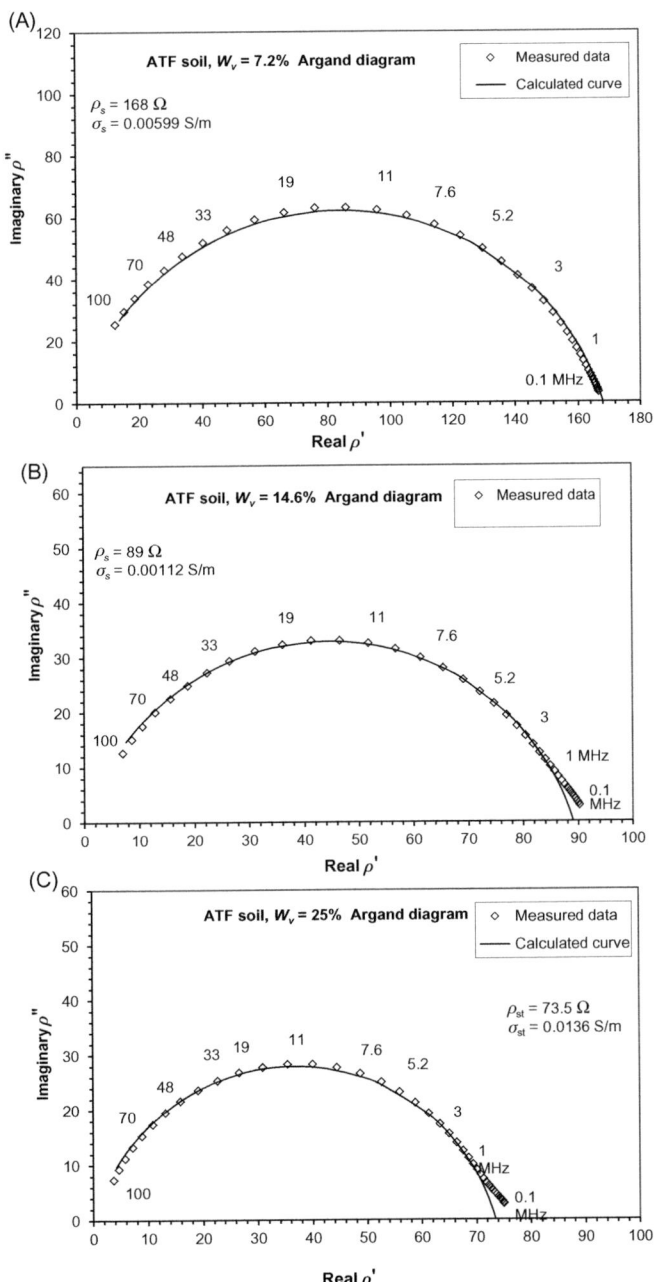

FIGURE 10.6

(A) Argand diagrams for ATF soil from Arizona with wetness $W_v = 7.2\%$. (B) Argand diagrams for ATF soil from Arizona with wetness $W_v = 14.6\%$. (C) Argand diagrams for ATF soil from Arizona with wetness $W_v = 25\%$. *ATF*, Antenna Test Facility.

Table 10.2 Argand Diagrams Parameters for Antenna Test Facility Samples With Various Water Contents W_v

Sample Wetness, W_v, %	ρ_s, Ω m	ρ_∞, Ω m	Distribution Parameter α	Relaxation Time τ, s	Static Conductivity σ_s, S/m
7.2	168	0	0.189	1.209×10^{-8}	0.0060
14.6	89	0	0.189	1.154×10^{-8}	0.0112
25.0	73.5	0	0.172	1.555×10^{-8}	0.0136

in Figs. 10.1 and 10.2, we defined also the real R_{el} and imaginary X_{el} components of the electrode polarization impedance, which were subtracted from the corresponding measured values. In Fig. 10.3, the dielectric permittivity is shown versus frequency for the ATF soil with wetness W_v. In addition to real ε' and imaginary ε'' parts of complex dielectric permittivity, the dielectric component of total electrical losses ε''_{diel} is also plotted. Analogous to soils at our other locations, these graphs of ε''_{diel} (f) also reveal some maximums at low frequencies, where the graphs for total ε'' (f) are approximated with a power function. The regression equations are displayed on the graphs. As seen in Fig. 10.4, power functions describe the imaginary conductivity σ'', mostly extending to higher frequencies. Fig. 10.5 shows the real ρ' and imaginary ρ'' components of resistivity versus frequency for samples with wetness W_v. Each sample reveals a dispersion region on the ρ' graph and a maximum on the ρ'' graph at the same characteristic, frequency that indicates the existence of a relaxation process. These relaxation processes are represented in a form of Argand diagrams in Fig. 10.6, where the calculated curves are also shown. The parameters of Argand diagrams are indicated in Table 10.2. Fig. 10.7 shows Cole–Cole diagrams, based on the data of dielectric losses ε''_{diel} (Fig. 10.3), along with the calculated curves. Cole–Cole parameters are summarized in Table 10.3.

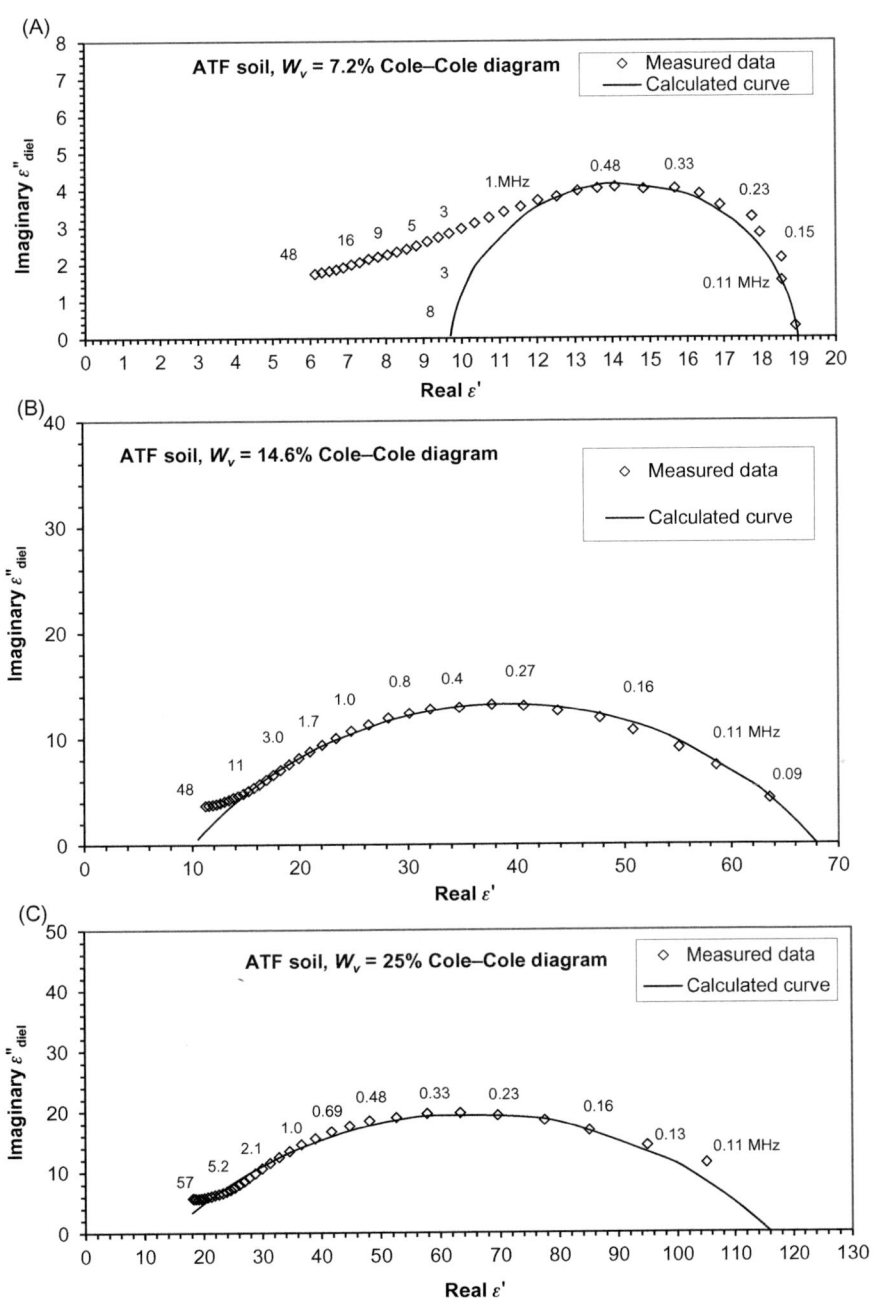

FIGURE 10.7

(A) Cole–Cole diagrams for ATF soil from Arizona with wetness $W_v = 7.2\%$. (B) Cole–Cole diagrams for ATF soil from Arizona with wetness $W_v = 14.6\%$. (C) Cole–Cole diagrams for ATF soil from Arizona with wetness $W_v = 25\%$. *ATF*, Antenna Test Facility.

Table 10.3 Cole–Cole Diagrams Parameters for Antenna Test Facility Samples With Various Water Contents W_v

Sample Wetness, W_v, %	ε_s	ε_∞	Distribution Parameter α	Relaxation Time τ, s
7.2	19	9.7	0.072	3.399×10^{-7}
14.6	68	10	0.456	5.43×10^{-7}
25.0	116	14	0.533	8.366×10^{-7}

Electrical properties of soils versus water content

Chapter Outline

Composition and water content are the main physical parameters that determine electrical properties of soil, such as dielectric permittivity ε and conductivity σ (or resistivity ρ). In this chapter, we will compare the data for four sites with varying compositions. A summary of the compositions is as follows:

1. Avra Valley (Tucson, AZ) soil contains 61% sand, 22% silt, and 17% clay;
2. Brookhaven (Long Island, NY) soil contains 96% clean sand, 1% silt, 3% clay;
3. Columbia University (City of New York) soil consists of washed quartz sand; and
4. Fort Huachuca Antenna Test Facility (ATF) soil 81% sand, 8% silt, and 11% clay.

Note that the clay measurement is a particle-size analysis. We do not have a mineralogy analysis of the clay fraction. We will also study the effect of water content on the soils from these four locations.

Figs. 11.1 and 11.2 represent the real dielectric permittivity ε' and the imaginary dielectric permittivity ε'' (dielectric loss factor) versus frequency for Avra Valley samples with various water contents W_v. These figures show that both components of the complex permittivity decrease with frequency and increase with water content. For comparison, data for an ovendried sample are also shown, which do not change significantly with frequency. Analogous behavior of the dielectric permittivity was observed for samples from other locations (Brookhaven, Columbia, and Ft. Huachuca). This is illustrated in Figs. 11.3–11.5 for the real dielectric permittivity ε' of these soil samples.

Electrical Spectroscopy of Earth Materials. DOI: https://doi.org/10.1016/B978-0-12-818603-9.00011-8

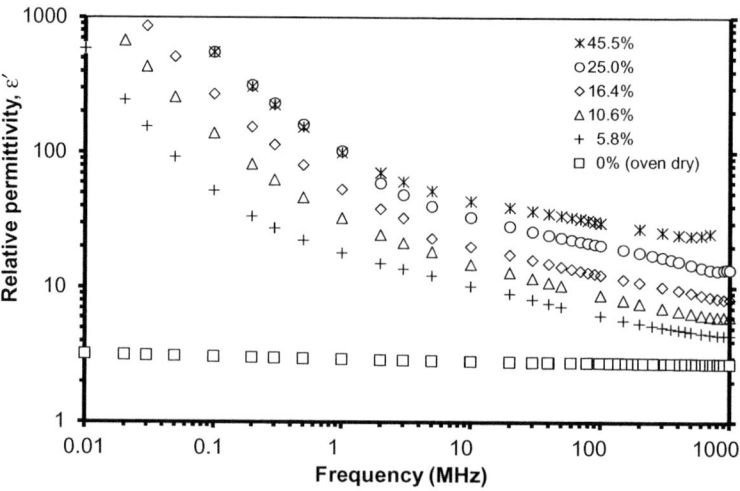

FIGURE 11.1

Real dielectric permittivity ε' versus frequency for soil samples from Avra Valley with various water contents W_v, %.

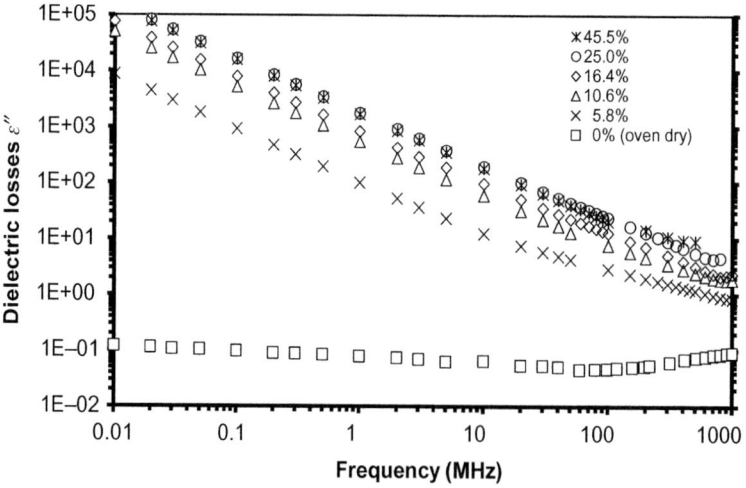

FIGURE 11.2

Imaginary dielectric permittivity ε'' (loss factor) versus frequency for soil samples from Avra Valley with various water contents W_v, %.

In Fig. 11.6, we show real conductivity σ' versus frequency for Avra Valley soils with various water contents. At lower frequencies, up to about 10 MHz, σ' increases slightly, while at higher frequencies, the conductivity

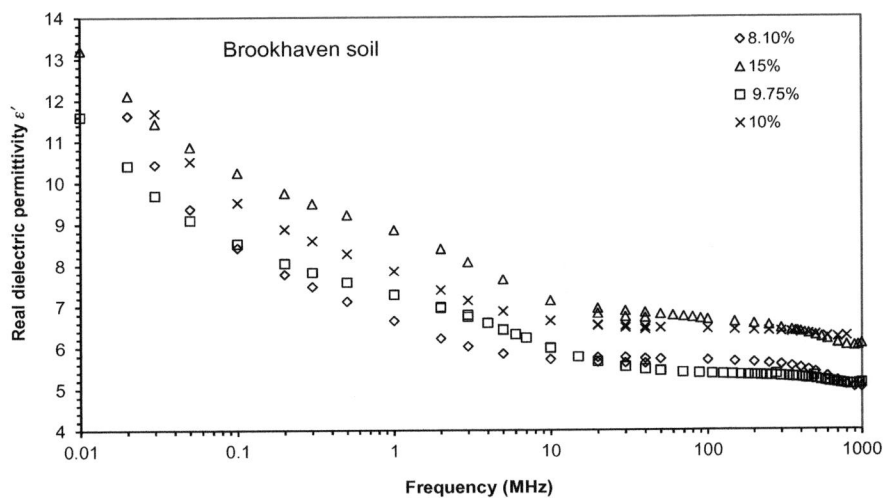

FIGURE 11.3

Real dielectric permittivity ε' versus frequency for soil samples from Brookhaven with various water contents W_v, %.

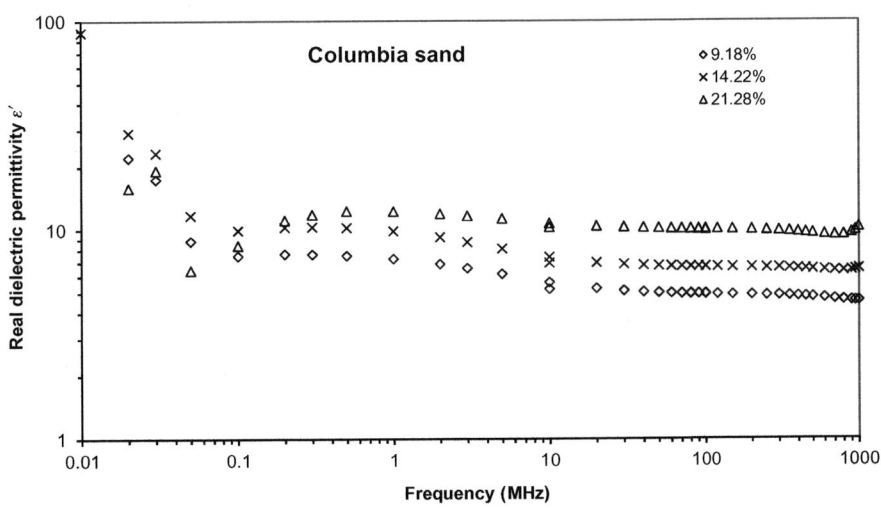

FIGURE 11.4

Real dielectric permittivity ε' versus frequency for soil samples from Columbia with various water contents W_v, %.

increases faster. For a dry sample, the real conductivity σ' increases in the entire frequency range.

Complex resistivity (real ρ' and imaginary ρ'' components) is shown in Fig. 11.7, for only two samples, with $W_v = 5.8\%$ and 16.4%, for clarity. It is seen

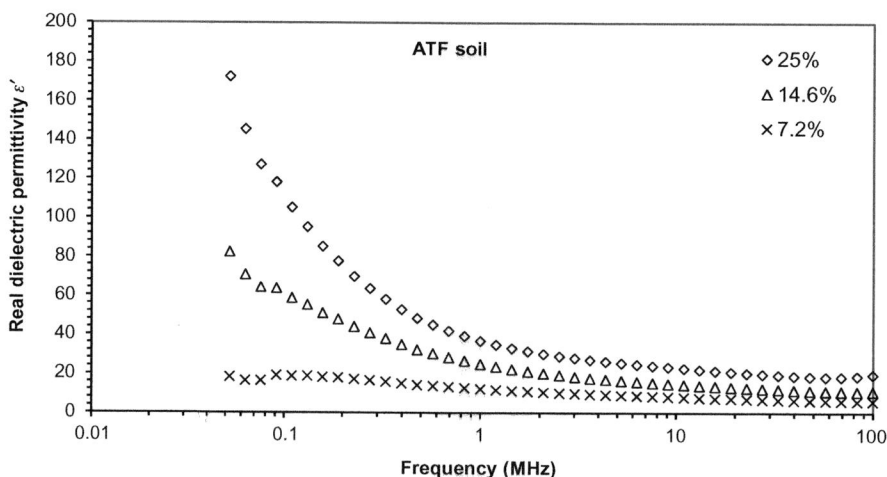

FIGURE 11.5

Real dielectric permittivity ε' versus frequency for ATF soil samples from Ft. Huachuca with various water contents W_v, %. *ATF*, Antenna Test Facility.

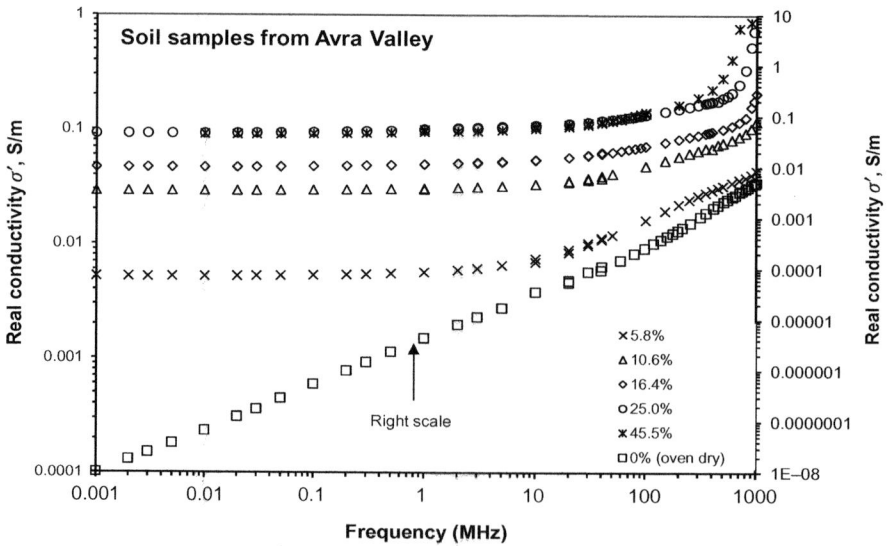

FIGURE 11.6

Real conductivity σ' versus frequency for soils rom Avra Valley with various W_v, %.

that ρ' decreases slightly with frequency, up to10 MHz. At higher frequencies, above 10 MHz, ρ' reveals dispersion, and the imaginary resistivity ρ'' has a maximum. Thus, unlike the complex dielectric permittivity, the complex resistivity

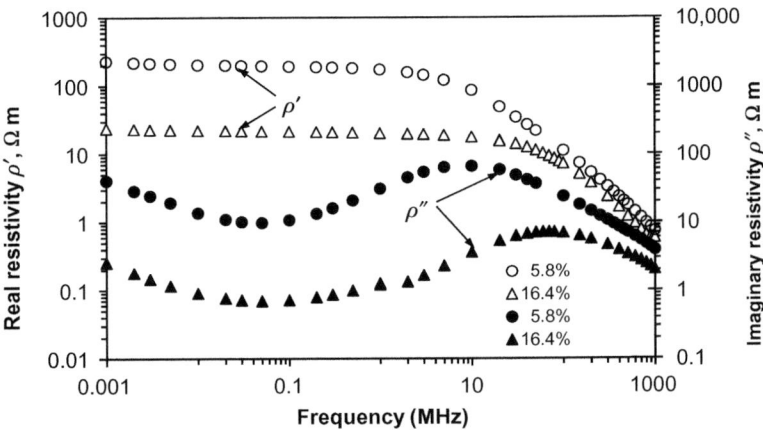

FIGURE 11.7

Complex resistivity (ρ' and ρ'') versus frequency for two soil samples from Avra Valley with various water contents W_v, %.

shows a relaxation process in wet soils. With higher water content in the sample, the relaxation shifts to higher frequencies. The Argand diagrams for these two samples are shown in Section 7.1, along with other samples from Avra Valley. The parameters derived from these diagrams are shown in Table 7.2. Knowing the parameters from the Argand diagrams (ρ'' vs ρ'), as well as from the Cole–Cole diagrams (ε''_{diel} vs ε'), we can investigate possible mechanisms of observed polarization processes in soils, which are dependent on their composition and water content. The observed relaxation processes may be caused by an interfacial (i.e., Maxwell–Wagner) polarization, which often occurs in inhomogeneous systems with a conductive component. Also, orientation polarization of polar groups and molecules may take place. See Section 2.2 and Debye (1945), Von Hippel (1954), Dukhin and Shilov (1974), and Levitskaya and Sternberg (1996).

11.1 Dielectric Permittivity Versus Water Content

In addition to interpretation of the physical process of polarization in soils, the data obtained may also have an important practical application for defining the soil water content, when the electrical parameters are known and vice versa. In particular, the dielectric permittivity is widely discussed in the literature in this respect (Hipp, 1974; Topp et al., 1980; Jackson, 1987; Campbell, 1990; O'Connor and Downing, 1999; Ferre, 1999).

In order to use these data for practical applications, it is convenient to consider the relationships of the real dielectric permittivity ε' versus volumetric water

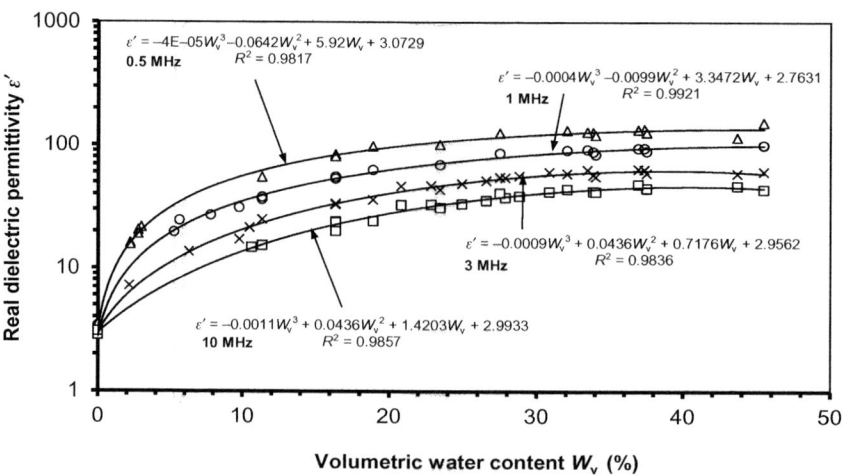

FIGURE 11.8

Graphs of real dielectric permittivity versus water content for Avra Valley soil at various low frequencies.

content W_v in soils at various frequencies. For example, such graphs for samples from Avra Valley are shown in Figs. 11.8 (lower frequencies) and 11.9 (higher frequencies). The regression equations for the graphs allow us to calculate the dielectric permittivity ε' at any water content in soil for the interval studied. These data show that the real dielectric permittivity is a function of, at least, two variables: soil water content W_v and the measurement frequency. The regression equations $\varepsilon' = F(W_v)$ are different for various frequencies; therefore, in order to use them for defining ε' values, the frequency of the equation used must be indicated. At frequencies, 0.5, 1.0 GHz, and above, when the real permittivity is virtually independent of frequency, a single equation for $\varepsilon' = F(W_v)$ may be derived. In Fig. 11.10, we show our data from Avra Valley at frequencies of 0.5–1 GHz with average values, which can be approximated with a second-degree polynomial. This regression equation appears to be more accurate, than the third-degree polynomial in Fig. 11.9, and it has another shape, closer to the graphs in other cases. Similar examples of a single equation ε' versus W_v for high frequencies (0.3–1.4 GHz), and also for soils of various textures, are given in Wang (1980) and Curtis (1994) and shown in Appendix F. In order to understand how much different the soil data are from our various locations, which have different lithology and texture, we made graphs for all of them on one plot, each plot for different frequencies (Figs. 11.11–11.13). These data show that soil from different locations may form separate graphs, except in the case of soils with similar lithology. For example, at 40 MHz, samples from Brookhaven and Columbia, which are mostly clean sand, are closer to each other, than, for example, Avra Valley samples, which contain more clay. The higher the frequency, the closer are the

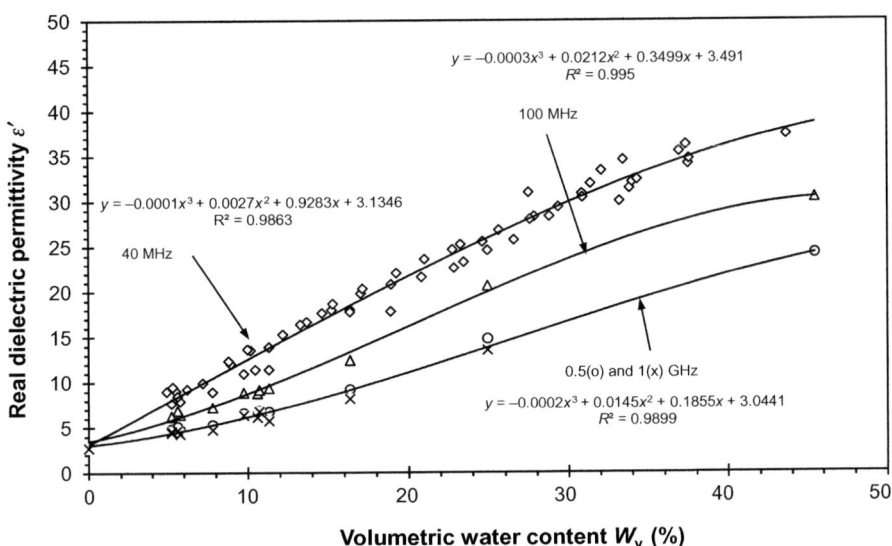

FIGURE 11.9

Graphs of real dielectric permittivity ε' versus water content for Avra Valley soil at various high frequencies.

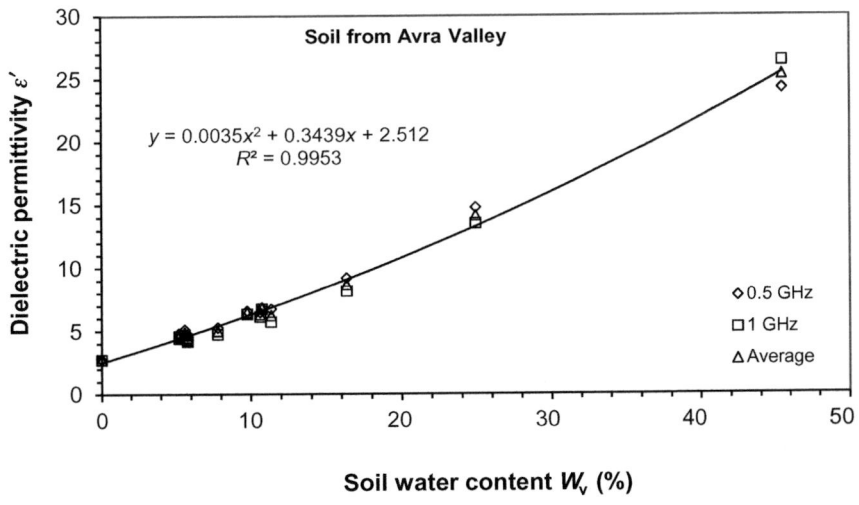

FIGURE 11.10

A single graph of ε' versus W_v at frequencies 0.5–1 GHz with average values.

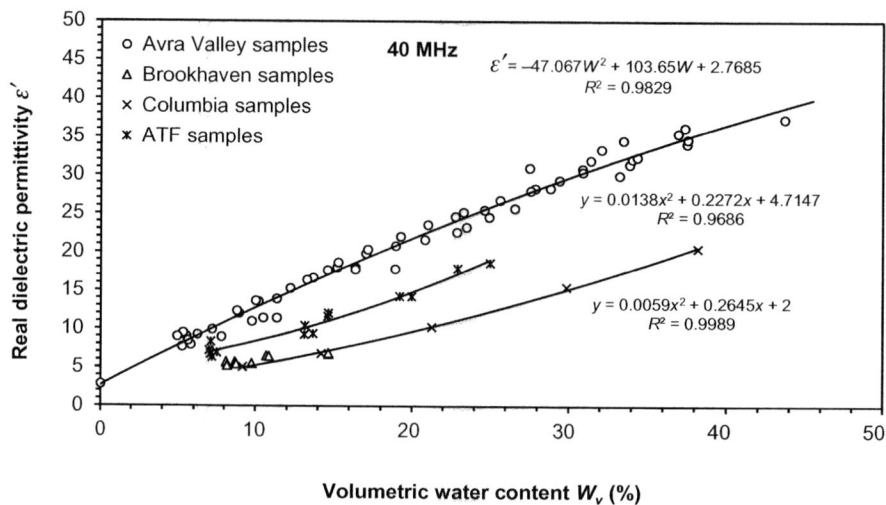

FIGURE 11.11

Real dielectric permittivity versus water content for all our locations at 40 MHz.

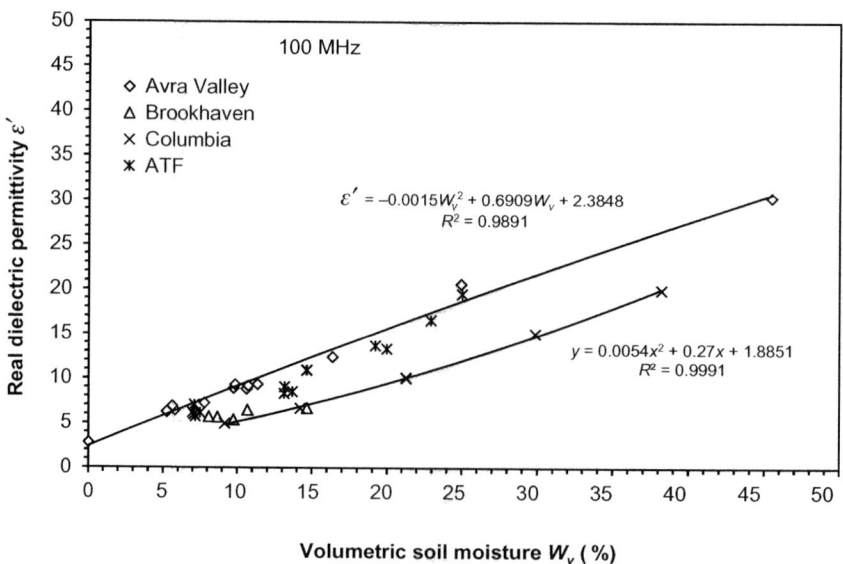

FIGURE 11.12

Real dielectric permittivity versus water content for all our locations at 100 MHz.

data from different locations and at frequencies of 0.5−1 GHz, and where all data form one plot with good repeatability (Fig. 11.13). This is in agreement with data from Wang (1980) and Curtis (1994).

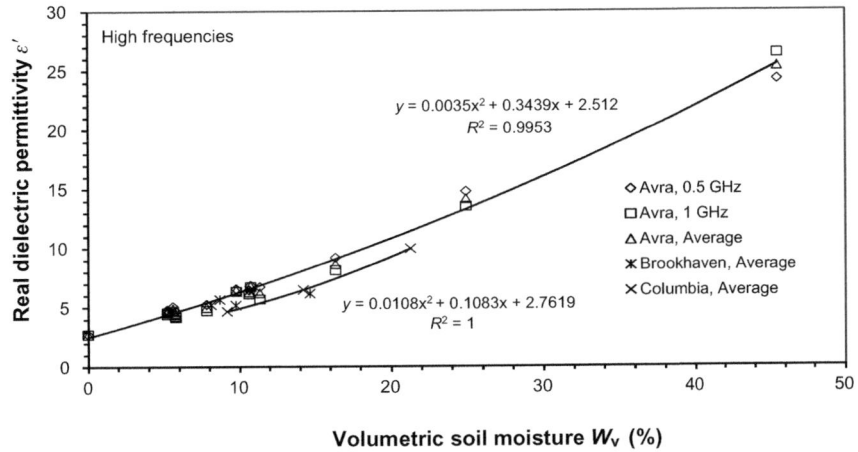

FIGURE 11.13

Real dielectric permittivity versus water content for all our locations at high frequencies.

Analysis of our data shows some special features in the shape of the graphs ε' versus W_v, which are more pronounced at lower frequencies and below 10 MHz. Our data are leveling off and no longer depending on water content (Fig. 11.8). This steady state, or stabilizing effect, is also visible in Figs. 11.1 and 11.2, where at higher water contents (25% and 45%), the graphs ε' and ε'' are nearly coinciding at frequencies below 1 and 100 MHz, respectively. This effect of ε' independence on soil water content is apparently caused by the conduction phenomena that prevail in such wet soils, and it is confirmed by Fig. 11.6 for the real conductivity σ'. Here, the steady-state graphs are also seen below 100 MHz, for soil with water contents of 25% and 45%.

In order to define the water content in soil when the real dielectric permittivity is known, we made the inverse graphs from the same experimental data and found the regression equations $W_v = F\ (\varepsilon')$, which are third-degree polynomials (Fig. 11.14). A more precise plot of W_v versus ε' for an average of high frequencies of 0.5 and 1 GHz (Fig. 11.15) can be well approximated with a second-degree polynomial, and it has a different shape. Analogous relationships were obtained by Wang (1980) and Curtis (1994).

11.2 Resistivity Versus Water Content

Analogous to dielectric permittivity, the resistivity of wet soils can also be used for defining the water content and wise versa. In Section 11.1, we presented our measurement results of Dielectric permittivity for Avra Valley

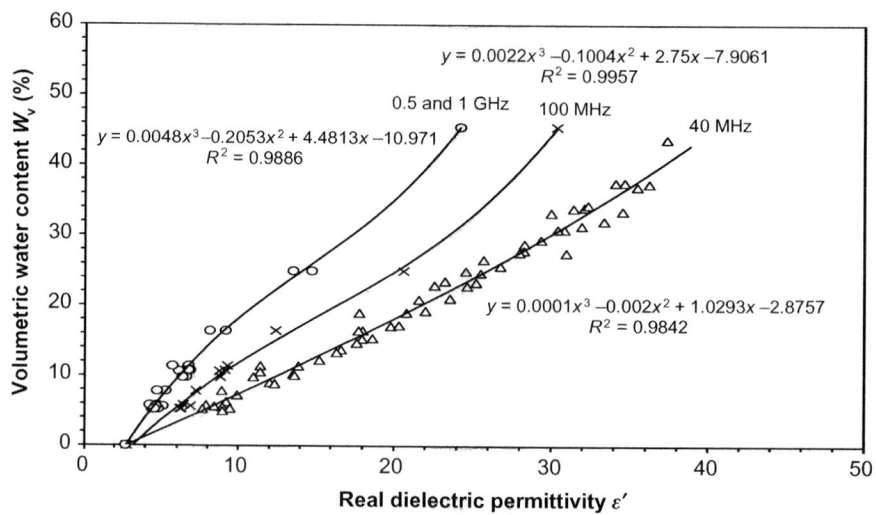

FIGURE 11.14

Soil water content versus dielectric permittivity at various frequencies for Avra Valley samples.

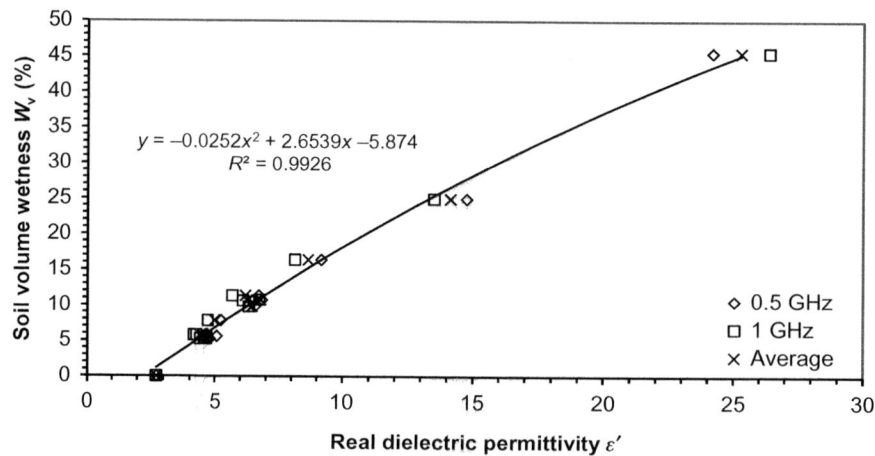

FIGURE 11.15

A single graph of W_v versus ε' at frequencies 0.5−1 GHz with average values for Avra Valley samples.

samples with various water contents, as well as for the other locations we studied. The measurements were done in a frequency range $f = 1\,kHz{-}1\,GHz$. Similarly, Figs. 11.16 and 11.17 in this chapter show frequency dependencies of resistivity for Avra Valley samples with various water contents. For other

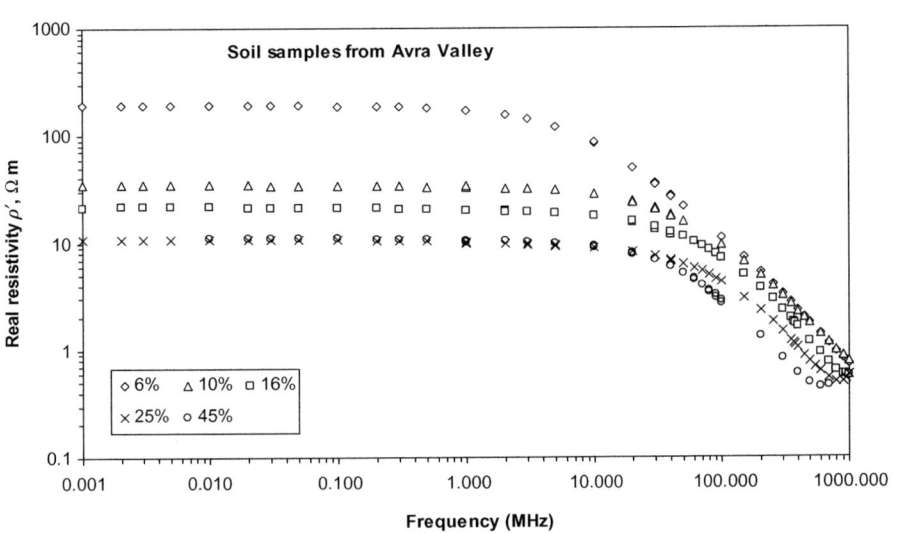

FIGURE 11.16

Real resistivity ρ' versus frequency for soil samples from Avra Valley with various water content W_v, %.

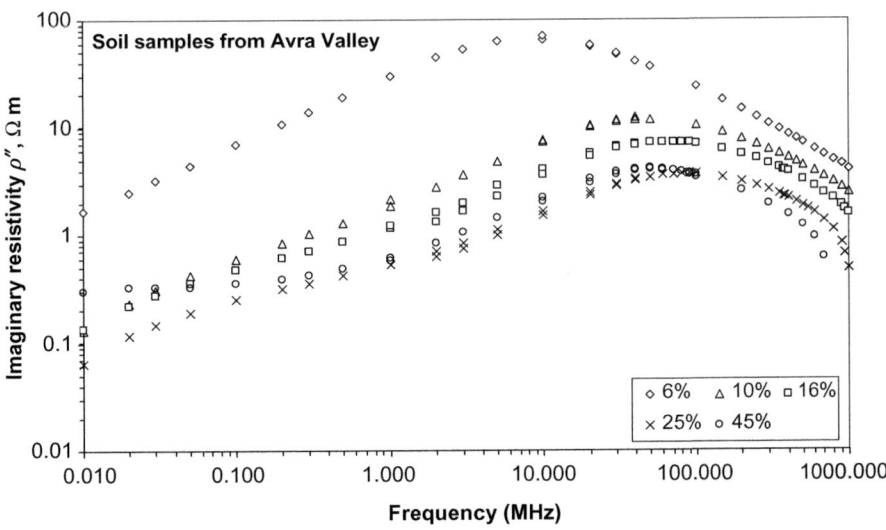

FIGURE 11.17

Imaginary resistivity ρ'' versus frequency for soil samples from Avra Valley with various water content W_v, %.

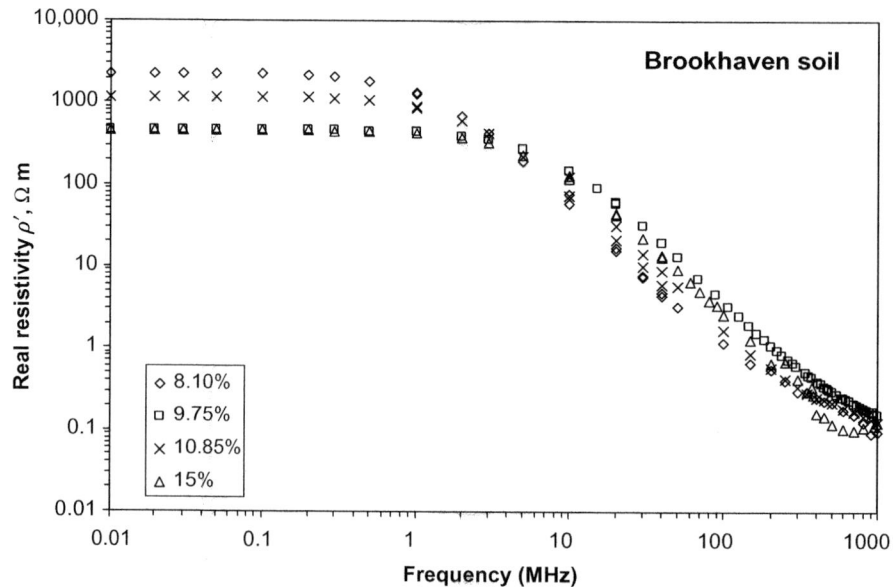

FIGURE 11.18

Real resistivity ρ' versus frequency for Brookhaven soil samples with various wetness W_v, %.

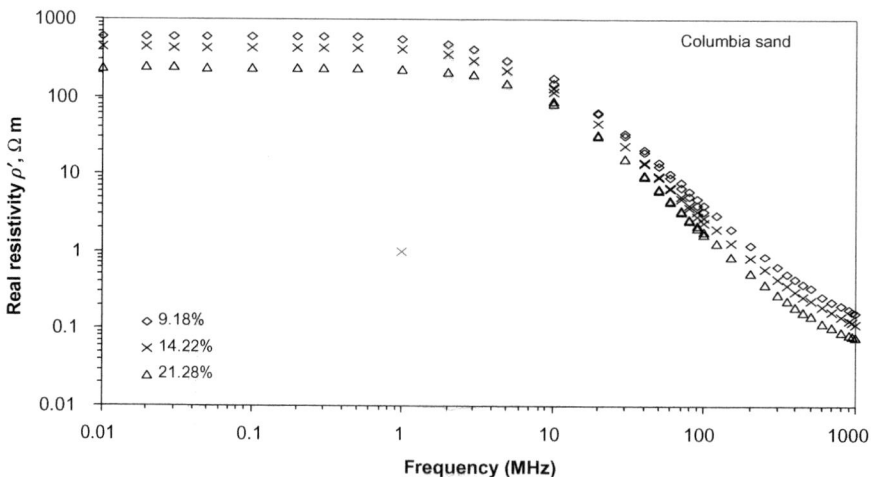

FIGURE 11.19

Real resistivity ρ' versus frequency for Columbia soil samples with various wetness W_v, %.

locations (Brookhaven, Columbia, and ATF) only real resistivity data ρ' are shown (Figs. 11.18–11.20). As seen, resistivity, as well as dielectric permittivity, depends on two variables, measurement frequency f and water content W_v.

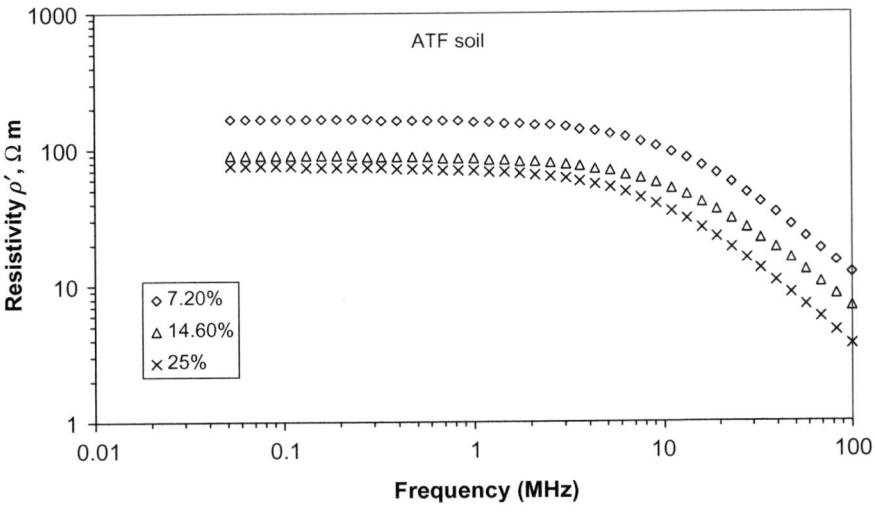

FIGURE 11.20

Real resistivity ρ' versus frequency for ATF soil with various water content W_v. ATF, Antenna Test Facility.

Each location shows a region of low frequencies, below 1 MHz, where the resistivity is independent of frequency. In order to use the resistivity data for defining water content in soils, we considered relationships of the real resistivity ρ' versus volumetric water content W_v in soils at various frequencies. Such data for Avra valley are shown in Fig. 11.21. The regression equation at each frequency, shown on the graphs, can be used for calculating the real resistivity at any water content inside the interval studied. In this figure, the graph of static resistivity is also shown. Static resistivity ρ_{st} is a limit at low frequency, when the frequency approaches zero, and can be considered as most informative. Its regression equation is close to relatively low-frequency data, when taken at 0.5−10 MHz.

Fig. 11.22 shows static values of real resistivity versus water content for soils from all the locations studied. The graphs for each location are clearly separated, except for soil from Brookhaven and Columbia, which are relatively clean sand, and they form essentially one graph. Avra Valley samples contain a significant amount of clay and are representative of most alluvial basins in southern Arizona. All graphs are well approximated with power functions, and the corresponding regression equations can be used for defining the resistivity at any water content in the studied interval.

In order to define the water content from known resistivity values, we made the inverse graphs from the same experimental data for each location, as shown in Fig. 11.23. The data are approximated with power functions, which can be used for calculating the water content in soil, when the static resistivity is known.

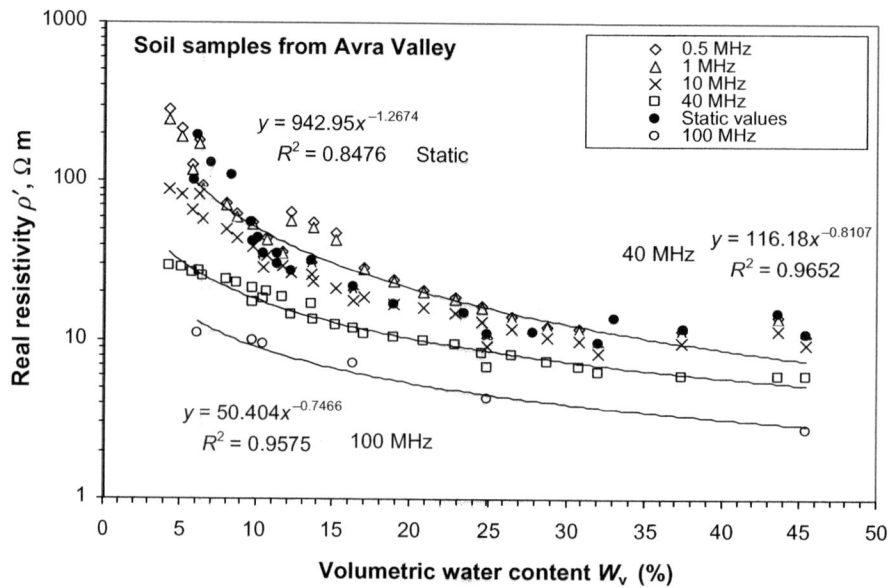

FIGURE 11.21

Real resistivity ρ' versus water content for Avra Valley soil at various frequencies.

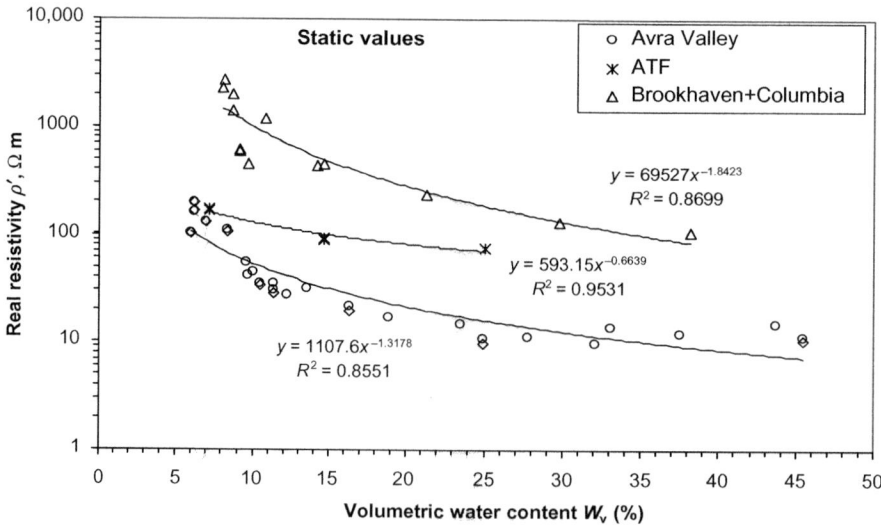

FIGURE 11.22

Static values of real resistivity ρ' versus water content W_v, % for soil from various locations.

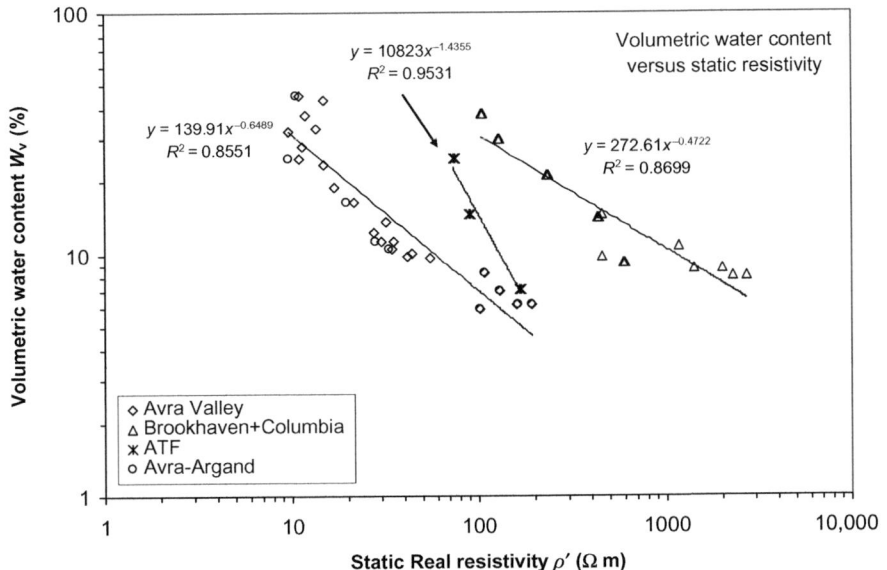

FIGURE 11.23

Water content versus static resistivity for soils from various locations.

11.3 Summary of Electrical Property Parameters for Geophysical Surveys at Various Locations and Varying Water Contents

In this section, we show a summary of key electrical property measurements for geophysical surveys, such as ground penetrating radar (GPR) surveys and electromagnetic (EM) surveys in the 1 kHz to 1 GHz frequency range. We incorporate data that are representative of a wide range of geographic locations and a wide range of environmental conditions.

Two key parameters for geophysical surveys are the attenuation constant α, and the phase velocity V_p, which are calculated from the measured electrical properties (ε' and ε''). Note that the symbol α in this chapter is the attenuation constant for EM fields in the material (as required by convention). In chapters 3, 7–10, and 13 we used the symbol α for the Cole–Cole and Argand diagram distribution parameter (as required by convention). From the dielectric parameters, we can define the complex wave-propagation constant $\gamma = \alpha + j\beta$ for the given material, where α is the attenuation constant, and β is the phase constant. The attenuation constant α and phase constant β are related to the dielectric parameters of a material as follows (Von Hippel, 1954):

$$\alpha = 8.686 \cdot \omega \cdot \sqrt{\frac{\varepsilon' \cdot \varepsilon_0 \cdot \mu' \cdot \mu_0}{2} \cdot \left(\sqrt{1 + \tan^2\delta} - 1\right)} \qquad (11.1)$$

$$\beta = \omega \cdot \sqrt{\frac{\varepsilon' \cdot \varepsilon_0 \cdot \mu' \cdot \mu_0}{2} \cdot \left(\sqrt{1 + \tan^2 \delta} + 1\right)} \qquad (11.2)$$

where α is in dB/m, β is in radian/m, $\mu' = 1$ (for a nonmagnetic material) is the real part of the relative magnetic permeability $\hat{\mu}/\mu_0$, and $\mu_0 = 1.257 \times 10^{-6}$ Henry/m is the magnetic permeability of vacuum. Phase velocity $V = \omega/\beta$ can be expressed as follows (in m/s):

$$V = \frac{1}{\sqrt{(\varepsilon' \cdot \varepsilon_0 \cdot \mu' \cdot \mu_0)/2 \cdot \left(\sqrt{1 + \tan^2 \delta} + 1\right)}} \qquad (11.3)$$

The attenuation constants α with various volume moisture contents are shown for Avra Valley samples in Fig. 11.24, and the phase velocity V_p with various volume moisture contents are shown for Avra Valley samples in Fig. 11.25. Tables with the data are in Appendix D and the spreadsheet with equations shown is in Appendix E.

The Avra Valley soil samples are representative of many soils throughout the southwestern US basins. The high attenuation rate (Fig. 11.24) for these soils explains the very limited depth of investigation of GPR surveys in this region (Sternberg and McGill, 1995). These high-loss soils are also representative of conditions at many other sites around the world.

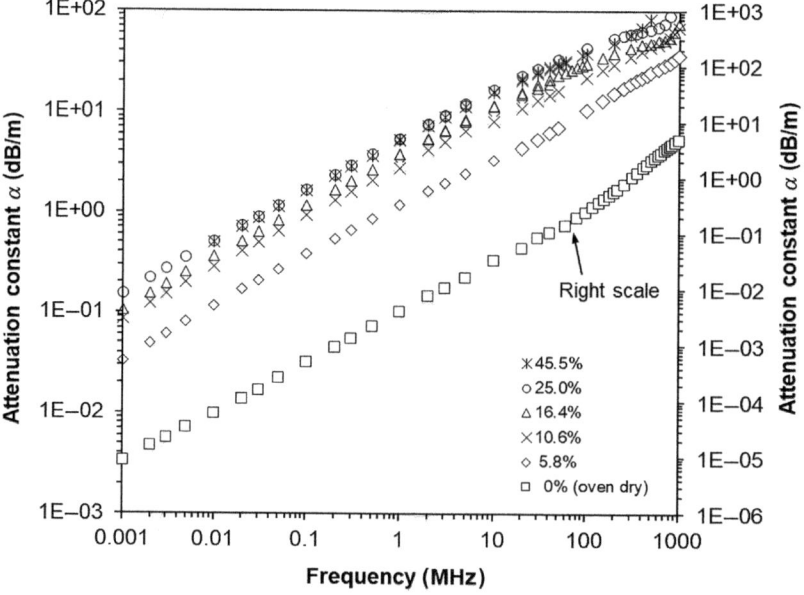

FIGURE 11.24

Attenuation constant α versus frequency for soils from Avra Valley with various moisture contents W_v.

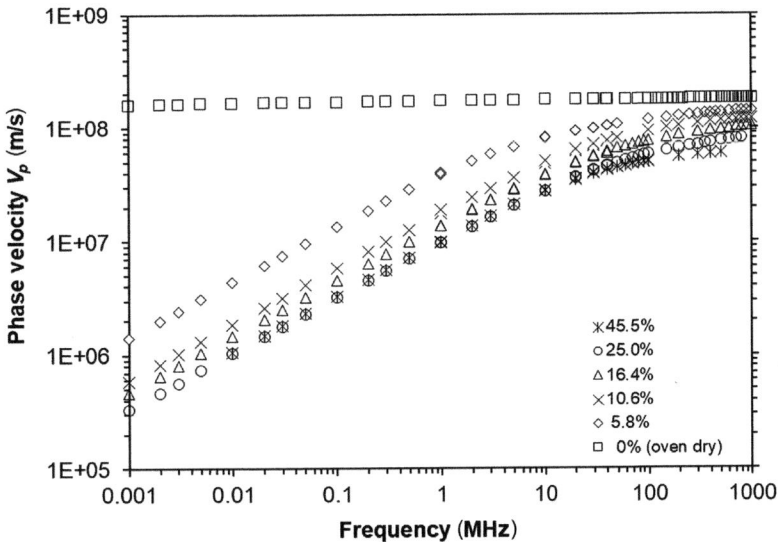

FIGURE 11.25

Phase velocity V_p versus frequency for soils from Avra Valley with various moisture contents W_v.

In contrast with the high-loss soil samples from Avra Valley, we have also studied low-loss soil samples from Brookhaven National Laboratory. Brookhaven National Laboratory is located on Long Island, New York. The soil samples from this location consist of relatively clean sand.

In Figs. 11.26–11.29, we show the electrical and propagation properties of a typical low-loss soil sample from Brookhaven with $W_v = 9.75\%$ in comparison with the data for a typical high-loss sample from Avra Valley with a very similar volume moisture content, i.e., $W_v = 10.6\%$. The dry bulk density values for these soil samples are also very close ($\rho_b = 1.50$ and 1.46, respectively). We have chosen to use a volumetric moisture content of $\sim 10\%$ since this is representative of soils at depths greater than a few centimeters, but above the water table.

As seen in Figs. 11.26 and 11.27, soils from Brookhaven, which contain very little clay, exhibit lower values of real dielectric permittivity ε' and dielectric losses ε'' than the Avra Valley samples, which contain a similar amount of water but $\sim 17\%$ clay. The soil samples from Brookhaven have a far lower attenuation factor α (Fig. 11.28), and higher phase velocity V_p (Fig. 11.29), for any given frequency than the Avra Valley soils.

In order to consider the complete range of electrical properties that may be encountered in the field, we have included in Figs. 11.26–11.29 the experimental results for the following samples in addition to the typical high-loss and low-loss soils: (1) Dry Brookhaven sample ($W_v = 0\%$). This condition is not typical in nature. Dry sand has, however, been frequently used in "sand box" experiments

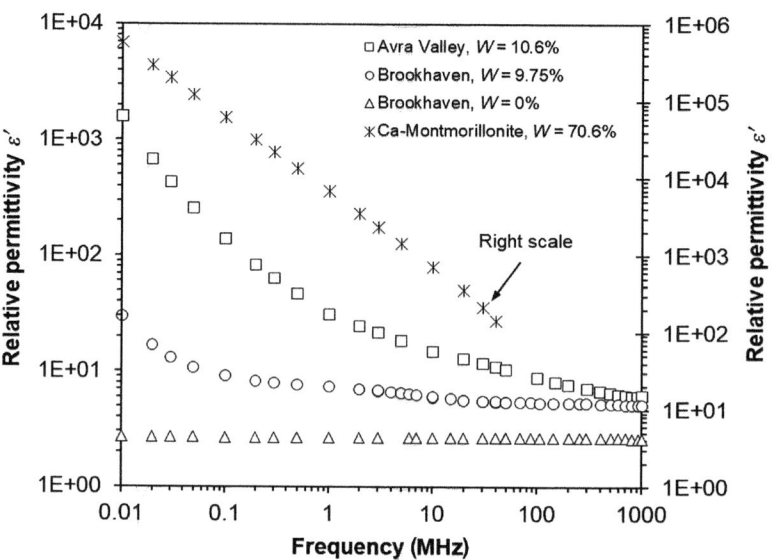

FIGURE 11.26

Relative permittivity ε' versus frequency for soils from various locations.

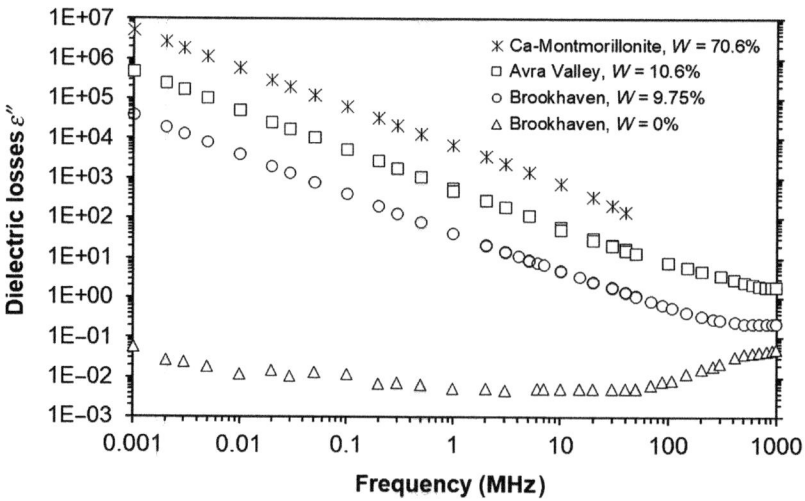

FIGURE 11.27

Dielectric losses ε'' versus frequency for soils from various locations.

for GPR and other EM survey tests. Our results may provide a useful comparison of these artificial conditions with a natural setting. (2) Wet Ca-Montmorillonite clay sample with water content $W_v = 70.6\%$. This sample is representative of a

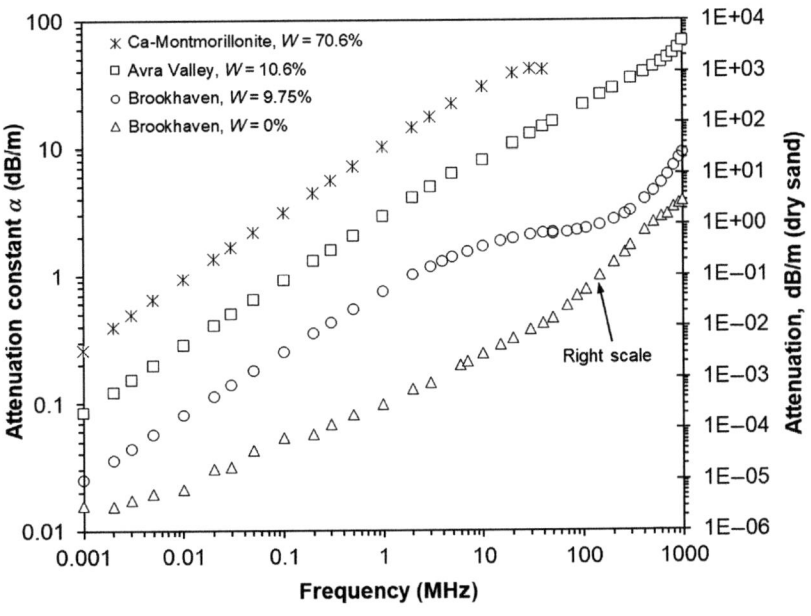

FIGURE 11.28

Attenuation constant α versus frequency for soils from various locations.

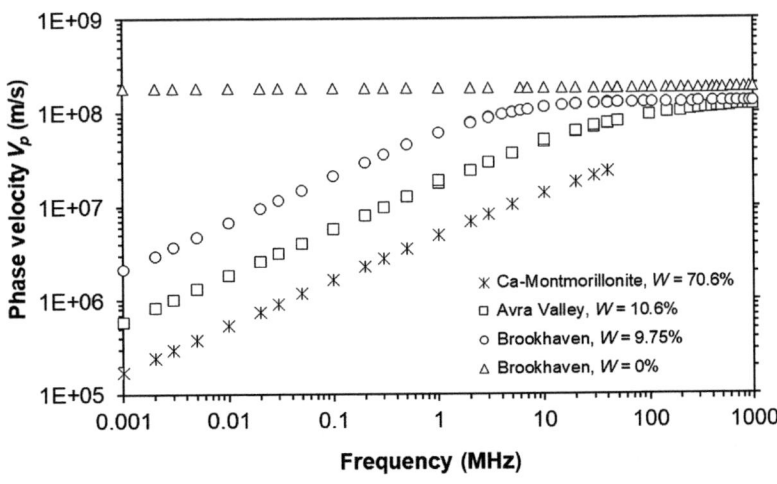

FIGURE 11.29

Phase velocity V_p versus frequency for soils from various locations.

very high loss soil, which is actually fairly common. For example, clay liners and clay caps often occur in environmental and engineering site investigations.

Data for the four samples displayed in Figs. 11.26–11.29 give a good idea of the range of electrical properties (including attenuation and depth of penetration) that may be encountered in GPR and other EM surveys over soils.

References

Campbell, J.E., 1990. Dielectric properties and influence of conductivity in soils at one to fifty Megahertz. Soil Sci. Soc. Am. 54, 332–341.

Curtis, J., 1994. Soil texture and broadband dielectric properties. In: Paper Presented at Unexploded Ordnance Detection and Range-Remediation Conference, Sponsored by U. S. Army Environ. Center and U.S. Army Yuma Proving Ground, Colorado, May 17–19, 1994.

Debye, P., 1945. Polar Molecules. Dover, Mineola, NY.

Dukhin, S.S., Shilov, V.N., 1974. Dielectric Phenomena and Double Layer in Disperse Systems and Electrolytes. Halsted, New York.

Ferre, T.P.A. Personal Communications, 1999.

Hipp, J.E., 1974. Soil electromagnetic parameters as function of frequency, soil density, and soil moisture. Proc. IEEE 62 (1), 98–103.

Jackson, T.J., 1987. Effects of soil properties on microwave dielectric constants. Transp. Res. Record 1119, 126–131.

Levitskaya, T.M., Sternberg, B.K., 1996. Polarization processes in rocks, part 1. Radio Sci. 31, 755–779.

O'Connor, K.M., Downing, C.H., 1999. Geo-Measurements by Pulsing TDR Cables and Probes. CRC Press, Boca Raton, FL.

Sternberg, B.K., McGill, J.W., 1995. Archaeology studies in southern Arizona using ground penetrating radar. J. Appl. Geophys. 33, 209–225.

Topp, G.C., Davis, J.L., Annan, A.P., 1980. Electromagnetic determination of soil water content: Measurements in coaxial transmission lines. Water Resour. Res. 16 (3), 574–582.

Von Hippel, A.R., 1954. Dielectrics and Waves. John Wiley, New York, Reprinted: Artech House, Boston, MA, 1995.

Wang, J.R., 1980. The dielectric properties of soil-water mixtures at microwave frequencies. Radio Sci. 15, 977–985.

Further Reading

Curtis, J.O., 2001. Moisture effects on the dielectric properties of soils. IEEE Trans. Geosci. Remote Sens. 39 (1), 125–128.

Kalinski, R.J., Kelly, W.E., 1993. Estimating water content of soils from electrical resistivity. Geotech. Test. J. 16 (3), 323–329.

Correlation between laboratory and in situ electrical measurements of soil

12

Chapter Outline

We used laboratory experimental resistivity results from Avra Valley soil with natural moisture, for comparing them with in situ well-logging measurements at the same location (Sternberg and Levitskaya, 1998). Our purpose was to define the necessary conditions for obtaining laboratory measurements that can be directly compared to field data.

Soil samples were collected at the University of Arizona, Avra Valley Test Site, located in the southern part of Avra Valley, about 20 mi. southwest of Tucson, Arizona. The average elevation of the area is 760 m above sea level. The climate is arid, characterized by low precipitation, high summer temperatures, and low humidity. The water table is at approximately 120 m depth (Sternberg et al., 1991).

We considered the moisture content as the main parameter, which when it changes, affects the electrical properties of the soil. Other parameters, such as temperature, contaminants, etc., are assumed stable. As indicated in Chapter 7, Soils from Avra Valley, Arizona, the soils are relatively uniform and consist of medium to fine—grained sands. Soil samples contained 61% sand, 22% silt, and 17% clay. The procedures for our laboratory measurements are described in Chapter 4, Measurements and analysis concept of distributed versus lumped parameters. Here, we show an example of electrical properties measurements for a sample with 10.5% water content versus frequency (Fig. 12.1A and B). The field measurements for comparing with our lab results were performed in wells with a Geonics EM-39 borehole conductivity meter at 39 kHz. Soil samples were collected in two ways: the "First Sampling" and the "Second Sampling."

Electrical Spectroscopy of Earth Materials. DOI: https://doi.org/10.1016/B978-0-12-818603-9.00012-X

175

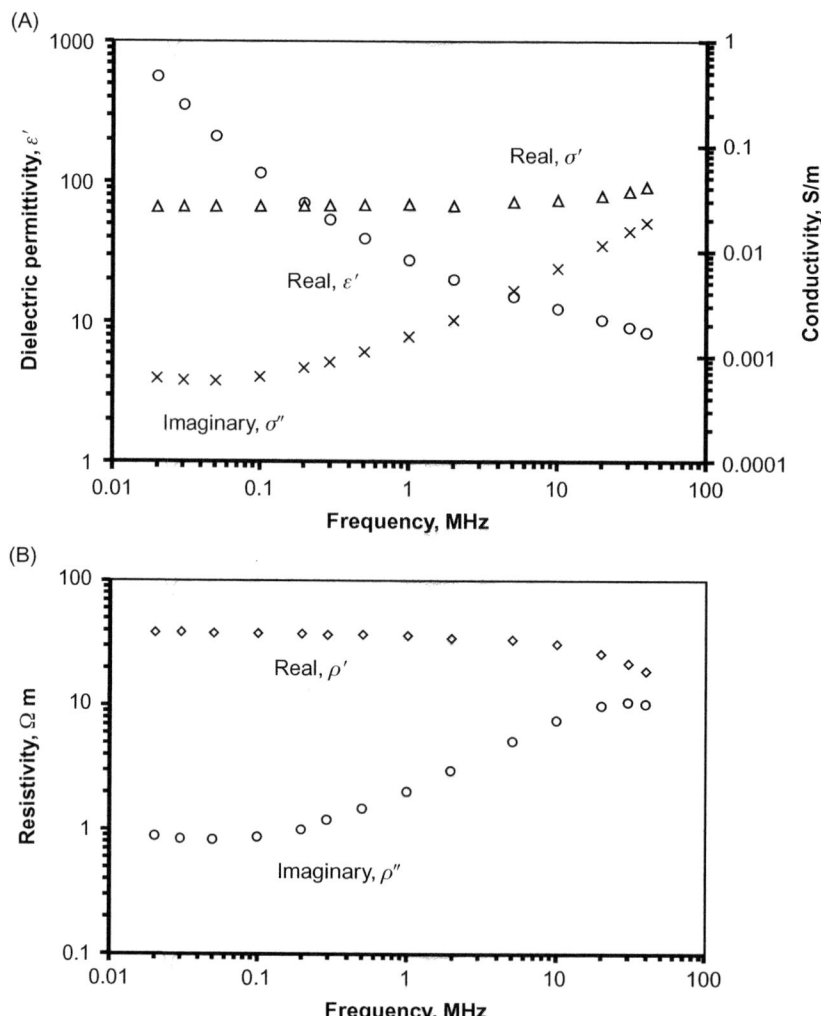

FIGURE 12.1

(A) Dielectric permittivity and conductivity versus frequency for a soil sample with volume wetness $W_v = 10.5\%$. (B) Real and imaginary resistivity versus frequency for a soil sample with volume wetness $W_v = 10.5\%$.

First Sampling

The "first sampling" was done at a relatively freshly excavated site in order to obtain undisturbed samples. The excavation area was approximately 10 m wide, 45 m long, and 4 m deep. Three sample locations referred to as locations I, II, and III were separated by 6 and 12 m along the length of the excavation, respectively.

This "first sampling" showed a poor correlation between laboratory and field data, with most of the laboratory data showing higher resistivities than the well-logging data (Fig. 12.2). We hypothesize that the cause for this poor agreement was that the samples were collected several hours after excavation, and therefore the soil could have dried out.

The highly scattered resistivities are likely caused by the variations in drying of the sample. This is demonstrated in Fig. 12.3, which shows the resistivity of these samples at 30 kHz, versus the volumetric water content.

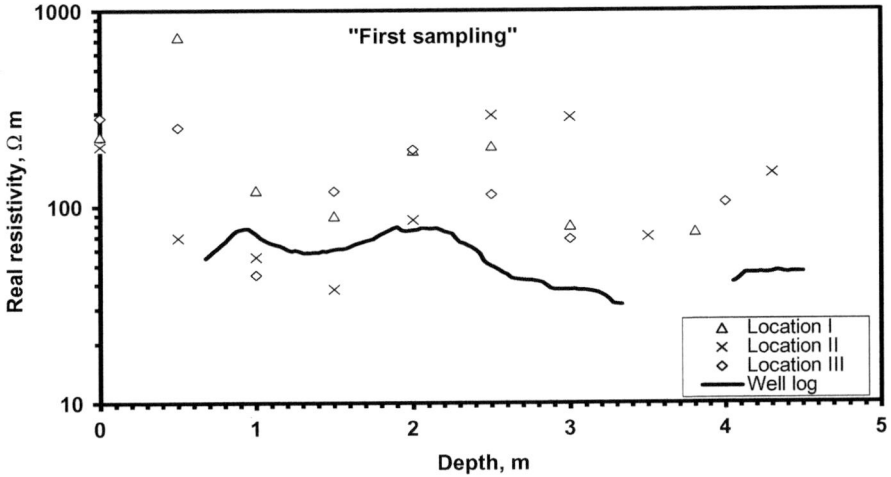

FIGURE 12.2

"First sampling." Resistivity of soils at 30 kHz (laboratory data) and well-logging data at 39 kHz versus depth.

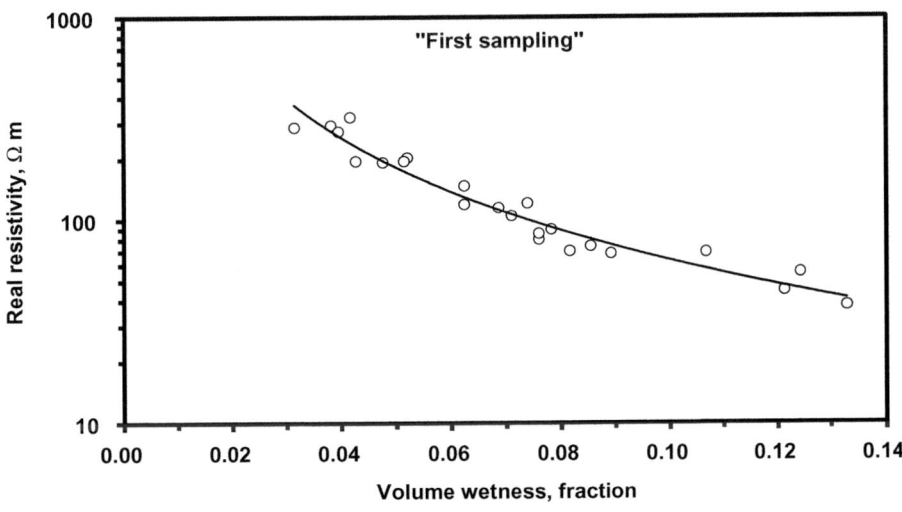

FIGURE 12.3

Resistivity of soil samples at 30 kHz versus volumetric water content.

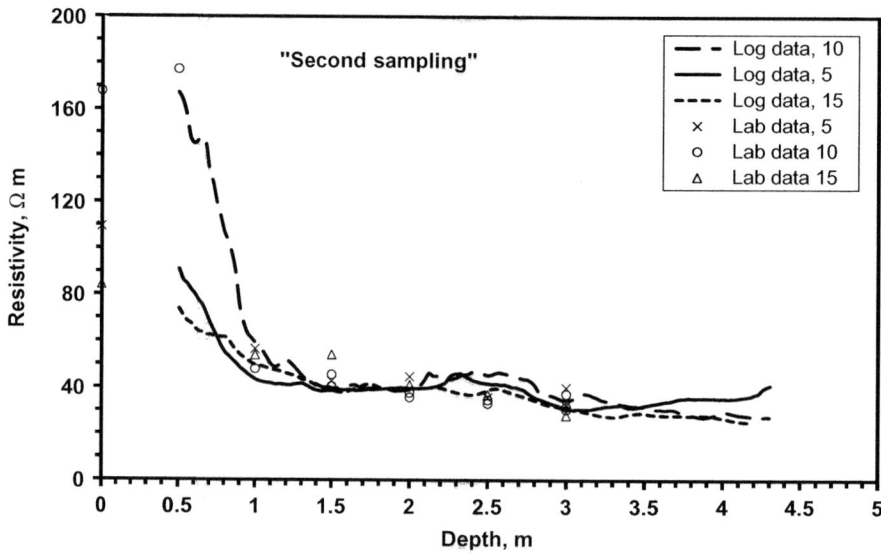

FIGURE 12.4

"Second sampling." Resistivity of soils at 30 kHz (laboratory data), and well-logging data at 39 kHz versus depth.

Second Sampling

For the "second sampling" excavations were made with a backhoe. From each depth, one or two samples were collected. The soil from the backhoe bucket was placed on the ground, and the loose soil was collected in tightly sealed jars or plastic bags within a few minutes after excavation. Samples were acquired to a depth of 3.0 m with 0.5 m intervals.

The laboratory-measured resistivity values at 30 kHz versus depth for three wells (one location) are shown in Fig. 12.4. The "second sampling" was performed with a very short time lag between excavation and sample collection, and excellent agreement with field measurements was obtained. The laboratory procedures of sample preparation are described in Chapter 4, Measurements and analysis concept of distributed versus lumped parameters; however, we repeat some details here for clarity.

12.1 Sample Preparation

For measuring electrical parameters in the laboratory, in both cases, the soil was sieved to a fraction less than 0.6 mm, in order to remove the clumps and stones. This was performed rather quickly (in 1 minute), so the water was not lost. We found that restricting the samples to a maximum size of 0.6 mm was essential for obtaining excellent agreement with the in situ results. The soil moisture is mostly located within the fine fraction, between the small grains, thus reducing the contact resistance of the grains. Consequently, most of the current in these soils flows through the damp fines rather than through the larger, low-porosity stones and hard calcite-cemented clumps. Therefore, the fine fraction is more representative for laboratory resistivity measurements.

Some of the sieved soil was then placed into a parallel-plate sample holder with disk electrodes. The electrodes, 5 cm in diameter, were made from brass, plated with silver. Three Teflon clamps allowed compression of the soil sample to densities of $\gamma = 1.5 - 1.6$ g/cm^3, which correspond to approximately in situ conditions (Sternberg, 1993).

The density γ for the wet sample was defined as the ratio of wet weight M_{wet} to its volume V. The sample density may serve as a measure of sample compaction in the sample holder. The volumetric water content of the sample was defined from its wet weight M_{wet}, dry weight M_{dry}, and its volume V, as follows (Hillel, 1982; Sternberg and Levitskaya, 2001):

$$W_v = \frac{M_{wet} - M_{dry}}{\gamma_w \cdot V} \tag{12.1}$$

where $\gamma_w = 1$ g/cm^3 is the water specific weight. The sample's dry weight was determined after the measurements by heating it in an oven at a temperature of 105°C until its weight became constant.

12.2 Laboratory Electrical Measurements

Earth materials, such as soils, behave in an electric field as both a dielectric and a conductor and can be described by a complex dielectric permittivity $\hat{\varepsilon}$ and a complex conductivity $\hat{\sigma}$, or a complex resistivity $\hat{\rho}$, which contain an imaginary component related to dielectric losses (Chapter 2: Parameters describing the material behavior in an electromagnetic field and Chapter 3: Methods of studying earth materials using alternating electric fields). The measurement procedures are described in Chapter 4, Measurements and analysis concept of distributed versus lumped parameters.

We measured the magnitude /Z/ and phase angle φ of the sample impedance Z which can be expressed through its resistance R and reactance X, related to a series equivalent-circuit model of the sample, as follows

$$\hat{Z} = /Z/ \cdot e^{j \cdot \varphi} = R + j \cdot X = /Z/ \cdot \cos\varphi + j \cdot /Z/ \cdot \sin\varphi \tag{12.2}$$

Using a parallel equivalent-circuit model, we may also define the admittance $Y = 1/Z$, which can be expressed through its conductance G and susceptance B as follows:

$$\hat{Y} = /Y/ \cdot e^{-j \cdot \varphi} = G - j \cdot B = /Y/ \cdot \cos\varphi - j \cdot /Y/ \cdot \sin\varphi \tag{12.3}$$

where $B = \omega \cdot C$ and C is the sample capacitance. From (12.2), and the geometry of the sample, we define a complex resistivity as $\hat{\rho} = \rho' + j \cdot \rho''$. In the same way, using Eq. (12.3), we may define the complex conductivity as $\hat{\sigma} = \sigma' - j \cdot \sigma'' = \sigma' - j \cdot \omega \cdot \varepsilon' \cdot \varepsilon_0$, where ε' is the real relative dielectric permittivity, and $\varepsilon_0 = 8.854 \cdot 10^{-12}$ F/m is the dielectric permittivity of vacuum.

The complex resistivity $\hat{\rho}$ is the reciprocal of the complex conductivity $\hat{\sigma}$ (Wait, 1959, 1982):

$$\rho' + j \cdot \rho'' = \frac{1}{\sigma' - j \cdot \omega \cdot \varepsilon' \cdot \varepsilon_0} \tag{12.4}$$

As seen from Eq. (12.4), the real resistivity $\rho' \neq 1/\sigma'$. In general, ρ' and real conductivity σ' are interrelated as follows (Chapter 4: Measurements and analysis concept of distributed versus lumped parameters):

$$\rho' = \frac{\sigma'}{\sigma'^2 + \sigma''^2} = \frac{\sigma'}{\sigma'^2 + (\omega \cdot \varepsilon' \cdot \varepsilon_0)^2} \tag{12.5}$$

Fig. 12.1A shows an example of frequency dependencies of real dielectric permittivity and complex conductivity. Fig. 12.1B shows the complex resistivity for a representative soil sample from the Avra Valley Test Site in Southern Arizona. This sample had a volumetric water content of $W_v = 10.5\%$ and a density of $\gamma = 1.53$ g/cm^3. The dielectric permittivity shows frequency dispersion. The real conductivity σ' remains almost constant with frequency, up to 1 MHz and then slowly increases. The imaginary conductivity σ'' increases with frequency, but in the range from 30 to 100 kHz, it is much smaller than the real σ' values. Analyzing the Eq. (12.5) with respect to our data from Fig. 12.1A, we found that at frequencies from 30 to 100 kHz, the second term in the denominator of Eq. (12.5) is 3.6×10^{-7} which is negligible in comparison with the first term $\sigma'^2 = 7.1 \times 10^{-4}$.

The real resistivity ρ' is also approximately constant below 1 MHz. The imaginary resistivity ρ'' increases with frequency, but it is very small over the frequency range from 30 to 100 kHz. Our subsequent analysis will use the real resistivity ρ' of the samples at frequencies from 30 to 50 kHz, where the resistivity may be considered as a real, and not a complex value. Therefore, for this frequency range, we used the real resistivity, which is simply the reciprocal of the real conductivity from the well logging.

The variation of real resistivity at 30 kHz with depth for the first sampling is shown in Fig. 12.2. The highly scattered data can be caused by changing the water content in soil with the depth and for three different locations. This is illustrated in Fig. 12.3, which shows the resistivity of these samples at 30 kHz versus volumetric water content can be described with the following regression equation:

$$\rho' = 1.90 \times (W_v)^{-1.52} \tag{12.6}$$

where W_v is the fractional volume wetness.

12.3 In Situ Well Logging

In order to provide a direct in situ field comparison with the laboratory data, we logged nearby boreholes with a Geonics EM-39 borehole conductivity meter. This instrument records real conductivity at a frequency of 39 kHz. The well-log measurements were made inside 5 cm ID PVC pipes. The data for the first sampling are shown in Fig. 12.2. Based on the comparison, we concluded that the soil collected for the laboratory first sampling had lost significant water during the several hours between the excavation and sample acquisition. That is why most of the laboratory ρ' values are much higher than the well-log results.

Well-log data for the second sampling are shown along with the laboratory-measured resistivity values at 30 kHz (for one location) in Fig. 12.4. As seen from the comparison, the agreement is excellent.

12.4 Summary

Results obtained from this study allow us to summarize the requirements for good agreement of laboratory and field measurements. Laboratory measurements may be directly comparable to field measurements, when special precautions are taken to preserve natural soil wetness. The rate of water evaporation from the soil sample depends on the ambient temperature and on the starting wetness. In hot and dry environmental conditions, the samples must be collected and sealed within a short time after excavation, i.e., within a few minutes. The higher the moister content of the soils, the shorter must be the time for sample acquisition. Careful sample preservation is crucial. They must be sealed in airtight containers with a minimal amount of air above the sample.

References

Hillel, D., 1982. Introduction to Soil Physics. Academic Press, New York.

McGill, J.W., 1990. Ground Penetrating Radar Investigations with Applications for Southern Arizona (M.S. thesis), Department of Mining and Geological Engineering, University of Arizona.

Sternberg, B.K., 1993. Construction of a lined basin for tests of a high-resolution subsurface imaging ellipticity system. EPRI Report, TR-103462. University of Arizona.

Sternberg, B.K., Levitskaya, T.M., 1998. Correlation between Laboratory and In Situ Electrical Resistivity Measurements of Soil. Journal Environmental and Engineering Geophysics 63–70.

Sternberg, B.K., Levitskaya, T.M., 2001. Electrical parameters of soils in the frequency range from 1 kHz to 1 GHz, using lumped-circuit methods. Radio Science 36 (4), 709–719.

Sternberg, B.K., Miletto, M.F., LaBrecque, D.J., Thomas, S.J., Poulton, M.M., 1991. The Avra Valley (Ajo Rd.) geophysical test site. LASI Report # 91-2. University of Arizona.

Wait, J.R. (Ed.), 1959. Overvoltage Research and Geophysical Applications. Pergamon Press, Oxford.

Wait, J.R., 1982. Geo-Electromagnetism. Academic Press, New York.

Further Reading

Hipp, J.E., 1974. Soil electromagnetic parameters as functions of frequency, soil density, and soil moisture. Proc. IEEE 62 (1), 98–103.

Olhoeft, G.R., 1985. Low-frequency electrical properties. Geophysics 50, 2492–2503.

Comparison of dielectric and conduction spectroscopy methods

13

Chapter Outline

In Chapter 2, Parameters describing the material behavior in an electromagnetic field, and Chapter 3, Methods of studying earth materials using alternating electric fields, we described the two main processes occurring in a material, placed in an alternating electric field, and the corresponding parameters that characterize the processes. The general term, impedance spectroscopy, includes all parameters, such as complex dielectric permittivity ε, which describes dielectric polarization processes (they produce displacement currents), and two transport, or conduction, parameters: complex conductivity σ, characterizing the ability of a material to conduct the electrical current, caused by free charges, and complex resistivity ρ, characterizing the material opposition to flowing current (Macdonald, 1987). Both of the conduction parameters can be used as spectroscopy methods and are called conduction spectroscopy, while the spectroscopy in terms of complex dielectric permittivity is called dielectric spectroscopy (Scott and Smith, 1986). We used the conduction spectroscopy in terms of complex resistivity (ρ). The spectroscopy methods are described in Chapter 3, Methods of studying earth materials using alternating electric fields. As seen from the above-mentioned chapter (Figs. 3.6 and 3.7), the frequency dependencies of real ε' and imaginary ε'' dielectric permittivity for Avra Valley samples with various water contents do not exhibit any polarization process in the studied frequency range. Because relaxation processes are highly probable in such heterogeneous materials; we assume that the conduction phenomena obscure them.

In contrast to the conductivity, the real ρ' and imaginary ρ'' components of complex resistivity reveal a region of polarization in the frequency range $f = 10-1000$ MHz for all soil samples, as shown in Chapter 7, Soils from Avra Valley, Arizona. The polarization processes in our soil samples, studied with resistivity spectroscopy, result in Argand circular diagrams with limit values ρ_s as the static resistivity. The limit infinity resistivity ρ_∞ is always approaching zero. The Argand diagrams are shown and analyzed for each sample from various locations in Chapter 7, Soils from Avra Valley, Arizona. The parameters defined are summarized in the corresponding tables.

Electrical Spectroscopy of Earth Materials. DOI: https://doi.org/10.1016/B978-0-12-818603-9.00013-1

The hidden relaxation process from the dielectric spectroscopy data can be revealed by subtracting the conduction component from the total energy losses ε''_{total}, thus obtaining the dielectric losses ε''_{diel} (Section 6.3). From these data, the Cole—Cole arcs $\varepsilon''_{diel} = F(\varepsilon')$ were obtained and the Cole—Cole parameters, ε_s, ε_∞, τ, and α, were defined. These results are also shown for each sample in Chapter 7, Soils from Avra Valley, Arizona, in the summary tables. Thus, in soils studied over the wide frequency range, both dielectric and resistivity spectroscopy methods reveal polarization processes, but in different frequency regions. From the dielectric losses (ε''_{diel} vs frequency) for each sample (Chapter 7: Soils from Avra Valley, Arizona), it is seen that the dielectric spectroscopy reveals a relaxation at lower frequencies ($f = 0.01 - 1$ MHz) than the resistivity. In addition, because the parameters $\hat{\varepsilon}$ and $\hat{\rho}$ have a different physical sense and behave differently in an alternating electric field, they may characterize different mechanisms of polarization processes in a material.

As shown in Sections 4.2 and 4.3, all parameters for both dielectric and resistivity spectroscopy methods, are obtained from the same measurement data, Z and ϕ, so it seems logical to expect both spectroscopy methods to describe the same polarization processes. In order to understand this, we compare the Argand diagrams and Cole—Cole diagrams for soil samples with various volume water contents W_v from all locations studied. We assume that if both methods describe the same polarization process then the relaxation parameters (τ and α) that are common for both Cole—Cole and Argand diagrams must have close values. The procedure of defining the relaxation parameters is described in Chapter 3, Methods of studying earth materials using alternating electric fields.

The comparisons of the common polarization parameters characterizing Argand and Cole—Cole diagrams for all the soil samples we studied are shown in Table 13.1. As seen from Table 13.1, some samples have rather close values of the relaxation time τ for both Argand and Cole—Cole diagrams, while in other samples, the τ-values differ by an order of magnitude. Distribution parameters α for the Cole—Cole and Argand diagrams are very different. The Cole—Cole diagrams appear to be broader than the Argand diagrams, and this may indicate that the two relaxation processes have a different origin.

For more clarity and convenience, we will represent the data from Table 13.1 graphically. But first, we show the limit parameters from Argand and Cole—Cole diagrams in Figs. 13.1—13.3. The static resistivity ρ_s (Fig. 13.1), defined with the Fricke—Lopatin (FL) method and from the Argand diagrams, has close values (the FL method was described in Chapter 5: Stray parameters of the measuring system and ways of defining them). They decrease with increasing water content in soil samples. The highest values of ρ_s belong to the clean sand (Brookhaven and Columbia locations), while the samples from Avra Valley have the lowest static resistivity. Data for Antenna Test Facility (ATF) samples from Ft. Huachuca are located in between, higher than Avra Valley, and lower than clean sand samples.

Static dielectric permittivity ε_s, calculated from the Cole—Cole diagrams, is shown in Fig. 13.2. These values increase with water content in the soil. Avra Valley samples have the highest values of ε_s, and the clean sand samples have the

Table 13.1 Comparison of the Common Polarization Parameters Characterizing Argand and Cole–Cole Diagrams for all the Soil Samples That We Measured

Volume Wetness W_v (%)	σ_{st}, S/m From		Relaxation Time τ (s)		Distribution Parameter α		Comments	
	FL Method	Argand Diagram	Argand Diagram	Cole–Cole Diagram	Argand Diagram	Cole–Cole Diagram	Measurement Arrangement	Resonance Presence
Soil From Avra Valley, Arizona. Content: 85% Sand, 10% Silt, and 5% Clay (by Weight)								
5.8	0.00519	0.00518	2.044×10^{-8}	3.367×10^{-6} 1.424×10^{-8}	0.228	0.467 0.444	HP4194A, fixture-disks HP4191A, coaxial. No resonance	
10.6	0.02895	0.0300	3.229×10^{-9}	8.686×10^{-8} 1.126×10^{-8}	0.194	0.439 0.433	HP4194A, fixture-disks HP419A, coaxial. No resonance	
16.4	0.04652	0.050761	2.460×10^{-9}	2.130×10^{-8}	0.194	0.467	HP4194A, fixture-coaxial HP419A, coaxial. Resonance: 970 MHz	
24.9	0.09171	0.10309	1.914×10^{-9}	1.369×10^{-9}	0.158	0.455	HP4194A, fixture-coaxial HP419A, coaxial. Resonance: 740 MHz	
45.5	0.09028	0.09434	1.51×10^{-9}	4.16×10^{-8}	0.172	0.478	HP4194A, Z-probe-coaxial HP4191A, coaxial. Resonance: 550 MHz	
Soil From Brookhaven, New York. Content: 96% Sand, 1% Silt, and 3% Clay Size								
8.1	0.000044	0.000435	1.340×10^{-7}	3.444×10^{-7}	0.0833	0.3555	HP4194A, fixture-coaxial HP419A, coaxial. No resonance	
9.75	0.002193	0.002193	2.525×10^{-8}	2.002×10^{-6} 4.188×10^{-8}	0.0833	0.494 0.272	HP4194A, fixture-coaxial HP4191A, coaxial. No resonance	
14.7	0.002200	0.002200	3.024×10^{-8}	1.00×10^{-6} 5.344×10^{-8}	0.0778	0.505 0.233	HP4194A, fixture-coaxial HP4191, coaxial. No resonance.	

(Continued)

Table 13.1 Comparison of the Common Polarization Parameters Characterizing Argand and Cole–Cole Diagrams for all the Soil Samples That We Measured *Continued*

Volume Wetness W_v (%)	σ_{st}, S/m From		Relaxation Time τ (s)		Distribution Parameter α		Comments
	FL Method	Argand Diagram	Argand Diagram	Cole–Cole Diagram	Argand Diagram	Cole–Cole Diagram	Measurement Arrangement Resonance Presence
Soil From Columbia, New York. Content: Washed Quartz Sand							
9.2	0.001682	0.001653	3.214×10^{-8}	3.473×10^{-8}	0.100	0.200	HP4194A, Z-probe-coaxial HP4191A, coaxial. No resonance
14.2	0.002323	0.002288	3.083×10^{-8}	4.180×10^{-8}	0.100	0.189	HP4194A, Z-probe-coaxial HP4191A, coaxial. No resonance
21.3	0.004271	0.004237	2.296×10^{-8}	4.180×10^{-8}	0.0555	0.289	HP4194A, Z-probe-coaxial HP4191A, coaxial. No resonance: 989 MHz
Soil From Ft. Huachuca (ATF-Subsurface), Arizona. Content: 81% Sand, 8% Silt, and 11% Clay							
7.2	0.0060	0.0060	1.209×10^{-8}	3.399×10^{-7}	0.189	0.072	HP4194A, Z-probe, disks
14.6	0.0110	0.0112	1.154×10^{-8}	5.430×10^{-7}	0.189	0.456	HP4194A, Z-probe, disks
25	0.0133	0.0136	1.55×10^{-8}	8.366×10^{-7}	0.172	0.533	HP4194A, Z-probe, disks

FL, *Fricke–Lopatin*.

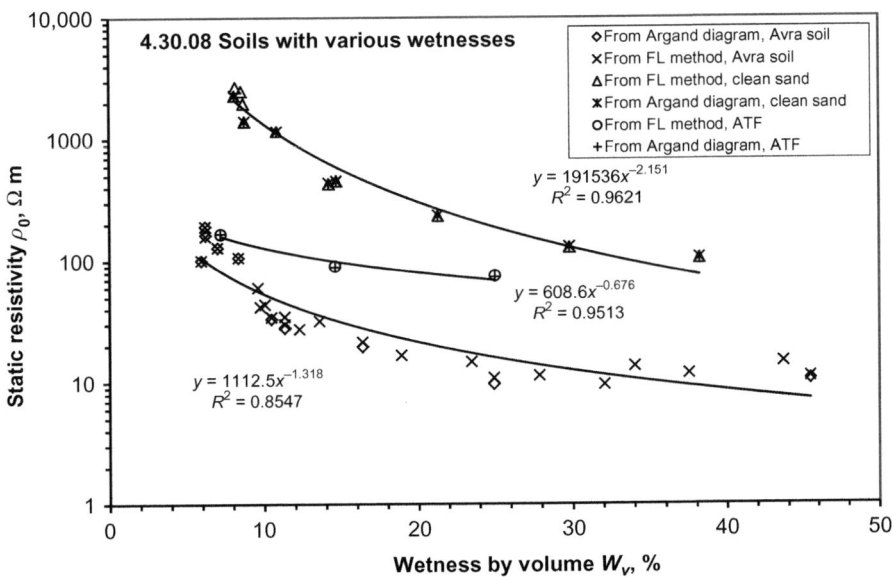

FIGURE 13.1

Static resistivity ρ_s, from Argand diagrams, for all locations.

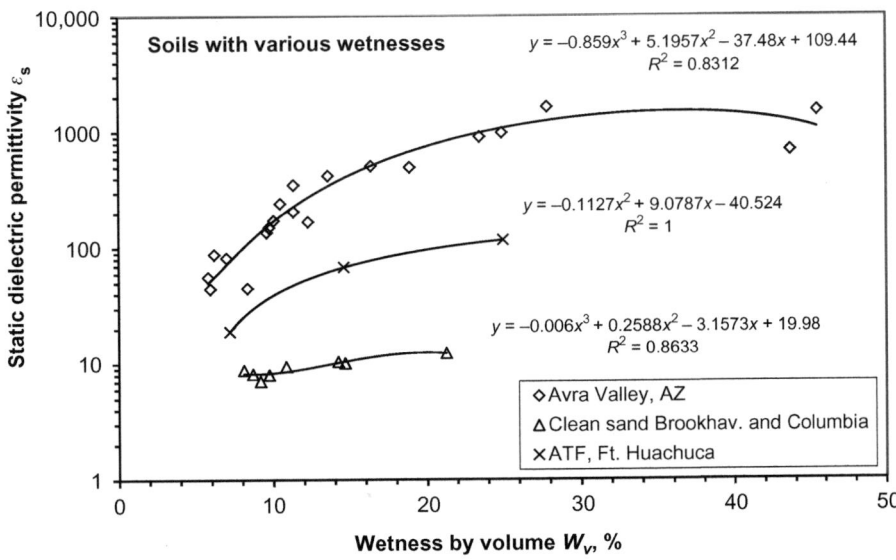

FIGURE 13.2

Static dielectric permittivity ε_s, from Cole–Cole diagrams, for all locations.

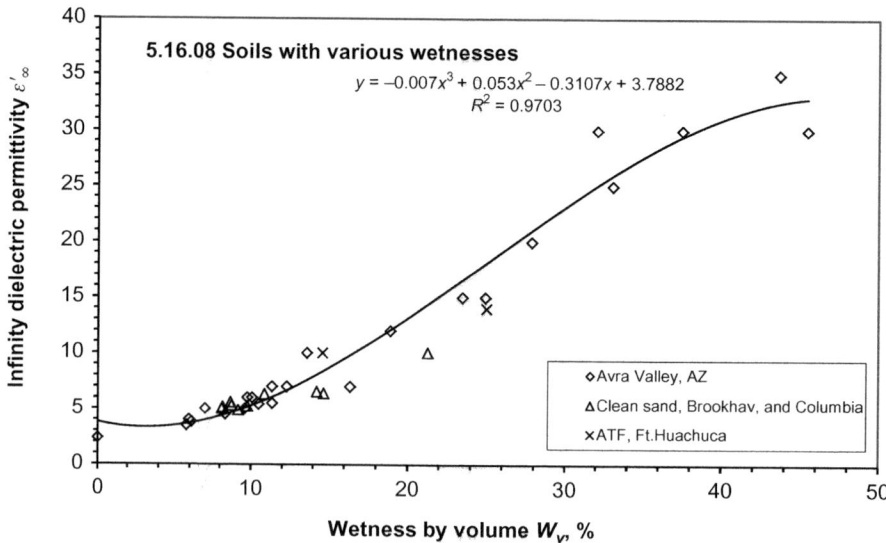

FIGURE 13.3

Infinity dielectric permittivity ε_{inf} from Cole–Cole diagrams for all locations.

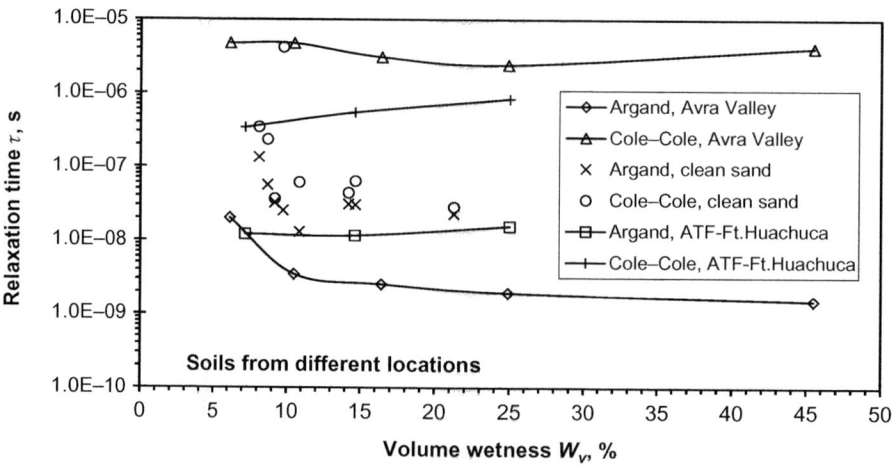

FIGURE 13.4

Relaxation time versus volume wetness for polarization processes, found with dielectric and resistivity spectroscopy methods in soils from all locations studied.

lowest ε_s. The infinity dielectric permittivity ε_∞, also defined from Cole–Cole diagrams, is shown to have close values for all locations studied, and are increasing with water content in the soil (Fig. 13.3).

Fig. 13.4 illustrates the relaxation time τ, obtained from Argand and Cole–Cole diagrams for all locations, versus volume water content. The values

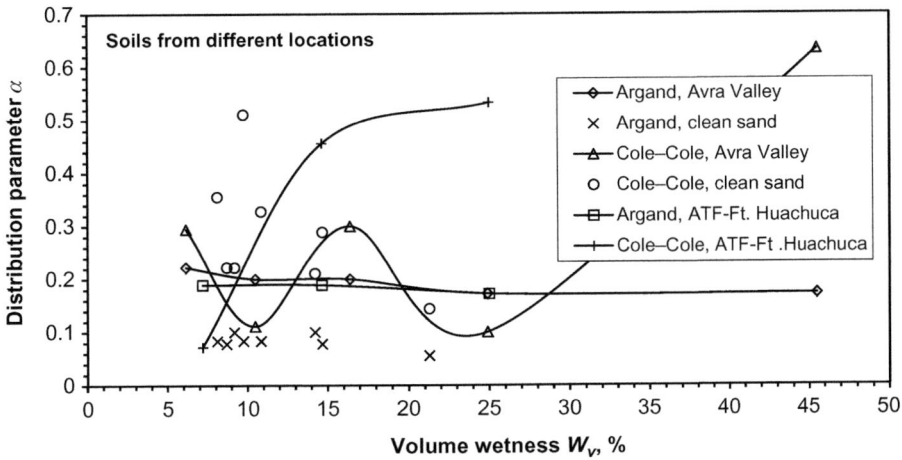

FIGURE 13.5

Distribution parameter versus volume wetness for polarization processes, found with dielectric and resistivity spectroscopy methods in soils from all locations studied.

of relaxation time are close for Argand and Cole–Cole diagrams only for the clean sand samples, from Brookhaven and Columbia. This indicates that both dielectric and resistivity spectroscopy methods describe the same relaxation process. Soils from other locations, Avra Valley and ATF, Ft. Huachuca, exhibit rather different values of τ for each spectroscopy method. Cole–Cole relaxation is characterized with higher τ values, of an order $10^{-7}–10^{-6}$ seconds, and Argand relaxation (in terms of resistivity) is described with lower τ, of an order $10^{-9}–10^{-8}$ seconds. This means that dielectric relaxation occurs at lower frequencies than the resistivity relaxation, and therefore, different mechanisms of these polarization processes are displayed. The distribution parameter α, shown in Fig. 13.5, is also seen to be different for Cole–Cole and Argand diagrams in soil from each location.

The types and mechanisms of polarization in materials are described and analyzed in Section 2.2. The slowly forming, relaxation types of polarization can be one of the following (Kobranova, 1989):

1. Orientation polarization of polar groups and bonds in the applied electric field with a relaxation time $\tau = 10^{-10}–10^{-7}$ secnds.
2. Interfacial, or migration polarization (Maxwell–Wagner effect) in materials, containing conducting components, such as wet rocks and soils. In this case, the range of relaxation times is $\tau = 10^{-6}–10^{-3}$ seconds.
3. Induced polarization, or electrochemical polarization, occurs in the electrical double layers (EDL) with the slowest relaxation times of the order $\tau > 10$ seconds.

Our data processing involves the removal of electrode polarization from measured data in the very beginning, prior to further calculations (Chapter 6: Corrections for stray parameters and error estimation). This procedure also eliminates other low-frequency processes, such as contact resistance on the border between electrodes and the sample, as well as the induced polarization in the EDL. All these processes behave similarly, decreasing with increasing frequency, and they cannot be distinguished. That is why our data do not reveal the induced polarization. The difference between τ-values for Cole—Cole and Argand diagrams means that the dielectric spectroscopy shows mostly the Maxwell—Wagner polarization mechanism, and the resistivity spectroscopy is more informative for the orientation mechanism of polarization. More descriptive in this respect are the frequency dependencies of dielectric permittivity ε and resistivity ρ for all samples (Chapter 7: Soils from Avra Valley, Arizona). These graphs show that the dielectric and resistivity spectroscopies reveal relaxation processes in different frequency ranges, and therefore their relaxation parameters, relaxation time τ, and distribution parameter α are different. Both spectroscopy approaches are complementary and together give a comprehensive spectrum for a material behavior in an alternative electric field.

References

Kobranova, V.N., 1989. Petrophysics,. Mir Publishers, Moscow, English translation.

Macdonald, J. Ross (Ed.), 1987. Impedance Spectroscopy. John Wiley & Sons, New York.

Scott, W.R.J., Smith, G.S., 1986. Dielectric spectroscopy. IEEE Trans. Antennas Propag. 34 (7), 919—929.

Conclusions

14

The book contains results from our studies of Earth Materials, in particular soils, by using Spectroscopy methods primarily in a frequency range of $f = 1$ kHz to 1 GHz. We used Dielectric Spectroscopy in terms of complex dielectric permittivity $\hat{\varepsilon}$, and Conduction Spectroscopy in terms of complex resistivity $\hat{\rho}$. For covering a wide range of frequencies, we used various Hewlett Packard (later Agilent and Keysight) Impedance Analyzers, which were described in Chapter 4, Measurements and analysis concept of distributed versus lumped parameters, along with sample holders for various frequencies.

From analyzing the electrical properties of soils versus water contents at various frequencies, we found that the high-frequency dependencies of dielectric permittivity ε' may be useful in defining the water content in soil from its ε' values and vice versa. Comparison of the two spectroscopy methods that we have used showed that they are complementary. The Dielectric Spectroscopy shows mostly the Maxwell–Wagner polarization mechanism, and the Resistivity Spectroscopy is more informative for the orientation mechanism of polarization. Together, both spectroscopy approaches give a comprehensive spectrum for a material behavior in an alternating electromagnetic field, and help to explain the polarization mechanisms in earth materials.

This book will be useful for researchers, as well as for students, studying the electrical properties of Earth materials from various locations, with different water contents, in a wide frequency range. This book will also be useful for geophysicists who apply electrical property measurements for geophysical surveys, such as Ground Penetrating Radar (GPR) surveys and Electromagnetic (EM) surveys in the 1 kHz to 1GHz frequency range. We incorporate data that are representative of a wide range of geographic locations and a wide range of environmental conditions.

Electrical Spectroscopy of Earth Materials. DOI: https://doi.org/10.1016/B978-0-12-818603-9.00014-3

Formulas for relaxation time

Defining the values of the relaxation time τ for Debye and Cole–Cole diagrams of circular and arc forms.

A.1 Derivation of the Expression for τ_0 in the Debye Equation

As discussed in Section 3.1, the circular diagram is described with Debye equations, shown on Fig. 3.2A, for a polarization process with a single relaxation time τ_0. The value τ_0 corresponds to the critical frequency f_c (or angular frequency $\omega_c = 2\pi f_c$), at which the maximum ε'' is attained, through the equation $2\pi f_c \tau_0 = 1$ (A.2).

In order to derive Eq. (A.2), we calculated the derivative of the Debye equation (A.1) for ε'' with respect to ω and equated it to 0 (Fig. A.1).

$$\text{The Debye equation:}\varepsilon'' = \frac{(\varepsilon_s - \varepsilon_\infty)\omega\tau_0}{1 + (\omega\tau_0)^2} \tag{A.1}$$

$$\text{Derivative:}\quad \frac{d(\varepsilon'')}{d(\omega)} = \frac{[(\varepsilon_s - \varepsilon_\infty)\cdot 1 \cdot \tau_0]\cdot[1 + (\omega\tau_0)^2] - 2\cdot\omega\tau_0\cdot\tau_0\cdot[(\varepsilon_s - \varepsilon_\infty)\cdot\omega\tau_0]}{[1 + (\omega\tau_0)^2]^2}$$

Replacing $(\varepsilon_s - \varepsilon_\infty) = \Delta\varepsilon$ for convenience, and setting the equation to 0, we obtain

$$\frac{d(\varepsilon'')}{d(\omega)} = \frac{[\Delta\varepsilon\cdot\tau_0]\cdot[1 + (\omega\tau_0)^2] - 2\cdot\omega\tau_0\cdot\tau_0\cdot[\Delta\varepsilon\cdot\omega\tau_0]}{[1 + (\omega\tau_0)^2]^2} = 0$$

For solving this equation, we have to equate the numerator to 0 as follows:

$$[\Delta\varepsilon\cdot\tau_0]\cdot[1 + (\omega\tau_0)^2] - 2\cdot\omega\tau_0\cdot\tau_0\cdot[\Delta\varepsilon\cdot\omega\tau_0] = 0$$

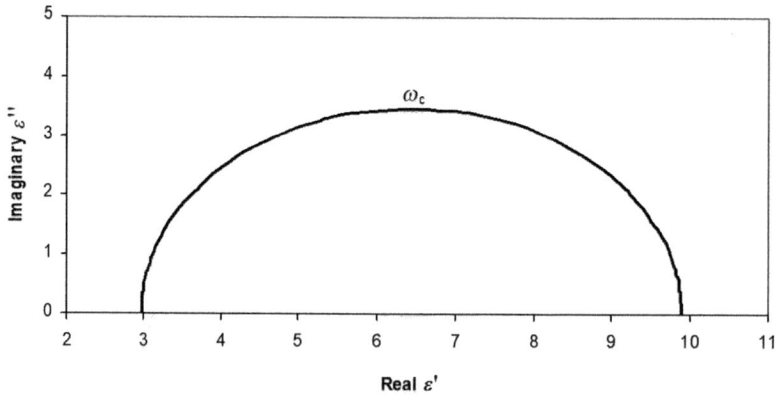

FIGURE A.1

Circular diagram.

After simplifying, we can find the expression for $\omega\tau_0$ as shown below:

$$\Delta\varepsilon\cdot\tau_0\cdot[1+(\omega\tau_0)^2]=2\cdot\omega\tau_0\cdot\tau_0\cdot[\Delta\varepsilon\cdot\omega\tau_0]$$
$$1+(\omega\tau_0)^2=2\cdot(\omega\tau_0)^2$$
$$1=2\cdot(\omega\tau_0)^2-(\omega\tau_0)^2$$
$$(\omega\tau_0)^2=1$$

Finally, the expression for τ_0 in the Debye equation is

$$\omega\tau_0=1 \qquad\qquad (A.2)$$

A.2 Derivation of the Expression for τ_0 in the Cole–Cole Equation

As discussed in Section 3.1, the heterogeneous materials reveal arc diagrams that are described by Cole–Cole with an empirical equation shown in Fig. 3.2A and B as a broad relaxation.

As seen from Fig. A.2, in this case, the maximum absorption is much smaller. Cole and Cole (1941) showed that the departure of such arcs from a semicircle can be measured with the angle between the ε' axis and the radius of the arc drawn to the point ε_∞. They estimated this angle as $\alpha\pi/2$, where α is the distribution parameter. For clarity, we repeat here the formulas for real ε' and imaginary ε'':

$$\varepsilon'=\varepsilon_\infty+\frac{(\varepsilon_s-\varepsilon_\infty)[1+(\omega\tau_0)^{1-\alpha}\sin(\alpha\pi/2)]}{1+2(\omega\tau_0)^{1-\alpha}\sin(\alpha\pi/2)+(\omega\tau_0)^{2(1-\alpha)}} \qquad\qquad (A.3)$$

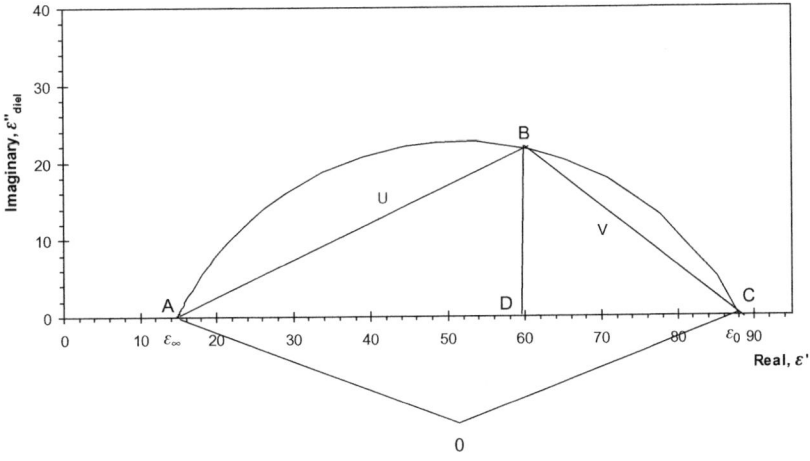

FIGURE A.2

Cole—Cole diagram for a heterogeneous material.

$$\varepsilon'' = \frac{(\varepsilon_s - \varepsilon_\infty)(\omega\tau_0)^{1-\alpha}\cos(\alpha\pi/2)}{1 + 2(\omega\tau_0)^{1-\alpha}\sin(\alpha\pi/2) + (\omega\tau_0)^{2(1-\alpha)}} \qquad (A.4)$$

As shown by Cole and Cole (1941) and Smyth (1955), the relaxation time τ_0 for the broad relaxation is related to the chords u and v as follows:

$$\frac{v}{u} = (\omega\tau_0)^{1-\alpha} \qquad (A.5)$$

In order to derive Eq. (A.5), we will use the following known theorems of a triangle:

1. The sum of three inner angles is equal to 180 degrees or π radian.
2. Each outer angle is equal to sum of two inner angles, not adjacent to it; for example, the outer angle at point B equals the sum of inner angles at points A and C.
3. Theorem of Sine. Lengths of sides related to the corresponding sine of the opposite angle are equal to each other.
4. The central angle (AOC) is measured with the arc (ABC) which subtends it.
5. The inscribed angle is measured with half of an arc, which subtends it.

Considering the part of Fig. A.2 below the axis ε', we see that the central angle
$\text{AOC} = \pi - \frac{\alpha\pi}{2} - \frac{\alpha\pi}{2} = \pi - \alpha\pi = \pi(1 - \alpha).$
Following the sine theorem, we write

$$\frac{U}{\sin C} = \frac{V}{\sin A} = \frac{\varepsilon_0 - \varepsilon_\infty}{\sin B}; \text{ then } \frac{V}{U} = \frac{\sin A}{\sin C} \qquad (A.6)$$

From the triangle ABD, we can find the following expression:

$$\tan A = \frac{\varepsilon''}{\varepsilon' - \varepsilon_\infty} \tag{A.7}$$

From the triangle BDC,

$$\tan C = \frac{\varepsilon''}{\varepsilon_0 - \varepsilon'} \tag{A.8}$$

Substituting the numerator and denominator in (A.7) and (A.8) with their expressions (A.3) and (A.4), we obtain

$$\tan A = \frac{(\varepsilon_s - \varepsilon_\infty)(\omega\tau_0)^{1-\alpha}\cos(\alpha\pi/2)}{(\varepsilon' - \varepsilon_\infty)[1 + 2(\omega\tau_0)^{1-\alpha}\sin(\alpha\pi/2) + (\omega\tau_0)^{2(1-\alpha)}]}$$

$$= \frac{(\varepsilon_s - \varepsilon_\infty)(\omega\tau_0)^{1-\alpha}\cos(\alpha\pi/2)\cdot[1 + 2(\omega\tau_0)^{1-\alpha}\sin(\alpha\pi/2) + (\omega\tau_0)^{2(1-\alpha)}]}{(\varepsilon_s - \varepsilon_\infty)[1 + (\omega\tau_0)^{1-\alpha}\sin(\alpha\pi/2)]\cdot[1 + 2(\omega\tau_0)^{1-\alpha}\sin(\alpha\pi/2) + (\omega\tau_0)^{2(1-\alpha)}]}$$

$$= \frac{(\omega\tau_0)^{1-\alpha}\cos(\alpha\pi/2)}{1 + (\omega\tau_0)^{1-\alpha}\sin(\alpha\pi/2)}$$

$$\tan A = \frac{(\omega\tau_0)^{1-\alpha}\cos(\alpha\pi/2)}{1 + (\omega\tau_0)^{1-\alpha}\sin(\alpha\pi/2)} \tag{A.9}$$

For tan C, we calculate first the denominator of Eq. (A.8):

$$\varepsilon_s - \varepsilon' = (\varepsilon_s - \varepsilon_\infty) - \frac{(\varepsilon_s - \varepsilon_\infty)[1 + (\omega\tau_0)^{1-\alpha}\sin(\alpha\pi/2)]}{1 + 2(\omega\tau_0)^{1-\alpha}\sin(\alpha\pi/2) + (\omega\tau_0)^{2(1-\alpha)}}$$

$$= (\varepsilon_s - \varepsilon_\infty)\left[\frac{1 + 2(\omega\tau_0)^{1-\alpha}\sin(\alpha\pi/2) + (\omega\tau_0)^{2(1-\alpha)} - 1 - (\omega\tau_0)^{1-\alpha}\sin(\alpha\pi/2)}{1 + 2(\omega\tau_0)^{1-\alpha}\sin(\alpha\pi/2) + (\omega\tau_0)^{2(1-\alpha)}}\right]$$

$$= \frac{(\varepsilon_s - \varepsilon_\infty)[(\omega\tau_0)^{1-\alpha}\sin(\alpha\pi/2) + (\omega\tau_0)^{2(1-\alpha)}]}{1 + 2(\omega\tau_0)^{1-\alpha}\sin(\alpha\pi/2) + (\omega\tau_0)^{2(1-\alpha)}}$$

$$= \frac{(\varepsilon_s - \varepsilon_\infty)[(\omega\tau_0)^{1-\alpha} + \sin(\alpha\pi/2)](\omega\tau_0)^{1-\alpha}}{1 + 2(\omega\tau_0)^{1-\alpha}\sin(\alpha\pi/2) + (\omega\tau_0)^{2(1-\alpha)}}$$

$$\varepsilon_s - \varepsilon' = \frac{(\varepsilon_s - \varepsilon_\infty)[(\omega\tau_0)^{1-\alpha} + \sin(\alpha\pi/2)](\omega\tau_0)^{1-\alpha}}{1 + 2(\omega\tau_0)^{1-\alpha}\sin(\alpha\pi/2) + (\omega\tau_0)^{2(1-\alpha)}} \tag{A.10}$$

Using Eqs. (A.4) and (A.10), we obtain the expression for tan C:

$$\tan C = \frac{\varepsilon''}{\varepsilon_s - \varepsilon'} = \frac{(\omega\tau_0)^{1-\alpha}\cdot\cos(\alpha\pi/2)}{[(\omega\tau_0)^{1-\alpha} + \sin(\alpha\pi/2)](\omega\tau_0)^{1-\alpha}} = \frac{\cos(\alpha\pi/2)}{(\omega\tau_0)^{1-\alpha} + \sin(\alpha\pi/2)} \tag{A.11}$$

Now, using Eqs. (A.9) and (A.11) for tan A and tan C, respectively, we can express the sin A and sin C as follows:

$$\sin A = \frac{\tan A}{\sqrt{1 + \tan^2 A}} = \frac{(\omega\tau_0)^{1-\alpha}\cos(\alpha\pi/2)}{\sqrt{1 + 2(\omega\tau_0)^{1-\alpha}\sin(\alpha\pi/2) + (\omega\tau_0)^{2(1-\alpha)}}} \tag{A.12}$$

$$\sin C = \frac{\tan C}{\sqrt{1 + \tan^2 C}} = \frac{\cos(\alpha\pi/2)}{\sqrt{1 + 2(\omega\tau_0)^{1-\alpha}\sin(\alpha\pi/2) + (\omega\tau_0)^{2(1-\alpha)}}} \tag{A.13}$$

Finally, the expression for the Cole−Cole equation is

$$\frac{V}{U} = \frac{\sin A}{\sin C} = (\omega\tau_0)^{1-\alpha} \tag{A.14}$$

References

Cole, K.S., Cole, R.H., 1941. Dispersion and absorption in dielectrics - I alternating current characteristics. J. Chem. Phys. 9, 341−352.

Smyth, C.P., 1955. Dielectric behavior and structure, McGraw-Hill, New York.

Measurement procedures with impedance analyzers

1. We used three experimental setups and two impedance analyzers:
   a. HP4194A that covered the frequency range $f = 1$ kHz-40 MHz with test fixture HP16047C.
   b. HP4194A that covered the frequency range $f = 10$ kHz-100 MHz with the impedance Z Probe HP41941A.
   c. HP4191A that covered a high-frequency interval from 1 to 1000 MHz.
2. Both instruments require calibration prior to measurements. The following are typical instrument setup parameters.
   a. HP4194A, HP16047C Fixture
      Select Function, Impedance $/Z/ - \vartheta$
      Select the initial and the ending frequencies: Parameter: Start and Stop
      Select Integration time, MED (medium)
      Select Averaging, 4, 8, or 16
      Select SWEEP: Log. Sweep
      Select Number of Points (N), Enter 51 for NOP.
      Select Compensation: Select Zero Open, Enter; Select Zero Short, Enter; For Short use a piece of copper foil.
   b. HP4194A, HP41941A Z Probe
      Select Function, Impedance with Z Probe, $/Z/ - \vartheta$
      The same preparation of the Instrument is used, and the Z Probe is calibrated: Select Compensation, More 1/3, More 2/3. The Instrument prompts:
      Connect 0 S—select (open), enter
      Connect 0 Ω—select (short), enter
      Connect 50 Ω—select (load), enter
      Calibration Complete
   c. HP4191A
      Calibration is done on an APC-7 connector.
      Before Calibration:
      Select Start: 1 MHz; Select Stop: 1000 MHz
      Select Blue Button followed by Log Sweep
      Select "Calibration". The Instrument prompts:
      Connect 0 Ω. Display R X (Short) start

199

Connect 0 S. Display G B (Open) start
Connect 50 Ω. Display Γx Γy (Load) start
Calibration End
Press "Calibration" in order to release the calibration function. The Instrument is automatically set to Z - ϑ (deg.) measurement mode of operation.

d. After compensation or calibration is complete, the sample holder with the sample is connected to the impedance analyzer. The impedance analyzers are connected to a computer via an HPIB interface, where a Lab View program controls the data acquisition. Each analyzer has its own version of Lab View program that indicates the address of the instrument, the amount of points, and the frequency range for collecting and recording the measured data in a table format. The program gives an option to save the data to a file with a chosen name to the C drive or to a removable memory device.

Derivation of the formulas for calculating sample parameters

Chapter Outline

These formulas are used in Chapter 6, Corrections for stray parameters and error estimation (Sections 6.1−6.4). They are derived on the base of series and parallel equivalent circuits shown in Chapter 5, Stray parameters of the measuring system and ways of defining them (Figs. C.1 and C.2).

C.1 Series Model—Sample Impedance Z

Measured impedance: $\hat{Z}_m = R_m + j \cdot X_m$

Sample impedance: $\hat{Z} = R + j \cdot X$ where $X = -1/(\omega \cdot C_{ser})$;

C_s is the stray capacitance of the measuring system. In parallel with the sample impedance, it composes the value $\hat{Z}1 = R1 + jX1$

$$\frac{1}{\hat{Z}1} = \frac{1}{\hat{Z}} + \frac{1}{\hat{Z}_S}; \ \hat{Z}_S = R_S + jX_S; \ \text{where} \ R_S = 0; \ X_S = -\frac{1}{\omega \cdot C_S}; \ \hat{Z}_S = -\frac{j}{\omega C_S};$$

$$\frac{1}{\hat{Z}_S} = -\frac{\omega C_S}{j} \cdot \frac{j}{j} = j\omega C_S$$

$$\frac{1}{\hat{Z}1} = \frac{1}{\hat{Z}} + \frac{1}{\hat{Z}_S} = \frac{1}{R + jX} + j\omega C_S;$$

$$\frac{1}{\hat{Z}} = \frac{1}{\hat{Z}1} - j\omega C_S = \frac{1 - j\omega C_S \cdot \hat{Z}1}{\hat{Z}1}; \quad \hat{Z} = \frac{\hat{Z}1}{1 - j\omega C_S \cdot \hat{Z}1}.$$

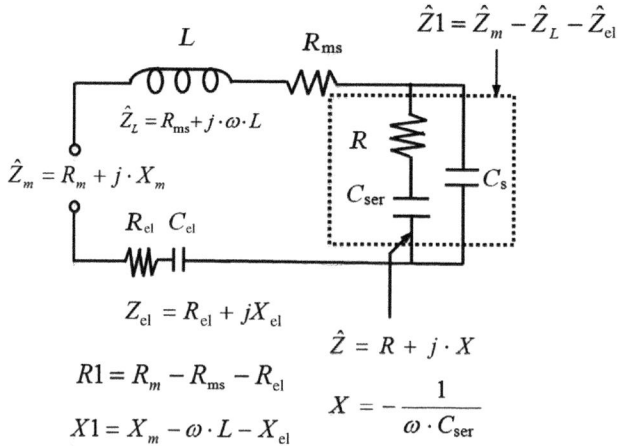

FIGURE C.1

Schematic for the sample series model.

Using the expressions for Z1, Z_m, Z_L, and Z_{el}, shown in Fig. C.1, we have

$$\hat{Z} = \frac{R_m + jX_m - R_{ms} - j\omega L - R_{el} - jX_{el}}{1 - j\omega C_S \cdot (R_m + jX_m - R_{ms} - j\omega L - R_{el} - jX_{el})}$$

$$= \frac{(R_m - R_{ms} - R_{el}) + j(X_m - \omega L - X_{el})}{1 - j\omega C_S \cdot R_m + \omega C_S \cdot X_m + j\omega C_S \cdot R_{ms} - \omega^2 C_S \cdot L + j\omega C_S \cdot R_{el} - \omega C_S \cdot X_{el}}$$

$$= \frac{R1 + jX1}{1 + \omega C_S(X_m - \omega L - X_{el}) - j\omega C_S(R_m - R_{ms} - R_{el})} = \frac{R1 + jX1}{1 + \omega C_S \cdot X1 - j\omega C_S \cdot R1}.$$

We can express the denominator in general as $A - jB$, where $A = 1 + \omega C_S \cdot X1$ and $B = \omega C_S \cdot R1$. In order to eliminate the number j from the denominator, we multiply and divide the last equation by conjugate denominator, such as $A + jB$. Then we obtain

$$\hat{Z} = \frac{(R1 + jX1) \cdot (1 + \omega C_S X1 + j\omega C_S R1)}{(1 + \omega C_S X1)^2 + (\omega C_S R1)^2}$$

$$= \frac{R1 + \omega C_S X1 R1 + j\omega C_S R1^2 + jX1 + j\omega C_S X1^2 - \omega C_S R1 X1}{1 + 2\omega C_S X1 + \omega^2 C_S^2 X1^2 + \omega^2 C_S^2 R1^2}$$

$$= \frac{R1 + j(X1 + \omega C_S R1^2 + \omega C_S X1^2)}{1 + 2\omega C_S X1 + \omega^2 C_S^2 (X1^2 + R1^2)}$$

Separating the real and imaginary components of the last expression, we receive following equations for R and X of the sample impedance $\hat{Z} = R + jX$:

$$R = \frac{R1}{1 + 2\omega C_S X1 + \omega^2 C_S^2 (X1^2 + R1^2)};$$

$$X = \frac{X1 + \omega C_S(R1^2 + X1^2)}{1 + 2\omega C_S X1 + \omega^2 C_S{}^2(X1^2 + R1^2)}.$$

C.2 Parallel Model—Sample Admittance *Y*

The parallel model is similar to the series model. In particular, their left parts, in terms of measured impedance $\hat{Z}_m$, stray parameters impedance $\hat{Z}_L$, and electrode impedance $\hat{Z}_{el}$, are the same.

The difference consists in presenting the sample as a parallel equivalent circuit of conductance G and capacitance C_{par} that compose the sample admittance $\hat{Y} = G - jB$, where $B = \omega C_{par}$ (Chapter 5: Stray parameters of the measuring system and ways of defining them, andChapter 6: Corrections for stray parameters and error estimation). The sample admittance $\hat{Y}$ is connected in parallel with the stray capacitance C_S.

The stray capacitance admittance $\hat{Y}_S = G_S - jB_S$, where $G_S = 0$ and $B_S = 1/X_S = -\omega C_S$.

Then, $\hat{Y}_S = -jB_S = j\omega C_S$.

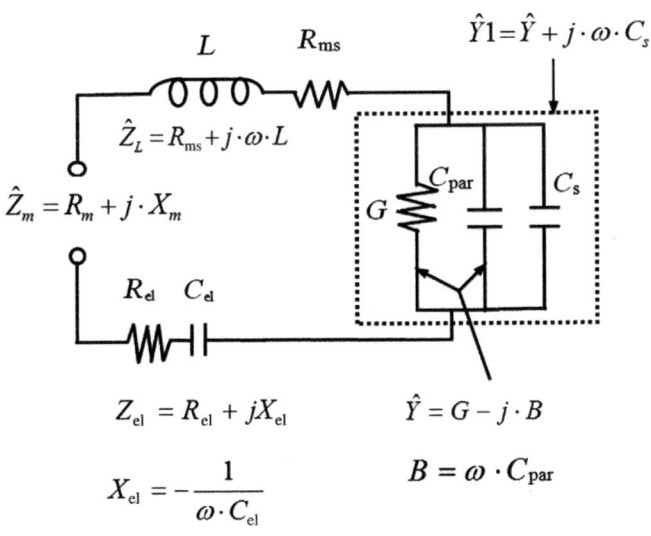

FIGURE C.2

Schematic for the sample parallel model.

$$\hat{Y}1 = \hat{Y} + \hat{Y}_S = G - j\omega C_{par} + j\omega C_S;$$

$$\frac{1}{\hat{Y}1} = \frac{1}{\hat{Y} + \hat{Y}_S} = \frac{1}{G - j\omega C_{par} + j\omega C_s};$$

$$\frac{1}{\hat{Y} + j\omega C_S} = R_m + jX_m - R_{ms} - j\omega L - R_{el} - jX_{el} = (R_m - R_{ms} - R_{el}) + j(X_m - \omega L - X_{el})$$

$$\frac{1}{\hat{Y} + j\omega C_S} = R1 + jX1; \quad (R1 + jX1)(\hat{Y} + j\omega C_S) = 1$$

$$R1 \cdot \hat{Y} + jR1 \cdot \omega C_S + jX1 \cdot \hat{Y} - X1 \cdot \omega C_S = 1$$

$$\hat{Y}(R1 + jX1) + j\omega C_S R1 - \omega C_S X1 = 1$$

$$\hat{Y} = \frac{1 - j\omega C_S R1 + \omega C_S X1}{R1 + jX1} \cdot \frac{(R1 - jX1)}{(R1 - jX1)}$$

$$= \frac{R1 - j\omega C_S R1^2 + \omega C_S X1 R1 - jX1 - \omega C_S R1 X1 - j\omega C_S X1^2}{R1^2 + X1^2}$$

Simplifying this equation, we obtain the following expression for sample admittance:

$$\hat{Y} = \frac{R1 - j(X1 + \omega C_S R1^2 + \omega C_S X1^2)}{R1^2 + X1^2}$$

Now, we have to separate the real and imaginary parts as follows:

$$\hat{Y} = \frac{R1}{R1^2 + X1^2} - j\left(\frac{X1}{R1^2 + X1^2} + \frac{\omega C_S(R1^2 + X1^2)}{R1^2 + X1^2}\right) = \frac{R1}{R1^2 + X1^2} - j\left(\frac{X1}{R1^2 + X1^2} + \omega C_S\right)$$

After simplifying,

$$\hat{Y} = G - jB = \frac{R1}{R1^2 + X1^2} - j\left(\frac{X1}{R1^2 + X1^2} + \omega C_S\right)$$

It follows that expressions for G and B are

$$G = \frac{R1}{R1^2 + X1^2}; \quad B = \frac{X1}{R1^2 + X1^2} + \omega C_S.$$

Spreadsheets for the measured data in this book

Chapter Outline

We have received many requests for our electrical-property laboratory-measurement data over the years. In order to preserve these data for future use, we are including all the measured and calculated data for the four sites, we have used in this book (Avra Valley, Brookhaven, Columbia, and Fort Huachuca).

One of the easiest ways to obtain machine-readable and editable data currently is to have a paper copy of a book or report, scan the data set you want, and then use optical character recognition. Because of this, we have chosen to have all the data on printed pages. A PDF copy can also be editable.

We have found that paper can still be a very safe means of storing data. We have many data sets on IBM 7-track tapes, 5-1/4 floppy drives, and many other storage devices. The only way we can access these data now is through our paper copies. We will see what the future holds for long-term storage.

Avra (6.15%)

	A	B	C	D	E	F	G	H	I	J	K	L	M
1	Avra - 6.15% Sample												
2	High Frequency - Coaxial Air Line												
3													
4				Sample holder, 3 cm, #154, spacers-1.5 mm (flat, h=0.324cm)									
5	Length	R, cm	r, cm	Weight	Dens.(wet)	Drying, weight in g.		Water cont.		Bulk density			
6	cm			M, g	g/cm³	Wet	Dry	W, %	Wv, %	g/cm³			
7	2.676	0.7144	0.3102	5.28	1.51652	5.21	5	4.200	6.15	1.46			
8	L, nH	Cs, pF			Volume					sc (from PL method)			
9	1.856	0.705			V, cm³					0.005186			
10					3.4816623								
11	f, MHz	Z, Ohm	φ, deg	tanδ	Res.meas. R(s), Ohm	Xmeas. Zsinφ	X_L=ωL Ohm	Rm-Rms R' = Rm	Xm-ωL X'	G, S	σ', S/m	B, S	tanδcorr G/B
12													
13	1.00E+00	9.89E+02	-1.28E+01	4.40E+00	9.64E+02	-2.19E+02	1.17E-02	9.64E+02	-2.19E+02	9.86E-04	4.89E-04	-2.20E-04	4.49E+00
14	1.00E+01	5.47E+02	-4.06E+01	1.17E+00	4.15E+02	-3.56E+02	1.17E-01	4.15E+02	-3.56E+02	1.39E-03	6.89E-03	-1.15E-03	1.21E+00
15	2.00E+01	3.70E+02	-5.12E+01	8.03E-01	2.32E+02	-2.89E+02	2.33E-01	2.32E+02	-2.89E+02	1.69E-03	8.38E-03	-2.02E-03	8.38E-01
16	3.00E+01	2.87E+02	-5.58E+01	6.79E-01	1.61E+02	-2.37E+02	3.50E-01	1.61E+02	-2.38E+02	1.95E-03	9.70E-03	-2.75E-03	7.11E-01
17	4.00E+01	2.37E+02	-5.85E+01	6.12E-01	1.24E+02	-2.02E+02	4.66E-01	1.24E+02	-2.03E+02	2.19E-03	1.09E-02	-3.42E-03	6.42E-01
18	5.00E+01	2.03E+02	-6.05E+01	5.65E-01	1.00E+02	-1.77E+02	5.83E-01	1.00E+02	-1.78E+02	2.41E-03	1.19E-02	-4.05E-03	5.94E-01
19	1.00E+02	1.23E+02	-6.61E+01	4.43E-01	4.96E+01	-1.12E+02	1.17E+00	4.96E+01	-1.13E+02	3.25E-03	1.61E-02	-6.96E-03	4.66E-01
20	1.50E+02	8.91E+01	-6.89E+01	3.85E-01	3.21E+01	-8.32E+01	1.75E+00	3.21E+01	-8.49E+01	3.89E-03	1.93E-02	-9.64E-03	4.04E-01
21	2.00E+02	7.02E+01	-7.08E+01	3.49E-01	2.31E+01	-6.62E+01	2.33E+00	2.31E+01	-6.86E+01	4.42E-03	2.19E-02	-1.22E-02	3.62E-01
22	2.50E+02	5.77E+01	-7.20E+01	3.25E-01	1.78E+01	-5.49E+01	2.92E+00	1.78E+01	-5.78E+01	4.87E-03	2.42E-02	-1.47E-02	3.32E-01
23	3.00E+02	4.88E+01	-7.30E+01	3.06E-01	1.43E+01	-4.66E+01	3.50E+00	1.43E+01	-5.01E+01	5.25E-03	2.61E-02	-1.71E-02	3.07E-01
24	3.50E+02	4.20E+01	-7.37E+01	2.93E-01	1.18E+01	-4.03E+01	4.08E+00	1.18E+01	-4.44E+01	5.60E-03	2.78E-02	-1.95E-02	2.87E-01
25	4.00E+02	3.66E+01	-7.42E+01	2.83E-01	9.98E+00	-3.52E+01	4.66E+00	9.98E+00	-3.99E+01	5.90E-03	2.93E-02	-2.18E-02	2.71E-01
26	4.50E+02	3.22E+01	-7.45E+01	2.78E-01	8.62E+00	-3.10E+01	5.25E+00	8.62E+00	-3.63E+01	6.21E-03	3.08E-02	-2.41E-02	2.57E-01
27	5.00E+02	2.85E+01	-7.46E+01	2.75E-01	7.57E+00	-2.75E+01	5.83E+00	7.57E+00	-3.33E+01	6.47E-03	3.21E-02	-2.63E-02	2.46E-01
28	6.00E+02	2.26E+01	-7.47E+01	2.74E-01	5.98E+00	-2.18E+01	7.00E+00	5.98E+00	-2.88E+01	6.90E-03	3.42E-02	-3.06E-02	2.25E-01
29	7.00E+02	1.79E+01	-7.42E+01	2.84E-01	4.89E+00	-1.72E+01	8.16E+00	4.89E+00	-2.54E+01	7.31E-03	3.63E-02	-3.49E-02	2.10E-01
30	8.00E+02	1.40E+01	-7.30E+01	3.05E-01	4.09E+00	-1.34E+01	9.33E+00	4.09E+00	-2.27E+01	7.66E-03	3.80E-02	-3.90E-02	1.96E-01
31	9.00E+02	1.06E+01	-7.07E+01	3.50E-01	3.52E+00	-1.00E+01	1.05E+01	3.52E+00	-2.05E+01	8.10E-03	4.02E-02	-4.33E-02	1.87E-01
32	1.00E+03	7.60E+00	-6.60E+01	4.46E-01	3.09E+00	-6.94E+00	1.17E+01	3.09E+00	-1.86E+01	8.70E-03	4.32E-02	-4.79E-02	1.82E-01

	N	O	P	Q	R	S	T	U	V	W	X	Y
1	Avra - 6.15% Sample											
2	High Frequency - Coaxial Air Line											
3												
4												
5												
6												
7												
8												
9												
10										Conduct. losses $\varepsilon''c$	Dielectric losses	
11	f, MHz	C, pF	ε'	$\phi corr$	$\varepsilon''tan\delta$	Zcorr	Rcorr	ρ', Ohm-m	ρ''		$\varepsilon''-\varepsilon''c$	σ', S/m
12		B/w			ε''	Ohm	Ohm	(Rcorr)		sc/we0		
13	1.00E+00	3.49E+01	1.96E+01	-1.26E+01	8.79E+01	9.90E+02	9.66E+02	1.95E+02	4.34E+01	9.32E+01	-5.28E+00	1.09E-03
14	1.00E+01	1.82E+01	1.02E+01	-3.95E+01	1.24E+01	5.56E+02	4.29E+02	8.64E+01	7.13E+01	9.32E+00	3.05E+00	5.68E-03
15	2.00E+01	1.61E+01	8.99E+00	-5.00E+01	7.54E+01	3.80E+02	2.44E+02	4.92E+01	5.87E+01	4.66E+00	2.87E+00	1.00E-02
16	3.00E+01	1.46E+01	8.17E+00	-5.46E+01	5.81E+00	2.96E+02	1.72E+02	3.46E+01	4.87E+01	3.11E+00	2.70E+00	1.36E-02
17	4.00E+01	1.36E+01	7.61E+00	-5.73E+01	4.89E+00	2.46E+02	1.33E+02	2.68E+01	4.18E+01	2.33E+00	2.56E+00	1.69E-02
18	5.00E+01	1.29E+01	7.23E+00	-5.93E+01	4.29E+00	2.12E+02	1.08E+02	2.18E+01	3.68E+01	1.86E+00	2.43E+00	2.01E-02
19	1.00E+02	1.11E+01	6.21E+00	-6.50E+01	2.90E+00	1.30E+02	5.50E+01	1.11E+01	2.38E+01	9.32E-01	1.96E+00	3.45E-02
20	1.50E+02	1.02E+01	5.73E+00	-6.80E+01	2.31E+00	9.62E+01	3.60E+01	7.25E+00	1.80E+01	6.21E-01	1.69E+00	4.78E-02
21	2.00E+02	9.71E+00	5.44E+00	-7.01E+01	1.97E+00	7.70E+01	2.62E+01	5.28E+00	1.46E+01	4.66E-01	1.50E+00	6.06E-02
22	2.50E+02	9.35E+00	5.24E+00	-7.17E+01	1.74E+00	6.46E+01	2.03E+01	4.10E+00	1.24E+01	3.73E-01	1.37E+00	7.29E-02
23	3.00E+02	9.09E+00	5.09E+00	-7.30E+01	1.56E+00	5.58E+01	1.64E+01	3.30E+00	1.08E+01	3.11E-01	1.25E+00	8.50E-02
24	3.50E+02	8.87E+00	4.97E+00	-7.40E+01	1.43E+00	4.93E+01	1.36E+01	2.74E+00	9.55E+00	2.66E-01	1.16E+00	9.67E-02
25	4.00E+02	8.68E+00	4.87E+00	-7.49E+01	1.32E+00	4.42E+01	1.15E+01	2.33E+00	8.60E+00	2.33E-01	1.08E+00	1.08E-01
26	4.50E+02	8.53E+00	4.78E+00	-7.56E+01	1.23E+00	4.02E+01	1.00E+01	2.02E+00	7.84E+00	2.07E-01	1.02E+00	1.20E-01
27	5.00E+02	8.37E+00	4.69E+00	-7.62E+01	1.15E+00	3.69E+01	8.82E+00	1.78E+00	7.23E+00	1.86E-01	9.68E-01	1.30E-01
28	6.00E+02	8.12E+00	4.55E+00	-7.73E+01	1.03E+00	3.19E+01	7.01E+00	1.41E+00	6.26E+00	1.55E-01	8.71E-01	1.52E-01
29	7.00E+02	7.93E+00	4.44E+00	-7.82E+01	9.31E-01	2.81E+01	5.76E+00	1.16E+00	5.54E+00	1.33E-01	7.98E-01	1.73E-01
30	8.00E+02	7.77E+00	4.35E+00	-7.89E+01	8.54E-01	2.51E+01	4.84E+00	9.76E-01	4.97E+00	1.17E-01	7.37E-01	1.94E-01
31	9.00E+02	7.66E+00	4.29E+00	-7.94E+01	8.03E-01	2.27E+01	4.17E+00	8.41E-01	4.50E+00	1.04E-01	7.00E-01	2.15E-01
32	1.00E+03	7.62E+00	4.27E+00	-7.97E+01	7.76E-01	2.05E+01	3.67E+00	7.40E-01	4.07E+00	9.32E-02	6.83E-01	2.38E-01

Avra - 6.15% Sample

Low Frequency - Disk Electrodes

Avra, # 16-2.5 (From the bag)

HP 4194A, Fixture.

	Diameter	Thickness	Volume	Weight	Dens.(wet)	Drying, weight in g.		Water cont.		Bulk density
	D, cm	h, cm	V, cm^3	M, g	g, g/cm^3	Wet	Dry	W, %	Wv, %	g/cm^3
	5	0.9233	18.12895	28.67	1.58145	28.61	27.49	4.074	6.150	1.52
	L, nH	Cs, pF		Cs, calc.		(Regression eqn. for Air meas. File "Methods"				
	50	2.1813		2.181303						

sc, S/m (from FL - method) 0.005186

									R'	X'	
f, MHz	Z, Ohm	ϕ, deg	ϵ'tanδ	Res.meas.	Xmeas.	$X_L=\omega L$	R_{el}	Xel	Rm-Rms Rel	Xm-ωL- Xel	G, S
				R(s), Ohm	Zsinϕ	Ohm	y=134.53x	y=-141.84x	Rel	Xel	
1.00E-03	1.09E+03	-1.00E+01	5.66E+00	1.07E+03	-1.90E+02	3.14E-04	1.70E+02	-1.79E+02	9.05E+02	-1.10E+01	1.11E-03
2.00E-03	1.04E+03	-7.49E+00	7.60E+00	1.03E+03	-1.35E+02	6.28E-04	1.20E+02	-1.27E+02	9.08E+02	-8.62E+00	1.10E-03
3.00E-03	1.01E+03	-6.49E+00	8.79E+00	1.01E+03	-1.15E+02	9.42E-04	9.80E+01	-1.03E+02	9.09E+02	-1.13E+01	1.10E-03
5.00E-03	9.88E+02	-5.29E+00	1.08E+01	9.84E+02	-9.11E+01	1.57E-03	7.59E+01	-8.00E+01	9.08E+02	-1.11E+01	1.10E-03
1.00E-02	9.62E+02	-3.86E+00	1.48E+01	9.60E+02	-6.47E+01	3.14E-03	5.37E+01	-5.66E+01	9.06E+02	-8.08E+00	1.10E-03
2.00E-02	9.44E+02	-3.16E+00	1.81E+01	9.43E+02	-5.20E+01	6.28E-03	3.80E+01	-4.00E+01	9.05E+02	-1.20E+01	1.11E-03
3.00E-02	9.32E+02	-2.97E+00	1.93E+01	9.31E+02	-4.82E+01	9.42E-03	3.10E+01	-3.27E+01	9.00E+02	-1.56E+01	1.11E-03
5.00E-02	9.21E+02	-2.92E+00	1.96E+01	9.20E+02	-4.69E+01	1.57E-02	2.40E+01	-2.53E+01	8.96E+02	-2.16E+01	1.12E-03
1.00E-01	9.07E+02	-3.27E+00	1.75E+01	9.05E+02	-5.17E+01	3.14E-02	1.70E+01	-1.79E+01	8.88E+02	-3.38E+01	1.12E-03
2.00E-01	8.90E+02	-4.19E+00	1.37E+01	8.88E+02	-6.50E+01	6.28E-02	1.20E+01	-1.27E+01	8.76E+02	-5.24E+01	1.14E-03
3.00E-01	8.79E+02	-5.12E+00	1.12E+01	8.76E+02	-7.85E+01	9.42E-02	9.80E+00	-1.03E+01	8.66E+02	-6.83E+01	1.15E-03
5.00E-01	8.62E+02	-6.87E+00	8.30E+00	8.56E+02	-1.03E+02	1.57E-01	7.59E+00	-8.00E+00	8.48E+02	-9.52E+01	1.16E-03
1.00E+00	8.28E+02	-1.08E+01	5.27E+00	8.13E+02	-1.54E+02	3.14E-01	5.37E+00	-5.66E+00	8.08E+02	-1.49E+02	1.20E-03
2.00E+00	7.71E+02	-1.71E+01	3.26E+00	7.37E+02	-2.26E+02	6.28E-01	3.80E+00	-4.00E+00	7.33E+02	-2.23E+02	1.25E-03
3.00E+00	7.19E+02	-2.21E+01	2.46E+00	6.66E+02	-2.71E+02	9.42E-01	3.10E+00	-3.27E+00	6.63E+02	-2.68E+02	1.30E-03
5.00E+00	6.31E+02	-2.97E+01	1.75E+00	5.48E+02	-3.13E+02	1.57E+00	2.40E+00	-2.53E+00	5.46E+02	-3.12E+02	1.38E-03
1.00E+01	4.79E+02	-4.10E+01	1.15E+00	3.62E+02	-3.14E+02	3.14E+00	1.70E+00	-1.79E+00	3.60E+02	-3.16E+02	1.57E-03
2.00E+01	3.26E+02	-5.09E+01	8.12E-01	2.05E+02	-2.53E+02	6.28E+00	1.20E+00	-1.27E+00	2.04E+02	-2.58E+02	1.89E-03
3.00E+01	2.48E+02	-5.55E+01	6.88E-01	1.41E+02	-2.05E+02	9.42E+00	9.80E-01	-1.03E+00	1.40E+02	-2.13E+02	2.15E-03
4.00E+01	2.00E+02	-5.80E+01	6.24E-01	1.06E+02	-1.70E+02	1.26E+01	8.49E-01	-8.95E-01	1.05E+02	-1.82E+02	2.39E-03

Avra - 6.15% Sample

Low Frequency - Disk Electrodes

f, MHz	σ', S/m	B, S	C, pF	ε'	tanδcorr	φcorr	ε''	Zcorr	Rcorr	ρ', Ohm-m	ρ''
			B/w		G/B		ε'tanδ	Ohm	Ohm	(Rcorr)	
1.00E-03	5.20E-03	-1.35E-05	2.15E+03	1.14E+03	8.20E+01	-6.99E-01	9.34E+04	9.05E+02	9.05E+02	1.92E+02	2.35E+00
2.00E-03	5.18E-03	-1.04E-05	8.31E+02	4.41E+02	1.06E+02	-5.43E-01	4.66E+04	9.08E+02	9.08E+02	1.93E+02	1.83E+02
3.00E-03	5.17E-03	-1.36E-05	7.23E+02	3.84E+02	8.07E+01	-7.10E-01	3.10E+04	9.09E+02	9.09E+02	1.93E+02	2.40E+00
5.00E-03	5.18E-03	-1.33E-05	4.24E+02	2.25E+02	8.25E+01	-6.94E-01	1.86E+04	9.08E+02	9.08E+02	1.93E+02	2.34E+00
1.00E-02	5.19E-03	-9.70E-06	1.54E+02	8.20E+01	1.14E+02	-5.04E-01	9.33E+03	9.06E+02	9.06E+02	1.93E+02	1.69E+00
2.00E-02	5.20E-03	-1.43E-05	1.14E+02	6.06E+01	7.70E+01	-7.44E-01	4.67E+03	9.05E+02	9.05E+02	1.92E+02	2.50E+00
3.00E-02	5.22E-03	-1.88E-05	9.97E+01	5.30E+01	5.91E+01	-9.70E-01	3.13E+03	9.00E+02	9.00E+02	1.91E+02	3.24E+00
5.00E-02	5.25E-03	-2.62E-05	8.34E+01	4.43E+01	4.26E+01	-1.35E+00	1.89E+03	8.96E+02	8.96E+02	1.91E+02	4.48E+00
1.00E-01	5.29E-03	-4.14E-05	6.60E+01	3.50E+01	2.71E+01	-2.11E+00	9.50E+02	8.89E+02	8.88E+02	1.89E+02	6.96E+00
2.00E-01	5.35E-03	-6.54E-05	5.20E+01	2.76E+01	1.74E+01	-3.29E+00	4.81E+02	8.78E+02	8.76E+02	1.86E+02	1.07E+01
3.00E-01	5.40E-03	-8.64E-05	4.58E+01	2.43E+01	1.33E+01	-4.31E+00	3.23E+02	8.69E+02	8.66E+02	1.84E+02	1.39E+01
5.00E-01	5.48E-03	-1.24E-04	3.94E+01	2.09E+01	9.40E+00	-6.07E+00	1.97E+02	8.54E+02	8.49E+02	1.81E+02	1.92E+01
1.00E+00	5.63E-03	-2.07E-04	3.30E+01	1.75E+01	5.78E+00	-9.82E+00	1.01E+02	8.23E+02	8.11E+02	1.73E+02	2.99E+01
2.00E+00	5.87E-03	-3.52E-04	2.80E+01	1.49E+01	3.55E+00	-1.57E+01	5.28E+01	7.70E+02	7.42E+02	1.58E+02	4.44E+01
3.00E+00	6.09E-03	-4.83E-04	2.56E+01	1.36E+01	2.68E+00	-2.04E+01	3.65E+01	7.23E+02	6.78E+02	1.44E+02	5.37E+01
5.00E+00	6.49E-03	-7.21E-04	2.30E+01	1.22E+01	1.91E+00	-2.76E+01	2.33E+01	6.42E+02	5.69E+02	1.21E+02	6.32E+01
1.00E+01	7.38E-03	-1.24E-03	1.97E+01	1.05E+01	1.27E+00	-3.83E+01	1.33E+01	5.00E+02	3.93E+02	8.35E+01	6.59E+01
2.00E+01	8.87E-03	-2.11E-03	1.68E+01	8.92E+00	8.94E-01	-4.82E+01	7.97E+00	3.53E+02	2.35E+02	5.01E+01	5.60E+01
3.00E+01	1.01E-02	-2.87E-03	1.52E+01	8.09E+00	7.50E-01	-5.31E+01	6.07E+00	2.79E+02	1.67E+02	3.56E+01	4.74E+01
4.00E+01	1.12E-02	-3.57E-03	1.42E+01	7.55E+00	6.68E-01	-5.63E+01	5.04E+00	2.33E+02	1.29E+02	2.75E+01	4.12E+01

Avra - 6.15% Sample

Low Frequency - Disk Electrodes

f, MHz	Conduct. losses,ε"c	Diel. losses ε"− ε"c	100/ω^0.5	Rms-Avg	Rms # 1	C(s), pF	σ", S/m	σ'diel	σ"diel	1/ω^0.5	ρ"
	sc/we0			Ohm	Used			σ'-σ'ohmic			
1.00E-03	9.32E+04	2.05E+02	1.26E+00	1.87E-02	1.51E-02	8.38E+05	6.34E-05	1.14E-05	1.39E-07	1.26E-02	2.35E+00
2.00E-03	4.66E+04	-4.99E+01	8.92E-01	1.87E-02	1.51E-02	5.89E+05	4.91E-05	-5.55E-06	-5.26E-08	8.92E-03	1.83E+00
3.00E-03	3.11E+04	-9.33E+01	7.28E-01	1.87E-02	1.51E-02	4.63E+05	6.41E-05	-1.56E-05	-1.93E-07	7.28E-03	2.40E+00
5.00E-03	1.86E+04	-3.54E+01	5.64E-01	1.87E-02	1.51E-02	3.49E+05	6.27E-05	-9.86E-06	-1.19E-07	5.64E-03	2.34E+00
1.00E-02	9.32E+03	8.64E+00	3.99E-01	1.87E-02	1.51E-02	2.46E+05	4.56E-05	4.81E-06	4.23E-08	3.99E-03	1.69E+00
2.00E-02	4.66E+03	1.01E+01	2.82E-01	1.87E-02	1.51E-02	1.53E+05	6.75E-05	1.13E-05	1.46E-07	2.82E-03	2.50E+00
3.00E-02	3.11E+03	2.17E+01	2.30E-01	1.87E-02	1.51E-02	1.10E+05	8.84E-05	3.61E-05	6.12E-07	2.30E-03	3.24E+00
5.00E-02	1.86E+03	2.15E+01	1.78E-01	1.87E-02	1.51E-02	6.79E+04	1.23E-04	5.98E-05	1.41E-06	1.78E-03	4.48E+00
1.00E-01	9.32E+02	1.81E+01	1.26E-01	1.87E-02	1.60E-02	3.08E+04	1.95E-04	1.00E-04	3.70E-06	1.26E-03	6.96E+00
2.00E-01	4.66E+02	1.48E+01	8.92E-02	2.20E-02	1.91E-02	1.22E+04	3.07E-04	1.64E-04	9.43E-06	8.92E-04	1.07E+01
3.00E-01	3.11E+02	1.27E+01	7.28E-02	2.40E-02	2.12E-02	6.76E+03	4.06E-04	2.12E-04	1.60E-05	7.28E-04	1.39E+01
5.00E-01	1.86E+02	1.04E+01	5.64E-02	2.80E-02	2.42E-02	3.09E+03	5.82E-04	2.90E-04	3.08E-05	5.64E-04	1.92E+01
1.00E+00	9.32E+01	7.95E+00	3.99E-02	3.30E-02	2.89E-02	1.03E+03	9.74E-04	4.42E-04	7.65E-05	3.99E-04	2.99E+01
2.00E+00	4.66E+01	6.19E+00	2.82E-02	3.90E-02	3.45E-02	3.52E+02	1.66E-03	6.89E-04	1.94E-04	2.82E-04	4.44E+01
3.00E+00	3.11E+01	5.43E+00	2.30E-02	4.40E-02	3.83E-02	1.96E+02	2.27E-03	9.06E-04	3.38E-04	2.30E-04	5.37E+01
5.00E+00	1.86E+01	4.70E+00	1.78E-02	5.00E-02	4.37E-02	1.02E+02	3.39E-03	1.31E-03	6.83E-04	1.78E-04	6.32E+01
1.00E+01	9.32E+00	3.95E+00	1.26E-02	5.90E-02	5.22E-02	5.06E+01	5.83E-03	2.20E-03	1.73E-03	1.26E-04	6.59E+01
2.00E+01	4.66E+00	3.31E+00	8.92E-03	7.10E-02	6.23E-02	3.15E+01	9.92E-03	3.68E-03	4.12E-03	8.92E-05	5.60E+01
3.00E+01	3.11E+00	2.96E+00	7.28E-03	7.80E-02	6.91E-02	2.59E+01	1.35E-02	4.94E-03	6.59E-03	7.28E-05	4.74E+01
4.00E+01	2.33E+00	2.71E+00	6.31E-03	8.40E-02	7.44E-02	2.34E+01	1.68E-02	6.03E-03	9.04E-03	6.31E-05	4.12E+01

Avra (10.48%)

	A	B	C	D	E	F	G	H	I	J	K	L	M
1	Avra - 10.48% Sample												
2	High Frequency - Disk Electrodes									Sample holder # 154, 3 cm Spacers are identical: 1.5 cm.(Avg:1.6217 mm			
3													
4	HP4194A		Avra, # 16-2.5. Water is added to completely dry soil. (Wet samples prepared for advance in a plastic can)										
5	Sample length: 3 cm - 0.324 - 0.1 = 2.576 cm						Stray capacitance Cs - from Air meas. with same Tef. Spacers. AVG: 0.705 pF.						
6	Length	R, cm	r, cm	Weight	Dens.(wet)	Drying, weight in glass.				Water cont.			Bulk dens.
7	cm			M, g	γ, g/cm³	Wet	Dry			W, %	W_v, %		g/cm³
8	2.58	0.71	0.31	5.44	1.62	5.42	5.07	Cair, pF		6.90	10.48		1.52
9	Induc. L	C_s, pF		σ'_c, S/m		Volume	C_s, pF		"CoaxStray-07Reverse",				
10	nH	Total		0.03		V, cm³	Air meas.		L - Calculated, AVG : 1.85633 nH (same Teflon spacers)				
11	1.86	0.77		From LFR		3.35	0.71	0.07	No Rms for coaxial				

f, MHz	Z, Ohm	ϕ, deg	tanδ	Res.meas. R(s), Ohm	Xmeas. Zsinϕ	$X_L=\omega L$ Ohm	Rm-Rms R' = Rms	Xm-ωL X'	G, S	σ', S/m	B, S	C(s), pF
1.00E+00	1.73E+02	-3.76E+00	1.52E+01	1.73E+02	-1.14E+01	1.17E-02	1.73E+02	-1.14E+01	5.76E-03	2.97E-02	-3.74E-04	1.40E+04
1.00E+01	1.50E+02	-1.55E+01	3.62E+00	1.45E+02	-4.00E+01	1.17E-01	1.45E+02	-4.02E+01	6.41E-03	3.30E-02	-1.73E-03	3.97E+02
2.00E+01	1.33E+02	-2.41E+01	2.24E+00	1.21E+02	-5.41E+01	2.33E-01	1.21E+02	-5.44E+01	6.87E-03	3.54E-02	-2.99E-03	1.47E+02
3.00E+01	1.19E+02	-2.99E+01	1.74E+00	1.03E+02	-5.93E+01	3.50E-01	1.03E+02	-5.96E+01	7.27E-03	3.75E-02	-4.06E-03	8.95E+01
4.00E+01	1.08E+02	-3.42E+01	1.47E+00	8.91E+01	-6.06E+01	4.66E-01	8.91E+01	-6.10E+01	7.64E-03	3.94E-02	-5.04E-03	6.57E+01
5.00E+01	9.87E+01	-3.77E+01	1.30E+00	7.81E+01	-6.03E+01	5.83E-01	7.81E+01	-6.09E+01	7.96E-03	4.10E-02	-5.96E-03	5.28E+01
1.00E+02	6.98E+01	-4.82E+01	8.94E-01	4.65E+01	-5.20E+01	1.17E+00	4.65E+01	-5.32E+01	9.31E-03	4.80E-02	-1.02E-02	3.06E+01
1.50E+02	5.42E+01	-5.38E+01	7.33E-01	3.20E+01	-4.37E+01	1.75E+00	3.20E+01	-4.54E+01	1.04E-02	5.34E-02	-1.40E-02	2.43E+01
2.00E+02	4.41E+01	-5.72E+01	6.43E-01	2.39E+01	-3.71E+01	2.33E+00	2.39E+01	-3.94E+01	1.12E-02	5.79E-02	-1.76E-02	2.15E+01
2.50E+02	3.70E+01	-5.96E+01	5.87E-01	1.87E+01	-3.19E+01	2.92E+00	1.87E+01	-3.48E+01	1.20E-02	6.18E-02	-2.11E-02	2.00E+01
3.00E+02	3.15E+01	-6.13E+01	5.48E-01	1.52E+01	-2.76E+01	3.50E+00	1.52E+01	-3.11E+01	1.26E-02	6.51E-02	-2.45E-02	1.92E+01
3.50E+02	2.72E+01	-6.24E+01	5.24E-01	1.26E+01	-2.41E+01	4.08E+00	1.26E+01	-2.82E+01	1.32E-02	6.82E-02	-2.78E-02	1.88E+01
4.00E+02	2.37E+01	-6.40E+01	4.87E-01	1.04E+01	-2.13E+01	4.66E+00	1.04E+01	-2.60E+01	1.33E-02	6.83E-02	-3.13E-02	1.87E+01
4.50E+02	2.07E+01	-6.33E+01	5.02E-01	9.30E+00	-1.85E+01	5.25E+00	9.30E+00	-2.38E+01	1.43E-02	7.36E-02	-3.43E-02	1.91E+01
5.00E+02	1.81E+01	-6.33E+01	5.04E-01	8.15E+00	-1.62E+01	5.83E+00	8.15E+00	-2.20E+01	1.48E-02	7.63E-02	-3.76E-02	1.97E+01
6.00E+02	1.39E+01	-6.21E+01	5.29E-01	6.50E+00	-1.23E+01	7.00E+00	6.50E+00	-1.93E+01	1.57E-02	8.09E-02	-4.37E-02	2.16E+01
7.00E+02	1.04E+01	-5.90E+01	6.02E-01	5.35E+00	-8.89E+00	8.16E+00	5.35E+00	-1.71E+01	1.67E-02	8.63E-02	-5.00E-02	2.56E+01
8.00E+02	7.45E+00	-5.24E+01	7.71E-01	4.55E+00	-5.90E+00	9.33E+00	4.55E+00	-1.52E+01	1.80E-02	9.28E-02	-5.64E-02	3.37E+01
9.00E+02	5.03E+00	-3.78E+01	1.29E+00	3.97E+00	-3.08E+00	1.05E+01	3.97E+00	-1.36E+01	1.99E-02	1.02E-01	-6.35E-02	5.74E+01
1.00E+03	3.60E+00	-5.90E+00	9.68E+00	3.58E+00	-3.70E-01	1.17E+01	3.58E+00	-1.20E+01	2.27E-02	1.17E-01	-7.15E-02	4.30E+02

	N	O	P	Q	R	S	T	U	V	W	X	Y	Z	AA
1		Avra - 10.48% Sample												
2		High Frequency - Disk Electrodes												
3														
4														
5														
6														
7														
8														
9														
10														
11														
12	σ'', S/m	f, MHz	C, pF	ε'	tanδcorr	φcorr	ε''	Zcorr	Rcorr	ρ', Ohm-m	ρ''	losses,ε''c	losses	$100/\omega^{0.5}$
13		B/w			G/B		e'tand	Ohm	Ohm	(Rcorr)		sc/we0	ε''-ε''c	
14	1.93E-03	1.00E+00	5.95E+01	3.47E+01	1.54E+01	-3.72E+00	5.34E+02	1.73E+02	1.73E+02	3.35E+01	2.18E+00	5.20E+02	1.34E+01	3.99E-02
15	8.91E-03	1.00E+01	2.75E+01	1.60E+01	3.71E+00	-1.51E+01	5.94E+01	1.51E+02	1.45E+02	2.82E+01	7.61E+00	5.20E+01	7.35E+00	1.26E-02
16	1.54E-02	2.00E+01	2.38E+01	1.39E+01	2.30E+00	-2.35E+01	3.18E+01	1.33E+02	1.22E+02	2.37E+01	1.03E+01	2.60E+01	5.82E+00	8.92E-03
17	2.09E-02	3.00E+01	2.15E+01	1.25E+01	1.79E+00	-2.92E+01	2.25E+01	1.20E+02	1.05E+02	2.03E+01	1.14E+01	1.73E+01	5.10E+00	7.28E-03
18	2.60E-02	4.00E+01	2.00E+01	1.17E+01	1.52E+00	-3.34E+01	1.77E+01	1.09E+02	9.12E+01	1.77E+01	1.17E+01	1.30E+01	4.68E+00	6.31E-03
19	3.07E-02	5.00E+01	1.90E+01	1.10E+01	1.34E+00	-3.68E+01	1.48E+01	1.01E+02	8.05E+01	1.56E+01	1.17E+01	1.04E+01	4.35E+00	5.64E-03
20	5.24E-02	1.00E+02	1.62E+01	9.42E+00	9.16E-01	-4.75E+01	8.63E+00	7.25E+01	4.90E+01	9.51E+00	1.04E+01	5.20E+00	3.42E+00	3.99E-03
21	7.21E-02	1.50E+02	1.48E+01	8.64E+00	7.42E-01	-5.34E+01	6.40E+00	5.75E+01	3.42E+01	6.64E+00	8.95E+00	3.47E+00	2.94E+00	3.26E-03
22	9.07E-02	2.00E+02	1.40E+01	8.15E+00	6.39E-01	-5.74E+01	5.21E+00	4.79E+01	2.58E+01	5.00E+00	7.83E+00	2.60E+00	2.61E+00	2.82E-03
23	1.09E-01	2.50E+02	1.34E+01	7.82E+00	5.69E-01	-6.04E+01	4.44E+00	4.12E+01	2.04E+01	3.95E+00	6.95E+00	2.08E+00	2.36E+00	2.52E-03
24	1.26E-01	3.00E+02	1.30E+01	7.57E+00	5.16E-01	-6.27E+01	3.90E+00	3.63E+01	1.66E+01	3.23E+00	6.25E+00	1.73E+00	2.17E+00	2.30E-03
25	1.43E-01	3.50E+02	1.27E+01	7.37E+00	4.75E-01	-6.46E+01	3.50E+00	3.25E+01	1.39E+01	2.70E+00	5.69E+00	1.49E+00	2.01E+00	2.13E-03
26	1.61E-01	4.00E+02	1.24E+01	7.24E+00	4.24E-01	-6.70E+01	3.07E+00	2.95E+01	1.15E+01	2.23E+00	5.26E+00	1.30E+00	1.77E+00	1.99E-03
27	1.77E-01	4.50E+02	1.21E+01	7.07E+00	4.16E-01	-6.74E+01	2.94E+00	2.69E+01	1.03E+01	2.01E+00	4.82E+00	1.16E+00	1.78E+00	1.88E-03
28	1.94E-01	5.00E+02	1.20E+01	6.96E+00	3.94E-01	-6.85E+01	2.74E+00	2.48E+01	9.09E+00	1.76E+00	4.47E+00	1.04E+00	1.70E+00	1.78E-03
29	2.25E-01	6.00E+02	1.16E+01	6.74E+00	3.60E-01	-7.02E+01	2.42E+00	2.16E+01	7.29E+00	1.42E+00	3.94E+00	8.67E-01	1.56E+00	1.63E-03
30	2.58E-01	7.00E+02	1.14E+01	6.61E+00	3.35E-01	-7.15E+01	2.22E+00	1.90E+01	6.03E+00	1.17E+00	3.49E+00	7.43E-01	1.47E+00	1.51E-03
31	2.91E-01	8.00E+02	1.12E+01	6.53E+00	3.19E-01	-7.23E+01	2.09E+00	1.69E+01	5.14E+00	9.97E-01	3.12E+00	6.50E-01	1.44E+00	1.41E-03
32	3.27E-01	9.00E+02	1.12E+01	6.53E+00	3.13E-01	-7.26E+01	2.04E+00	1.50E+01	4.49E+00	8.71E-01	2.78E+00	5.78E-01	1.47E+00	1.33E-03
33	3.69E-01	1.00E+03	1.14E+01	6.63E+00	3.18E-01	-7.24E+01	2.10E+00	1.33E+01	4.03E+00	7.83E-01	2.46E+00	5.20E-01	1.58E+00	1.26E-03

Avra - 10.48% Sample

Low Frequency - Disk Electrodes

Avra, # 16-2.5. Water is added to completely dry soil.

HP4194A

Subtracting the Impedance of Electrode Polarization after Lopatin.

	A	B	C	D	E	F	G	H	I	J	K	L
7	Diameter	Thickness	Volume	Weight	Dens.(wet)	Drying, weight in g.		Water cont.		Bulk density		
8	D, cm	h, cm	V, cm³	M, g	γ, g/cm³	Wet	Dry	W, %	W_v, %	g/cm³		
9	5	0.924967	18.161685	29.96	1.6496267	25.65	24.02	6.7860117	10.483008	1.5447966		
10	L, nH	Cs, pF (calcul.)			Cs, calc	(Regression eqn. for Air. "MethodsP")			sc, S/m	from FL - method		
11	50	2.18			2.1798316	L - AVG from Air & Tefl. "MethodsP"			0.02895	R'	X'	
12	f, MHz	Z, Ohm	ϕ, deg	tanδ	Res.meas.	Xmeas.	$X_L=\omega L$	R_{el}	X_{el}	Rm-Rms-	Xm-ωL-	G, S
13					R(s), Ohm	Zsinϕ	Ohm	y=10.455*x	y=-11.591*x	R_{el}	X_{el}	
14	1.00E-03	1.77E+02	-5.08E+00	1.13E+01	1.76E+02	-1.56E+02	3.14E-04	1.32E+01	-1.46E+01	1.63E+02	-9.99E-01	6.15E-03
15	2.00E-03	1.72E+02	-3.57E+00	1.60E+01	1.72E+02	-1.07E+01	6.28E-04	9.33E+00	-1.03E+01	1.63E+02	-3.94E-01	6.14E-03
16	3.00E-03	1.71E+02	-2.94E+00	1.95E+01	1.70E+02	-8.74E+00	9.42E-04	7.62E+00	-8.44E+00	1.63E+02	-2.99E-01	6.14E-03
17	5.00E-03	1.69E+02	-2.34E+00	2.45E+01	1.69E+02	-6.90E+00	1.57E-03	5.90E+00	-6.54E+00	1.63E+02	-3.60E-01	6.14E-03
18	1.00E-02	1.67E+02	-1.80E+00	3.18E+01	1.67E+02	-5.25E+00	3.14E-03	4.17E+00	-4.62E+00	1.63E+02	-6.27E-01	6.15E-03
19	2.00E-02	1.65E+02	-1.51E+00	3.79E+01	1.65E+02	-4.35E+00	6.28E-03	2.95E+00	-3.27E+00	1.62E+02	-1.09E+00	6.16E-03
20	3.00E-02	1.64E+02	-1.45E+00	3.96E+01	1.64E+02	-4.15E+00	9.42E-03	2.41E+00	-2.67E+00	1.62E+02	-1.49E+00	6.18E-03
21	5.00E-02	1.63E+02	-1.42E+00	4.03E+01	1.63E+02	-4.05E+00	1.57E-02	1.87E+00	-2.07E+00	1.61E+02	-2.00E+00	6.20E-03
22	1.00E-01	1.62E+02	-1.52E+00	3.77E+01	1.62E+02	-4.29E+00	3.14E-02	1.32E+00	-1.46E+00	1.60E+02	-2.86E+00	6.23E-03
23	2.00E-01	1.60E+02	-1.78E+00	3.22E+01	1.60E+02	-4.98E+00	6.28E-02	9.33E-01	-1.03E+00	1.59E+02	-4.01E+00	6.28E-03
24	3.00E-01	1.59E+02	-2.03E+00	2.82E+01	1.59E+02	-5.65E+00	9.42E-02	7.62E-01	-8.44E-01	1.58E+02	-4.90E+00	6.31E-03
25	5.00E-01	1.58E+02	-2.48E+00	2.31E+01	1.58E+02	-6.83E+00	1.57E-01	5.90E-01	-6.54E-01	1.57E+02	-6.34E+00	6.36E-03
26	1.00E+00	1.55E+02	-3.45E+00	1.66E+01	1.55E+02	-9.35E+00	3.14E-01	4.17E-01	-4.62E-01	1.55E+02	-9.20E+00	6.44E-03
27	2.00E+00	1.52E+02	-5.08E+00	1.12E+01	1.52E+02	-1.35E+01	6.28E-01	2.95E-01	-3.27E-01	1.51E+02	-1.38E+01	6.55E-03
28	3.00E+00	1.50E+02	-6.53E+00	8.73E+00	1.49E+02	-1.70E+01	9.42E-01	2.41E-01	-2.67E-01	1.49E+02	-1.77E+01	6.64E-03
29	5.00E+00	1.46E+02	-9.09E+00	6.25E+00	1.44E+02	-2.30E+01	1.57E+00	1.87E-01	-2.07E-01	1.44E+02	-2.44E+01	6.76E-03
30	1.00E+01	1.37E+02	-1.44E+01	3.89E+00	1.32E+02	-3.40E+01	3.14E+00	1.32E-01	-1.46E-01	1.32E+02	-3.70E+01	7.02E-03
31	2.00E+01	1.20E+02	-2.20E+01	2.47E+00	1.11E+02	-4.51E+01	6.28E+00	9.33E-02	-1.03E-01	1.11E+02	-5.13E+01	7.41E-03
32	3.00E+01	1.06E+02	-2.72E+01	1.95E+00	9.42E+01	-4.84E+01	9.42E+00	7.62E-02	-8.44E-02	9.40E+01	-5.77E+01	7.73E-03
33	4.00E+01	9.36E+01	-3.06E+01	1.69E+00	8.05E+01	-4.76E+01	1.26E+01	6.59E-02	-7.31E-02	8.04E+01	-6.01E+01	7.98E-03

	M	N	O	P	Q	R	S	T	U	V	W	X
1	Avra - 10.48% Sample											
2	Low Frequency - Disk Electrodes											
3												
4												
5												
6												
7												
8												
9												
10												
11												
12	f, MHz	C, pF	σ', S/m	B, S	ε'	tandcorr	ϕcorr	ε''	Zcorr	Rcorr	ρ', Ohm-m	ρ''
13		B/w				G/B		ε'tand	Ohm	Ohm	(Rcorr)	
14	1.00E-03	6.01E+03	2.90E-02	-3.77E-05	3.20E+03	1.63E+02	-3.52E-01	5.21E+05	1.63E+02	1.63E+02	3.45E+01	2.12E-01
15	2.00E-03	1.18E+03	2.89E-02	-1.48E-05	6.29E+02	4.14E+02	-1.38E-01	2.60E+05	1.63E+02	1.63E+02	3.46E+01	8.35E-02
16	3.00E-03	5.96E+02	2.89E-02	-1.12E-05	3.17E+02	5.46E+02	-1.05E-01	1.73E+05	1.63E+02	1.63E+02	3.46E+01	6.33E-02
17	5.00E-03	4.30E+02	2.89E-02	-1.35E-05	2.29E+02	4.55E+02	-1.26E-01	1.04E+05	1.63E+02	1.63E+02	3.46E+01	7.60E-02
18	1.00E-02	3.76E+02	2.90E-02	-2.36E-05	2.00E+02	2.61E+02	-2.20E-01	5.21E+04	1.63E+02	1.63E+02	3.45E+01	1.32E-01
19	2.00E-02	3.28E+02	2.90E-02	-4.12E-05	1.74E+02	1.50E+02	-3.83E-01	2.61E+04	1.62E+02	1.62E+02	3.44E+01	2.30E-01
20	3.00E-02	2.99E+02	2.91E-02	-5.64E-05	1.59E+02	1.10E+02	-5.23E-01	1.74E+04	1.62E+02	1.62E+02	3.44E+01	3.14E-01
21	5.00E-02	2.42E+02	2.92E-02	-7.60E-05	1.29E+02	8.15E+01	-7.03E-01	1.05E+04	1.61E+02	1.61E+02	3.43E+01	4.20E-01
22	1.00E-01	1.75E+02	2.93E-02	-1.10E-04	9.30E+01	5.68E+01	-1.01E+00	5.28E+03	1.60E+02	1.60E+02	3.41E+01	6.00E-01
23	2.00E-01	1.24E+02	2.96E-02	-1.55E-04	6.58E+01	4.04E+01	-1.42E+00	2.66E+03	1.59E+02	1.59E+02	3.38E+01	8.36E-01
24	3.00E-01	1.01E+02	2.97E-02	-1.91E-04	5.39E+01	3.30E+01	-1.73E+00	1.78E+03	1.58E+02	1.58E+02	3.36E+01	1.02E+00
25	5.00E-01	7.95E+01	3.00E-02	-2.50E-04	4.23E+01	2.54E+01	-2.25E+00	1.08E+03	1.57E+02	1.57E+02	3.33E+01	1.31E+00
26	1.00E+00	5.89E+01	3.04E-02	-3.70E-04	3.13E+01	1.74E+01	-3.29E+00	5.46E+02	1.55E+02	1.55E+02	3.28E+01	1.88E+00
27	2.00E+00	4.54E+01	3.09E-02	-5.70E-04	2.41E+01	1.15E+01	-4.97E+00	2.77E+02	1.52E+02	1.51E+02	3.21E+01	2.80E+00
28	3.00E+00	3.98E+01	3.13E-02	-7.50E-04	2.12E+01	8.84E+00	-6.45E+00	1.87E+02	1.50E+02	1.49E+02	3.16E+01	3.57E+00
29	5.00E+00	3.44E+01	3.19E-02	-1.08E-03	1.83E+01	6.26E+00	-9.07E+00	1.15E+02	1.46E+02	1.44E+02	3.06E+01	4.89E+00
30	1.00E+01	2.91E+01	3.31E-02	-1.83E-03	1.55E+01	3.84E+00	-1.46E+01	5.94E+01	1.38E+02	1.33E+02	2.83E+01	7.38E+00
31	2.00E+01	2.50E+01	3.49E-02	-3.14E-03	1.33E+01	2.36E+00	-2.30E+01	3.14E+01	1.24E+02	1.14E+02	2.43E+01	1.03E+01
32	3.00E+01	2.30E+01	3.64E-02	-4.33E-03	1.22E+01	1.78E+00	-2.93E+01	2.18E+01	1.13E+02	9.85E+01	2.09E+01	1.17E+01
33	4.00E+01	2.16E+01	3.76E-02	-5.42E-03	1.15E+01	1.47E+00	-3.42E+01	1.69E+01	1.04E+02	8.58E+01	1.82E+01	1.24E+01

Avra - 10.48% Sample

Low Frequency - Disk Electrodes

	Conduct.	Dielectric					5.28.98			
f, MHz	losses, ε''_c	losses	ε'	ε''	sc/we0	$\varepsilon''-\varepsilon''_c$	Rms-Avg	Rms # 1	σ'', S/m	$100/\omega^{0.5}$
	sc/we0	$\varepsilon''-\varepsilon''_c$	$32.781 \cdot f^{-0.4259}$	$541.9788 \cdot f^{-0.9901}$			Ohm	Used		
1.00E-03			3.20E-03	5.21E+05	5.20E+05	2.60E+02	1.87E-02	1.51E-02	1.78E-04	1.26E+00
2.00E-03			6.29E-02	2.60E+05	2.60E+05	-7.33E+01	1.87E-02	1.51E-02	7.00E-05	8.92E-01
3.00E-03	5.20E+05	2.49E+02	3.17E+02	1.73E+05	1.73E+05	-1.03E+02	1.87E-02	1.51E-02	5.29E-05	7.28E-01
3.00E-03	2.60E+05	-7.90E+01	3.89E+02	1.71E+05	1.73E+05	-2.88E+03	1.87E-02	1.51E-02	6.36E-05	5.64E-01
4.00E-03	1.73E+05	-1.06E+02	3.44E+02	1.28E+05	1.30E+05	-1.80E+03	1.87E-02	1.51E-02	1.11E-04	3.99E-01
5.00E-03	1.04E+05	-3.55E+01	3.13E+02	1.03E+05	1.04E+05	-1.21E+03	1.87E-02	1.51E-02	1.94E-04	2.82E-01
6.00E-03	5.20E+04	4.28E+01	2.90E+02	8.59E+04	8.67E+04	-8.54E+02	1.87E-02	1.51E-02	2.66E-04	2.30E-01
7.00E-03	2.60E+04	7.31E+01	2.71E+02	7.37E+04	7.43E+04	-6.20E+02	1.87E-02	1.51E-02	3.58E-04	1.78E-01
8.00E-03	1.73E+04	8.79E+01	2.56E+02	6.46E+04	6.50E+04	-4.57E+02	1.87E-02	1.60E-02	5.17E-04	1.26E-01
9.00E-03	1.04E+04	8.55E+01	2.44E+02	5.75E+04	5.78E+04	-3.40E+02	2.18E-02	1.91E-02	7.32E-04	8.92E-02
1.00E-02	5.20E+03	7.19E+01	2.33E+02	5.18E+04	5.20E+04	-2.52E+02	2.43E-02	2.12E-02	9.00E-04	7.28E-02
1.00E-02	2.60E+03	5.55E+01	2.00E+02	5.21E+04	5.20E+04	4.40E+01	2.77E-02	2.42E-02	1.18E-03	5.64E-02
1.50E-02	1.73E+03	4.64E+01	1.96E+02	3.47E+04	3.47E+04	-2.93E+01	3.33E-02	2.89E-02	1.74E-03	3.99E-02
2.00E-02	1.04E+03	3.63E+01	1.73E+02	2.61E+04	2.60E+04	5.00E+01	4.00E-02	3.45E-02	2.69E-03	2.82E-02
2.00E-02	5.20E+02	2.53E+01	1.74E+02	2.61E+04	2.60E+04	7.37E+01	4.45E-02	3.83E-02	3.54E-03	2.30E-02
2.50E-02	2.60E+02	1.73E+01	1.58E+02	2.09E+04	2.08E+04	8.76E+01	5.09E-02	4.37E-02	5.09E-03	1.78E-02
3.00E-02	1.73E+02	1.38E+01	1.46E+02	1.75E+04	1.73E+04	1.04E+02	6.10E-02	5.22E-02	8.61E-03	1.26E-02
4.00E-02	1.04E+02	1.05E+01	1.29E+02	1.31E+04	1.30E+04	1.16E+02	7.33E-02	6.23E-02	1.48E-02	8.92E-03
5.00E-02	5.20E+01	7.39E+00	1.17E+02	1.05E+04	1.04E+04	1.16E+02	8.15E-02	6.91E-02	2.04E-02	7.28E-03
6.00E-02	2.60E+01	5.36E+00	1.09E+02	8.79E+03	8.67E+03	1.12E+02	8.79E-02	7.44E-02	2.55E-02	6.31E-03
7.00E-02	1.73E+01	4.46E+00	1.02E+02	7.54E+03	7.43E+03	1.08E+02				
8.00E-02	1.30E+01	3.88E+00	9.61E+01	6.61E+03	6.50E+03	1.03E+02				

Avra (16.4%)

Avra - 16.4 Sample — Avra Valley, # 5-3.0, with 10 % water by weight.

High Frequency - Coaxial Fixture

HP 4191A — Coaxial sample holder #154, patterned teflon spacers (2h = 0.4135 cm)

	Length,cm	R, cm	Volume V, cm³	Weight M, g	Dens.(wet) γ, g/cm³	Drying, weight in g. Wet	Dry	Water content. W, %	W_v, %	Bulk dens. g/cm³	method: sc', S/m
	2.5865	0.7144	3.365217	5.56	1.65220	13.22	11.91	10.999	16.372	1.49	0.046520

L, nH = 1.36 C_s, pF = 0.563 (stray capac. C_s: AVG from Air-6.3.98 & 9.15.98 with same Teflon's 2h); "CoaxStray-07Reverse" $X_m = \omega L$

f, MHz	Z, Ohm	φ, deg	tanδ	Res.meas. R(s), Ohm	Xmeas. Zsinφ	$X_L=\omega L$ Ohm	Rm-Rms R'=Rm	Xm-ωL X'	G, S	σ, S/m	B, S	tanocorr G/B	C(s), pF
1.00E+00	1.04E+02	-3.44E+00	1.66E+01	1.03E+02	-6.21E+00	8.55E-03	1.03E+02	-6.22E+00	9.64E-03	4.95E-02	-5.77E-04	1.67E+01	2.56E+04
2.00E+00	1.01E+02	-3.92E+00	1.46E+01	1.01E+02	-6.91E+00	1.71E-02	1.01E+02	-6.93E+00	9.86E-03	5.06E-02	-6.71E-04	1.47E+01	1.15E+04
3.00E+00	9.95E+01	-5.00E+00	1.14E+01	9.91E+01	-8.67E+00	2.56E-02	9.91E+01	-8.70E+00	1.00E-02	5.14E-02	-8.68E-04	1.15E+01	6.12E+03
5.00E+00	9.72E+01	-7.09E+00	8.04E+00	9.65E+01	-1.20E+01	4.27E-02	9.65E+01	-1.20E+01	1.02E-02	5.24E-02	-1.26E-03	8.12E+00	2.65E+03
1.00E+01	9.29E+01	-1.18E+01	4.78E+00	9.09E+01	-1.90E+01	8.55E-02	9.09E+01	-1.91E+01	1.05E-02	5.41E-02	-2.18E-03	4.83E+00	8.36E+02
2.00E+01	8.53E+01	-1.92E+01	2.88E+00	8.06E+01	-2.80E+01	1.71E-01	8.06E+01	-2.82E+01	1.11E-02	5.68E-02	-3.80E-03	2.91E+00	2.84E+02
3.00E+01	7.88E+01	-2.47E+01	2.18E+00	7.16E+01	-3.29E+01	2.56E-01	7.16E+01	-3.31E+01	1.15E-02	5.91E-02	-5.22E-03	2.21E+00	1.61E+02
4.00E+01	7.29E+01	-2.90E+01	1.80E+00	6.37E+01	-3.53E+01	3.42E-01	6.37E+01	-3.57E+01	1.19E-02	6.13E-02	-6.54E-03	1.83E+00	1.13E+02
5.00E+01	6.79E+01	-3.25E+01	1.57E+00	5.73E+01	-3.65E+01	4.27E-01	5.73E+01	-3.69E+01	1.23E-02	6.33E-02	-7.77E-03	1.59E+00	8.72E+01
6.00E+01	6.35E+01	-3.56E+01	1.40E+00	5.16E+01	-3.70E+01	5.13E-01	5.16E+01	-3.75E+01	1.27E-02	6.51E-02	-9.01E-03	1.41E+00	7.17E+01
7.00E+01	5.96E+01	-3.82E+01	1.27E+00	4.68E+01	-3.69E+01	5.98E-01	4.68E+01	-3.75E+01	1.30E-02	6.69E-02	-1.02E-02	1.28E+00	6.17E+01
8.00E+01	5.62E+01	-4.05E+01	1.17E+00	4.27E+01	-3.65E+01	6.84E-01	4.27E+01	-3.72E+01	1.33E-02	6.84E-02	-1.13E-02	1.18E+00	5.45E+01
9.00E+01	5.30E+01	-4.26E+01	1.09E+00	3.90E+01	-3.59E+01	7.69E-01	3.90E+01	-3.67E+01	1.36E-02	6.99E-02	-1.25E-02	1.09E+00	4.93E+01
1.00E+02	5.03E+01	-4.44E+01	1.02E+00	3.59E+01	-3.52E+01	8.55E-01	3.59E+01	-3.60E+01	1.39E-02	7.13E-02	-1.36E-02	1.02E+00	4.53E+01
1.50E+02	3.97E+01	-5.11E+01	8.06E-01	2.49E+01	-3.09E+01	1.28E+00	2.49E+01	-3.22E+01	1.50E-02	7.72E-02	-1.89E-02	7.96E-01	3.43E+01
2.00E+02	3.26E+01	-5.56E+01	6.86E-01	1.84E+01	-2.69E+01	1.71E+00	1.84E+01	-2.86E+01	1.59E-02	8.18E-02	-2.40E-02	6.64E-01	2.96E+01
2.50E+02	2.74E+01	-5.85E+01	6.14E-01	1.43E+01	-2.33E+01	2.14E+00	1.43E+01	-2.55E+01	1.68E-02	8.61E-02	-2.89E-02	5.79E-01	2.73E+01
3.00E+02	2.34E+01	-6.09E+01	5.58E-01	1.14E+01	-2.04E+01	2.56E+00	1.14E+01	-2.30E+01	1.73E-02	8.89E-02	-3.39E-02	5.11E-01	2.60E+01
3.50E+02	2.02E+01	-6.22E+01	5.28E-01	9.41E+00	-1.78E+01	2.99E+00	9.41E+00	-2.08E+01	1.80E-02	9.25E-02	-3.86E-02	4.66E-01	2.55E+01
3.70E+02	1.91E+01	-6.26E+01	5.19E-01	8.78E+00	-1.69E+01	3.16E+00	8.78E+00	-2.01E+01	1.83E-02	9.38E-02	-4.05E-02	4.51E-01	2.54E+01
3.80E+02	1.85E+01	-6.28E+01	5.13E-01	8.45E+00	-1.65E+01	3.25E+00	8.45E+00	-1.97E+01	1.84E-02	9.43E-02	-4.15E-02	4.42E-01	2.54E+01
4.00E+02	1.75E+01	-6.30E+01	5.09E-01	7.94E+00	-1.56E+01	3.42E+00	7.94E+00	-1.90E+01	1.87E-02	9.59E-02	-4.34E-02	4.31E-01	2.55E+01
5.00E+02	1.32E+01	-6.42E+01	4.84E-01	5.77E+00	-1.19E+01	4.27E+00	5.77E+00	-1.62E+01	1.95E-02	1.00E-01	-5.30E-02	3.68E-01	2.67E+01
6.00E+02	9.98E+00	-6.30E+01	5.09E-01	4.53E+00	-8.89E+00	5.13E+00	4.53E+00	-1.40E+01	2.09E-02	1.07E-01	-6.25E-02	3.34E-01	2.98E+01
7.00E+02	7.26E+00	-5.96E+01	5.86E-01	3.67E+00	-6.26E+00	5.98E+00	3.67E+00	-1.22E+01	2.25E-02	1.15E-01	-7.25E-02	3.10E-01	3.63E+01
8.00E+02	4.93E+00	-5.18E+01	7.87E-01	3.05E+00	-3.87E+00	6.84E+00	3.05E+00	-1.07E+01	2.46E-02	1.26E-01	-8.36E-02	2.94E-01	5.14E+01
9.00E+02	3.28E+00	-2.89E+01	1.81E+00	2.87E+00	-1.58E+00	7.69E+00	2.87E+00	-9.27E+00	3.04E-02	1.56E-01	-9.52E-02	3.20E-01	1.12E+02
9.50E+02	2.84E+00	-9.00E+00	6.31E+00	2.80E+00	-4.44E-01	8.12E+00	2.80E+00	-8.56E+00	3.45E-02	1.77E-01	-1.02E-01	3.38E-01	3.77E+02
1.00E+03	2.85E+00	1.41E+01	-3.98E+00	2.76E+00	6.93E-01	8.55E+00	2.76E+00	-7.85E+00	3.98E-02	2.05E-01	-1.10E-01	3.63E-01	-2.30E+02

Avra - 16.4 Sample
High Frequency - Coaxial Fixture

f, MHz	C, pF	ε	φcorr	ε"	Zcorr	Rcorr	ρ', Ohm-m	ρ	losses,ε"c	losses	10U/ω^0.5	σ", S/m
	B/w			ε'tanδ	Ohm	Ohm	(Rcorr)		sc/we0	ε"-ε"c		
1.00E+00	9.18E+01	5.32E+01	-3.42E+00	8.90E+02	1.04E+02	1.03E+02	2.01E+01	1.20E+00	8.36E+02	5.35E+01	3.99E-02	2.96E-03
2.00E+00	5.34E+01	3.09E+01	-3.89E+00	4.55E+02	1.01E+02	1.01E+02	1.97E+01	1.34E+00	4.18E+02	3.70E+01	2.82E-02	3.44E-03
3.00E+00	4.60E+01	2.67E+01	-4.95E+00	3.08E+02	9.95E+01	9.92E+01	1.93E+01	1.67E+00	2.79E+02	2.91E+01	2.30E-02	4.45E-03
5.00E+00	4.00E+01	2.32E+01	-7.02E+00	1.88E+02	9.73E+01	9.65E+01	1.88E+01	2.31E+00	1.67E+02	2.11E+01	1.78E-02	6.45E-03
1.00E+01	3.47E+01	2.01E+01	-1.17E+01	9.72E+01	9.30E+01	9.11E+01	1.77E+01	3.67E+00	8.36E+01	1.35E+01	1.26E-02	1.12E-02
2.00E+01	3.02E+01	1.75E+01	-1.90E+01	5.10E+01	8.55E+01	8.09E+01	1.58E+01	5.41E+00	4.18E+01	9.22E+00	8.92E-03	1.95E-02
3.00E+01	2.77E+01	1.60E+01	-2.44E+01	3.54E+01	7.92E+01	7.21E+01	1.40E+01	6.37E+00	2.79E+01	7.51E+00	7.28E-03	2.68E-02
4.00E+01	2.60E+01	1.51E+01	-2.87E+01	2.76E+01	7.34E+01	6.44E+01	1.25E+01	6.87E+00	2.09E+01	6.66E+00	6.31E-03	3.36E-02
5.00E+01	2.47E+01	1.43E+01	-3.22E+01	2.28E+01	6.86E+01	5.81E+01	1.13E+01	7.12E+00	1.67E+01	6.04E+00	5.64E-03	3.99E-02
6.00E+01	2.39E+01	1.39E+01	-3.54E+01	1.95E+01	6.43E+01	5.24E+01	1.02E+01	7.25E+00	1.39E+01	5.57E+00	5.15E-03	4.62E-02
7.00E+01	2.31E+01	1.34E+01	-3.80E+01	1.72E+01	6.05E+01	4.77E+01	9.29E+00	7.26E+00	1.19E+01	5.22E+00	4.77E-03	5.22E-02
8.00E+01	2.25E+01	1.30E+01	-4.03E+01	1.54E+01	5.72E+01	4.36E+01	8.50E+00	7.22E+00	1.05E+01	4.91E+00	4.46E-03	5.81E-02
9.00E+01	2.21E+01	1.28E+01	-4.25E+01	1.40E+01	5.42E+01	3.99E+01	7.78E+00	7.13E+00	9.29E+00	4.66E+00	4.21E-03	6.40E-02
1.00E+02	2.16E+01	1.25E+01	-4.43E+01	1.28E+01	5.15E+01	3.68E+01	7.18E+00	7.01E+00	8.36E+00	4.45E+00	3.99E-03	6.96E-02
1.50E+02	2.00E+01	1.16E+01	-5.15E+01	9.25E+00	4.14E+01	2.58E+01	5.03E+00	6.31E+00	5.57E+00	3.67E+00	3.26E-03	9.70E-02
2.00E+02	1.91E+01	1.11E+01	-5.64E+01	7.36E+00	3.47E+01	1.92E+01	3.74E+00	5.63E+00	4.18E+00	3.18E+00	2.82E-03	1.23E-01
2.50E+02	1.84E+01	1.07E+01	-5.99E+01	6.19E+00	2.99E+01	1.50E+01	2.92E+00	5.04E+00	3.34E+00	2.84E+00	2.52E-03	1.49E-01
3.00E+02	1.80E+01	1.04E+01	-6.29E+01	5.32E+00	2.63E+01	1.20E+01	2.33E+00	4.56E+00	2.79E+00	2.54E+00	2.30E-03	1.74E-01
3.50E+02	1.76E+01	1.02E+01	-6.50E+01	4.75E+00	2.35E+01	9.91E+00	1.93E+00	4.14E+00	2.39E+00	2.36E+00	2.13E-03	1.98E-01
3.70E+02	1.74E+01	1.01E+01	-6.57E+01	4.56E+00	2.25E+01	9.26E+00	1.80E+00	4.00E+00	2.26E+00	2.30E+00	2.07E-03	2.08E-01
3.80E+02	1.74E+01	1.01E+01	-6.61E+01	4.46E+00	2.20E+01	8.91E+00	1.74E+00	3.93E+00	2.20E+00	2.26E+00	2.05E-03	2.13E-01
4.00E+02	1.72E+01	1.00E+01	-6.67E+01	4.31E+00	2.12E+01	8.38E+00	1.63E+00	3.79E+00	2.09E+00	2.22E+00	1.99E-03	2.23E-01
5.00E+02	1.69E+01	9.79E+00	-6.98E+01	3.60E+00	1.77E+01	6.11E+00	1.19E+00	3.23E+00	1.67E+00	1.93E+00	1.78E-03	2.72E-01
6.00E+02	1.66E+01	9.61E+00	-7.15E+01	3.21E+00	1.52E+01	4.81E+00	9.37E-01	2.81E+00	1.39E+00	1.81E+00	1.63E-03	3.21E-01
7.00E+02	1.65E+01	9.55E+00	-7.28E+01	2.96E+00	1.32E+01	3.90E+00	7.60E-01	2.45E+00	1.19E+00	1.77E+00	1.51E-03	3.72E-01
8.00E+02	1.66E+01	9.64E+00	-7.36E+01	2.84E+00	1.15E+01	3.24E+00	6.31E-01	2.15E+00	1.05E+00	1.79E+00	1.41E-03	4.29E-01
9.00E+02	1.68E+01	9.76E+00	-7.23E+01	3.12E+00	1.00E+01	3.04E+00	5.93E-01	1.86E+00	9.29E-01	2.19E+00	1.33E-03	4.89E-01
9.50E+02	1.71E+01	9.92E+00	-7.13E+01	3.35E+00	9.27E+00	2.97E+00	5.79E-01	1.71E+00	8.80E-01	2.47E+00	1.29E-03	5.24E-01
1.00E+03	1.75E+01	1.01E+01	-7.01E+01	3.68E+00	8.56E+00	2.92E+00	5.69E-01	1.57E+00	8.36E-01	2.84E+00	1.26E-03	5.64E-01

Avra - 16.4 % Sample

Low Frequency - Coaxial Air Line

Subtracting the Impedance of Electrode Polarization after Lopatin.

HP 4194A, Fixture — Coaxial sample holder #154, patterned teflon spacers (2h = 0.4135 cm)

Avra Valley, # 5-3.0, with 10 % water by weight.

Not using Rms for coaxial

	Length	R, cm	r, cm	Volume	Weight	Dens.(wet)	Drying, weight in g.		Water cont.		Bulk dens.	method:
	cm			V, cm³	M, g	γ, g/cm³	Wet	Dry	W, %	W_v, %	g/cm³	sc', S/m
	2.5865	0.7144	0.3102	3.3652166	5.56	1.65220	13.22	11.91	10.999	16.372	1.49	0.04652

L, nH = 14.7 C_s, pF = 0.586

Stray capacitance Cs - from 4.4.97, Air measurements with same Teflon spacers.
L - from shorted coaxial, brass disk
CoaxStray-07Reverse

f, MHz	Z, Ohm	ϕ, deg	$\tan\delta$	Res.meas.	Xmeas.	$X_L=\omega L$	R_{el}	Xel	Rm-Rms-	Xm-ωL-	G, S
				R(s), Ohm	Zsinϕ	Ohm	$y=9.7925*x$	$y=-9.936*x$	Rel	Xel	
1.00E-03	1.23E+02	-6.13E+00	9.32E+00	1.23E+02	-1.32E+01	9.24E-05	1.24E+01	-1.25E+01	1.10E+02	-6.21E-01	9.07E-03
2.00E-03	1.19E+02	-4.41E+00	1.30E+01	1.19E+02	-9.18E+00	1.85E-04	8.74E+00	-8.86E+00	1.10E+02	-3.17E-01	9.06E-03
3.00E-03	1.18E+02	-3.66E+00	1.56E+01	1.18E+02	-7.52E+00	2.77E-04	7.13E+00	-7.24E+00	1.11E+02	-2.86E-01	9.05E-03
5.00E-03	1.16E+02	-2.94E+00	1.94E+01	1.16E+02	-5.97E+00	4.62E-04	5.52E+00	-5.61E+00	1.11E+02	-3.62E-01	9.05E-03
1.00E-02	1.14E+02	-2.34E+00	2.45E+01	1.14E+02	-4.66E+00	9.24E-04	3.91E+00	-3.96E+00	1.10E+02	-6.99E-01	9.06E-03
2.00E-02	1.13E+02	-1.99E+00	2.87E+01	1.13E+02	-3.93E+00	1.85E-03	2.76E+00	-2.80E+00	1.10E+02	-1.13E+00	9.08E-03
3.00E-02	1.12E+02	-1.89E+00	3.03E+01	1.12E+02	-3.70E+00	2.77E-03	2.26E+00	-2.29E+00	1.10E+02	-1.42E+00	9.11E-03
5.00E-02	1.11E+02	-1.85E+00	3.10E+01	1.11E+02	-3.58E+00	4.62E-03	1.75E+00	-1.77E+00	1.09E+02	-1.81E+00	9.15E-03
1.00E-01	1.10E+02	-1.92E+00	2.98E+01	1.10E+02	-3.68E+00	9.24E-03	1.24E+00	-1.25E+00	1.08E+02	-2.43E+00	9.22E-03
2.00E-01	1.08E+02	-2.14E+00	2.68E+01	1.08E+02	-4.04E+00	1.85E-02	8.74E-01	-8.86E-01	1.07E+02	-3.17E+00	9.31E-03
3.00E-01	1.07E+02	-2.34E+00	2.45E+01	1.07E+02	-4.39E+00	2.77E-02	7.13E-01	-7.24E-01	1.07E+02	-3.69E+00	9.37E-03
5.00E-01	1.06E+02	-2.70E+00	2.12E+01	1.06E+02	-5.00E+00	4.62E-02	5.52E-01	-5.61E-01	1.06E+02	-4.48E+00	9.46E-03
1.00E+00	1.04E+02	-3.46E+00	1.66E+01	1.04E+02	-6.29E+00	9.24E-02	3.91E-01	-3.96E-01	1.04E+02	-5.99E+00	9.61E-03
2.00E+00	1.02E+02	-4.75E+00	1.20E+01	1.02E+02	-8.46E+00	1.85E-01	2.76E-01	-2.80E-01	1.02E+02	-8.36E+00	9.78E-03
3.00E+00	1.01E+02	-5.92E+00	9.64E+00	1.00E+02	-1.04E+01	2.77E-01	2.26E-01	-2.29E-01	9.99E+01	-1.04E+01	9.90E-03
5.00E+00	9.82E+01	-8.70E+00	6.54E+00	9.71E+01	-1.49E+01	4.62E-01	1.75E-01	-1.77E-01	9.69E+01	-1.51E+01	1.01E-02
1.00E+01	9.32E+01	-1.27E+01	4.43E+00	9.09E+01	-2.05E+01	9.24E-01	1.24E-01	-1.25E-01	9.08E+01	-2.13E+01	1.04E-02
2.00E+01	8.39E+01	-1.98E+01	2.78E+00	7.89E+01	-2.84E+01	1.85E+00	8.74E-02	-8.86E-02	7.89E+01	-3.01E+01	1.11E-02
3.00E+01	7.54E+01	-2.49E+01	2.15E+00	6.84E+01	-3.17E+01	2.77E+00	7.13E-02	-7.24E-02	6.83E+01	-3.44E+01	1.17E-02
4.00E+01	6.78E+01	-2.86E+01	1.83E+00	5.95E+01	-3.25E+01	3.69E+00	6.18E-02	-6.27E-02	5.94E+01	-3.61E+01	1.23E-02

R' X'

Avra - 16.4 % Sample

Low Frequency - Coaxial Air Line

f, MHz	C, pF	σ', S/m	ε'	tanδcorr	φcorr	ε''	Zcorr	Rcorr	ρ', Ohm-m	ρ''	Conduct. losses,ε''c
	B/w			G/B		e'tand	Ohm	Ohm	(Rcorr)		sc/we0
1.00E-03	8.13E+03	4.66E-02	4.71E+03	1.78E+02	-3.23E-01	8.37E+05	1.10E+02	1.10E+02	2.15E+01	1.21E-01	8.36E+05
2.00E-03	2.07E+03	4.65E-02	1.20E+03	3.48E+02	-1.64E-01	4.18E+05	1.10E+02	1.10E+02	2.15E+01	6.17E-02	4.18E+05
3.00E-03	1.24E+03	4.65E-02	7.20E+02	3.87E+02	-1.48E-01	2.78E+05	1.11E+02	1.11E+02	2.15E+01	5.57E-02	2.79E+05
5.00E-03	9.44E+02	4.64E-02	5.47E+02	3.05E+02	-1.88E-01	1.67E+05	1.11E+02	1.11E+02	2.15E+01	7.05E-02	1.67E+05
1.00E-02	9.12E+02	4.65E-02	5.29E+02	1.58E+02	-3.62E-01	8.36E+04	1.10E+02	1.10E+02	2.15E+01	1.36E-01	8.36E+04
2.00E-02	7.39E+02	4.66E-02	4.28E+02	9.78E+01	-5.86E-01	4.19E+04	1.10E+02	1.10E+02	2.14E+01	2.19E-01	4.18E+04
3.00E-02	6.22E+02	4.68E-02	3.61E+02	7.76E+01	-7.38E-01	2.80E+04	1.10E+02	1.10E+02	2.14E+01	2.75E-01	2.79E+04
5.00E-02	4.82E+02	4.69E-02	2.79E+02	6.04E+01	-9.48E-01	1.69E+04	1.09E+02	1.09E+02	2.13E+01	3.52E-01	1.67E+04
1.00E-01	3.28E+02	4.73E-02	1.90E+02	4.47E+01	-1.28E+00	8.50E+03	1.08E+02	1.08E+02	2.11E+01	4.73E-01	8.36E+03
2.00E-01	2.18E+02	4.78E-02	1.27E+02	3.39E+01	-1.69E+00	4.29E+03	1.07E+02	1.07E+02	2.09E+01	6.16E-01	4.18E+03
3.00E-01	1.72E+02	4.81E-02	9.95E+01	2.90E+01	-1.98E+00	2.88E+03	1.07E+02	1.07E+02	2.08E+01	7.17E-01	2.79E+03
5.00E-01	1.27E+02	4.86E-02	7.38E+01	2.37E+01	-2.42E+00	1.75E+03	1.06E+02	1.06E+02	2.06E+01	8.69E-01	1.67E+03
1.00E+00	8.77E+01	4.93E-02	5.08E+01	1.74E+01	-3.28E+00	8.87E+02	1.04E+02	1.04E+02	2.02E+01	1.16E+00	8.36E+02
2.00E+00	6.35E+01	5.02E-02	3.68E+01	1.23E+01	-4.66E+00	4.51E+02	1.02E+02	1.02E+02	1.98E+01	1.61E+00	4.18E+02
3.00E+00	5.43E+01	5.08E-02	3.15E+01	9.68E+00	-5.90E+00	3.05E+02	1.00E+02	9.99E+01	1.95E+01	2.01E+00	2.79E+02
5.00E+00	4.95E+01	5.17E-02	2.87E+01	6.48E+00	-8.78E+00	1.86E+02	9.81E+01	9.70E+01	1.89E+01	2.92E+00	1.67E+02
1.00E+01	3.84E+01	5.36E-02	2.23E+01	4.32E+00	-1.30E+01	9.64E+01	9.33E+01	9.09E+01	1.77E+01	4.10E+00	8.36E+01
2.00E+01	3.31E+01	5.68E-02	1.92E+01	2.66E+00	-2.06E+01	5.11E+01	8.46E+01	7.92E+01	1.54E+01	5.79E+00	4.18E+01
3.00E+01	3.07E+01	5.99E-02	1.78E+01	2.02E+00	-2.63E+01	3.59E+01	7.68E+01	6.88E+01	1.34E+01	6.63E+00	2.79E+01
4.00E+01	2.91E+01	6.31E-02	1.69E+01	1.68E+00	-3.08E+01	2.84E+01	6.99E+01	6.01E+01	1.17E+01	6.97E+00	2.09E+01

	Y	Z	AA	AB	AC	AD
1	Avra - 16.4 % Sample					
2	Low Frequency - Coaxial Air Line					
3						
4						
5						
6						
7						
8						
9						
10		Dieletric				
11	f, MHz	losses	$100/\omega^{0.5}$	B, S	C(s), pF	σ'', S/m
12		$\varepsilon''-\varepsilon''c$				
13	1.00E-03	9.64E+02	1.26E+00	-5.11E-05	1.21E+07	2.62E-04
14	2.00E-03	-1.34E+02	8.92E-01	-2.60E-05	8.67E+06	1.33E-04
15	3.00E-03	-3.97E+02	7.28E-01	-2.34E-05	7.05E+06	1.20E-04
16	5.00E-03	-2.52E+02	5.64E-01	-2.96E-05	5.33E+06	1.52E-04
17	1.00E-02	-6.14E+01	3.99E-01	-5.73E-05	3.41E+06	2.94E-04
18	2.00E-02	9.30E+01	2.82E-01	-9.28E-05	2.03E+06	4.77E-04
19	3.00E-02	1.38E+02	2.30E-01	-1.17E-04	1.43E+06	6.02E-04
20	5.00E-02	1.54E+02	1.78E-01	-1.51E-04	8.89E+05	7.77E-04
21	1.00E-01	1.41E+02	1.26E-01	-2.06E-04	4.33E+05	1.06E-03
22	2.00E-01	1.13E+02	8.92E-02	-2.74E-04	1.97E+05	1.41E-03
23	3.00E-01	9.46E+01	7.28E-02	-3.24E-04	1.21E+05	1.66E-03
24	5.00E-01	7.36E+01	5.64E-02	-4.00E-04	6.37E+04	2.05E-03
25	1.00E+00	5.04E+01	3.99E-02	-5.51E-04	2.53E+04	2.83E-03
26	2.00E+00	3.32E+01	2.82E-02	-7.98E-04	9.41E+03	4.10E-03
27	3.00E+00	2.58E+01	2.30E-02	-1.02E-03	5.11E+03	5.25E-03
28	5.00E+00	1.86E+01	1.78E-02	-1.56E-03	2.14E+03	7.98E-03
29	1.00E+01	1.27E+01	1.26E-02	-2.41E-03	7.76E+02	1.24E-02
30	2.00E+01	9.25E+00	8.92E-03	-4.15E-03	2.81E+02	2.13E-02
31	3.00E+01	8.04E+00	7.28E-03	-5.78E-03	1.67E+02	2.97E-02
32	4.00E+01	7.45E+00	6.31E-03	-7.32E-03	1.23E+02	3.76E-02

Avra (25%)

A sample of Avra # 5-2.5 with 15% water (by weight)

	A	B	C	D	E	F	G	H	I	J	K	L	M
1	Avra - 25% Sample												
2	High Frequency - Coaxial												
3	Coaxial sample holder #154, patterned teflon spacers: 0.4135 cm								**Not using Rms for coaxial**				
4	HP 4194A								Subtracting the Impedance of Electrode Polarization after Lopatin.				
5													
6	Length,cm	R, cm	r, cm	Volume	Weight	Dens.(wet)	Drying, weight in g.		Water cont.		Bulk dens.		
7	2.5865	0.7144	0.3102	V, cm^3	M, g	γ, g/cm^3	Wet	Dry	W, %	Wv, %	g/cm^3	**sc', S/m**	
8	L, nH	Cs, pF		3.365217	6.44	1.91370	21.19	18.43	14.976	24.926	1.66	**0.092**	
9	1.39	0.563		L - from resonance;	(Cs: AVG from Air, 6.3.98 & 9.15.98 w/same Teflon spacers)							(PL)	ε'
10	f, MHz	Z, Ohm	ϕ, deg	tanδ	Res.meas.	Xmeas.	$X_L=\omega L$	Rm-Rms	$Xm-\omega L$	G, S	σ', S/m	B, S	
11					R(s), Ohm	Zsinϕ	Ohm	R'=Rm	X'				
12	1	51.18	-3.1	18.46447	51.11	-2.767754	0.0087	51.11	-2.7765	0.01951	1.00E-01	-0.001056	97.478
13	10	46.83	-9.69	5.85640	46.16	-7.882301	0.0873	46.16	-7.9696	0.02104	1.08E-01	-0.003596	33.185
14	20	43.97	-15.75	3.54573	42.32	-11.93524	0.1747	42.32	-12.1099	0.02184	1.12E-01	-0.006179	28.509
15	30	41.41	-20.55	2.66752	38.77	-14.53593	0.2620	38.77	-14.7979	0.02251	1.16E-01	-0.008485	26.098
16	40	39.01	-24.56	2.18822	35.48	-16.21435	0.3493	35.48	-16.5637	0.02314	1.19E-01	-0.010662	24.594
17	50	36.82	-28.01	1.87993	32.51	-17.29162	0.4367	32.51	-17.7283	0.02371	1.22E-01	-0.012754	23.537
18	100	28.33	-39.85	1.19811	21.75	-18.15329	0.8734	21.75	-19.0267	0.02605	1.34E-01	-0.022431	20.698
19	150	22.62	-46.69	0.94268	15.52	-16.45951	1.3100	15.52	-17.7696	0.02788	1.43E-01	-0.0314	19.316
20	200	18.55	-50.97	0.81065	11.68	-14.40994	1.7467	11.68	-16.1567	0.02939	1.51E-01	-0.039939	18.426
21	250	15.46	-53.69	0.73484	9.15	-12.45805	2.1834	9.15	-14.6415	0.03070	1.58E-01	-0.048218	17.797
22	300	13.01	-55.37	0.69063	7.39	-10.70513	2.6201	7.39	-13.3252	0.03184	1.63E-01	-0.05632	17.323
23	350	11	-56.10	0.67197	6.14	-9.130135	3.0568	6.14	-12.1869	0.03296	1.69E-01	-0.064226	16.932
24	400	9.3	-56.05	0.67324	5.19	-7.714585	3.4935	5.19	-11.2080	0.03404	1.75E-01	-0.072034	16.617
25	450	7.87	-55.84	0.67858	4.42	-6.512211	3.9301	4.42	-10.4423	0.03437	1.76E-01	-0.079627	16.327
26	500	6.52	-52.99	0.75383	3.92	-5.206419	4.3668	3.92	-9.5732	0.03666	1.88E-01	-0.087659	16.177
27	550	5.365	-49.53	0.85318	3.48	-4.081402	4.8035	3.48	-8.8849	0.03824	1.96E-01	-0.095619	16.042
28	600	4.34	-43.4	1.05747	3.15	-2.98196	5.2402	3.15	-8.2221	0.04066	2.09E-01	-0.103905	15.979
29	700	2.839	-17.9	3.09606	2.70	-0.872585	6.1135	2.70	-6.9861	0.04815	2.47E-01	-0.122044	16.088
30	800	2.807	26.9	-1.97111	2.50	1.269984	6.9869	2.50	-5.7169	0.06427	3.30E-01	-0.143948	16.603
31	900	4.392	54.6	-0.71066	2.54	3.580041	7.8603	2.54	-4.2802	0.10262	5.27E-01	-0.169453	17.373
32	1000	6.885	64.55	-0.47590	2.96	6.216884	8.7336	2.96	-2.5167	0.19610	1.0E+00	-0.163272	15.065

f, MHz	tanδcorr	C, pF	φcorr	ε"	Zcorr	Rcorr	ρ', Ohm-m	ρ"	Conduct. losses,ε"c	Dielectric losses	100/ω^0.5	σ", S/m
	G/B	B/w		ε'tanδ	Ohm	Ohm	(Rcorr)		sc/we0	ε"-ε"c		
1	18.46802	168.134	-3.0994	1800.23	51.18097	51.10611	9.9559	0.5391	1653.745	146.484	0.039894	5.42E-03
10	5.84919	57.238	-9.7017	194.103	46.85793	46.18779	8.9978	1.5383	165.3745	28.729	0.012616	1.85E-02
20	3.5346	49.174	-15.7971	100.768	44.05528	42.39138	8.2582	2.3364	82.68723	18.081	0.008921	3.17E-02
30	2.653065	45.014	-20.6526	69.239	41.56762	38.89634	7.5774	2.8561	55.12482	14.114	0.007284	4.36E-02
40	2.170501	42.421	-24.7366	53.382	39.24798	35.64663	6.9443	3.1994	41.34362	12.038	0.006308	5.47E-02
50	1.859057	40.597	-28.2761	43.756	37.14293	32.71085	6.3724	3.4277	33.07489	10.681	0.005642	6.55E-02
100	1.161141	35.700	-40.7357	24.033	29.09235	22.04407	4.2944	3.6984	16.53745	7.495	0.003989	1.15E-01
150	0.887939	33.316	-48.3969	17.151	23.8141	15.81178	3.0803	3.4690	11.02496	6.126	0.003257	1.61E-01
200	0.735818	31.782	-53.6537	13.558	20.16708	11.95231	2.3284	3.1644	8.268723	5.290	0.002821	2.05E-01
250	0.636726	30.697	-57.5140	11.332	17.49385	9.395827	1.8304	2.8747	6.614978	4.717	0.002523	2.48E-01
300	0.565286	29.879	-60.5211	9.792	15.45692	7.60639	1.4818	2.6213	5.512482	4.280	0.002303	2.89E-01
350	0.51313	29.205	-62.8363	8.688	13.85271	6.32424	1.2320	2.4010	4.724985	3.963	0.002132	3.30E-01
400	0.472499	28.662	-64.7093	7.852	12.55166	5.362205	1.0446	2.2108	4.134362	3.717	0.001995	3.70E-01
450	0.431646	28.162	-66.6528	7.048	11.53027	4.569475	0.8902	2.0623	3.674988	3.373	0.001881	4.09E-01
500	0.418243	27.903	-67.3032	6.766	10.52446	4.060904	0.7911	1.8915	3.307489	3.458	0.001784	4.50E-01
550	0.399893	27.670	-68.2039	6.415	9.710524	3.605563	0.7024	1.7565	3.006808	3.408	0.001701	4.91E-01
600	0.391352	27.562	-68.6270	6.254	8.962278	3.266191	0.6363	1.6259	2.756241	3.497	0.001629	5.33E-01
700	0.394552	27.748	-68.4682	6.347	7.62197	2.797399	0.5450	1.3812	2.362492	3.985	0.001508	6.26E-01
800	0.44648	28.637	-65.9402	7.413	6.343422	2.586149	0.5038	1.1284	2.067181	5.346	0.00141	7.39E-01
900	0.605577	29.966	-58.8019	10.521	5.047905	2.614809	0.5094	0.8412	1.837494	8.683	0.00133	8.70E-01
1000	1.201055	25.985	-39.7808	18.094	3.918946	3.011701	0.5867	0.4885	1.653745	16.441	0.001262	8.38E-01

Avra - 25% Sample

Low Frequency - Coaxial

A sample of Avra # 5-2.5 with 15% water (by weight)

Not using Rms for coaxial

HP 4194A — Coaxial sample holder #154, patterned teflon spacers: 0.4135 cm — Subtracting the Impedance of Electrode Polarization after Lopatin.

	Length,cm	R, cm	r, cm	Volume V, cm^3	Weight M, g	Dens.(wet) $\gamma, g/cm^3$	Water cont. Wet	Dry	W_v, %	Bulk dens. g/cm^3	method: sc', S/m
	2.5865	0.7144	0.3102				Drying, weight in g.		W, %		
L, nH	C_s, pF			3.365217	6.44	1.91370	21.19	18.43	14.976	1.66	0.092
14.7	0.586								24.926		

From Air, 4.4.97 w/same Teflon spacers — L=14.7 from shorted coaxial, brass disk.

f, MHz	Z, Ohm	ϕ, deg	$\tan\delta$	Res.meas. $R(s)$, Ohm	Xmeas. $Z\sin\phi$	$X_L=\omega L$ Ohm	R_{el} $\gamma=3.346*x$	X_{el} $\gamma=-3.98*x$	R' Rm-Rms- R_{el}	X' Xm-ωL- X_{el}	G, S	σ', S/m	B, S
1.00E-03	6.04E+01	-5.04E+00	1.13E+01	6.01E+01	-5.31E+01	9.24E-05	4.22E+00	-5.02E+00	5.59E+01	-2.83E-01	1.79E-02	9.18E-02	-9.04E-05
2.00E-03	5.91E+01	-3.47E+00	1.65E+01	5.90E+01	-3.57E+00	1.85E-04	2.99E+00	-3.55E+00	5.60E+01	-2.30E-02	1.78E-02	9.16E-02	-7.31E-06
3.00E-03	5.85E+01	-2.85E+00	2.01E+01	5.85E+01	-2.91E+00	2.77E-04	2.44E+00	-2.90E+00	5.60E+01	-1.36E-02	1.78E-02	9.16E-02	-4.32E-06
5.00E-03	5.80E+01	-2.32E+00	2.47E+01	5.79E+01	-2.34E+00	4.62E-04	1.89E+00	-2.25E+00	5.60E+01	-9.85E-02	1.78E-02	9.16E-02	-3.14E-05
1.00E-02	5.73E+01	-1.92E+00	2.99E+01	5.72E+01	-1.92E+00	9.24E-04	1.33E+00	-1.59E+00	5.59E+01	-3.29E-01	1.79E-02	9.18E-02	-1.05E-04
2.00E-02	5.66E+01	-1.74E+00	3.30E+01	5.66E+01	-1.71E+00	1.85E-03	9.44E-01	-1.12E+00	5.57E+01	-5.93E-01	1.80E-02	9.22E-02	-1.91E-04
3.00E-02	5.62E+01	-1.71E+00	3.35E+01	5.62E+01	-1.68E+00	2.77E-03	7.71E-01	-9.17E-01	5.55E+01	-7.66E-01	1.80E-02	9.26E-02	-2.49E-04
5.00E-02	5.58E+01	-1.74E+00	3.29E+01	5.57E+01	-1.69E+00	4.62E-03	5.97E-01	-7.10E-01	5.51E+01	-9.87E-01	1.81E-02	9.31E-02	-3.25E-04
1.00E-01	5.51E+01	-1.87E+00	3.07E+01	5.50E+01	-1.79E+00	9.24E-03	4.22E-01	-5.02E-01	5.46E+01	-1.30E+00	1.83E-02	9.39E-02	-4.35E-04
2.00E-01	5.43E+01	-2.08E+00	2.76E+01	5.43E+01	-1.97E+00	1.85E-02	2.99E-01	-3.55E-01	5.40E+01	-1.63E+00	1.85E-02	9.50E-02	-5.58E-04
3.00E-01	5.38E+01	-2.24E+00	2.55E+01	5.38E+01	-2.11E+00	2.77E-02	2.44E-01	-2.90E-01	5.36E+01	-1.84E+00	1.86E-02	9.57E-02	-6.41E-04
5.00E-01	5.32E+01	-2.52E+00	2.28E+01	5.32E+01	-2.34E+00	4.62E-02	1.89E-01	-2.25E-01	5.30E+01	-2.16E+00	1.88E-02	9.67E-02	-7.66E-04
1.00E+00	5.23E+01	-3.05E+00	1.87E+01	5.22E+01	-2.79E+00	9.24E-02	1.33E-01	-1.59E-01	5.21E+01	-2.72E+00	1.91E-02	9.83E-02	-9.96E-04
2.00E+00	5.13E+01	-3.96E+00	1.44E+01	5.12E+01	-3.54E+00	1.85E-01	9.44E-02	-1.12E-01	5.11E+01	-3.61E+00	1.95E-02	1.00E-01	-1.37E-03
3.00E+00	5.06E+01	-4.77E+00	1.20E+01	5.04E+01	-4.21E+00	2.77E-01	7.71E-02	-9.17E-02	5.03E+01	-4.40E+00	1.97E-02	1.01E-01	-1.71E-03
5.00E+00	4.96E+01	-6.27E+00	9.10E+00	4.93E+01	-5.41E+00	4.62E-01	5.97E-02	-7.10E-02	4.92E+01	-5.80E+00	2.00E-02	1.03E-01	-2.35E-03
1.00E+01	4.75E+01	-9.51E+00	5.97E+00	4.69E+01	-7.85E+00	9.24E-01	4.22E-02	-5.02E-02	4.68E+01	-8.72E+00	2.06E-02	1.06E-01	-3.81E-03
2.00E+01	4.39E+01	-1.45E+01	3.87E+00	4.25E+01	-1.10E+01	1.85E+00	2.99E-02	-3.55E-02	4.25E+01	-1.28E+01	2.16E-02	1.11E-01	-6.43E-03
3.00E+01	4.04E+01	-1.84E+01	3.01E+00	3.84E+01	-1.28E+01	2.77E+00	2.44E-02	-2.90E-02	3.83E+01	-1.55E+01	2.24E-02	1.15E-01	-8.95E-03
4.00E+01	3.71E+01	-2.11E+01	2.59E+00	3.46E+01	-1.34E+01	3.69E+00	2.11E-02	-2.51E-02	3.46E+01	-1.70E+01	2.33E-02	1.20E-01	-1.13E-02

	O	P	Q	R	S	T	U	V	W	X	Y	Z
1	Avra - 25% Sample											
2	Low Frequency - Coaxial											
3												
4												
5												
6												
7												
8												
9												Conduct.	Dielectric
10	f, MHz	C, pF	ε'	tanδcorr	φcorr	ε''	Zcorr	Rcorr	ρ', Ohm-m	ρ''	losses,ε''c	losses
11		B/w		G/B		ε'tand	Ohm	Ohm	(Rcorr)		sc/we0	ε''-ε''c
12	1.00E-03	1.44E+04	8.34E+03	1.98E+02	-2.90E-01	1.65E+06	5.59E+01	5.59E+01	1.09E+01	5.51E-02	1.65E+06	-3.75E+03
13	2.00E-03	5.81E+02	3.37E+02	2.44E+03	-2.35E-02	8.23E+05	5.60E+01	5.60E+01	1.09E+01	4.47E-03	8.27E+05	-3.43E+03
14	3.00E-03	2.29E+02	1.33E+02	4.14E+03	-1.39E-02	5.49E+05	5.60E+01	5.60E+01	1.09E+01	2.64E-03	5.51E+05	-2.37E+03
15	5.00E-03	9.99E+02	5.79E+02	5.69E+02	-1.01E-01	3.29E+05	5.60E+01	5.60E+01	1.09E+01	1.92E-02	3.31E+05	-1.35E+03
16	1.00E-02	1.68E+03	9.71E+02	1.70E+02	-3.37E-01	1.65E+05	5.59E+01	5.59E+01	1.09E+01	6.40E-02	1.65E+05	-2.75E+02
17	2.00E-02	1.52E+03	8.83E+02	9.38E+01	-6.11E-01	8.29E+04	5.57E+01	5.57E+01	1.08E+01	1.16E-01	8.27E+04	1.90E+02
18	3.00E-02	1.32E+03	7.66E+02	7.24E+01	-7.91E-01	5.55E+04	5.55E+01	5.55E+01	1.08E+01	1.49E-01	5.51E+04	3.33E+02
19	5.00E-02	1.03E+03	5.99E+02	5.59E+01	-1.03E+00	3.35E+04	5.51E+01	5.51E+01	1.07E+01	1.92E-01	3.31E+04	3.83E+02
20	1.00E-01	6.93E+02	4.02E+02	4.20E+01	-1.36E+00	1.69E+04	5.46E+01	5.46E+01	1.06E+01	2.53E-01	1.65E+04	3.46E+02
21	2.00E-01	4.44E+02	2.57E+02	3.32E+01	-1.73E+00	8.54E+03	5.40E+01	5.40E+01	1.05E+01	3.17E-01	8.27E+03	2.70E+02
22	3.00E-01	3.40E+02	1.97E+02	2.91E+01	-1.97E+00	5.74E+03	5.36E+01	5.36E+01	1.04E+01	3.59E-01	5.51E+03	2.24E+02
23	5.00E-01	2.44E+02	1.41E+02	2.46E+01	-2.33E+00	3.48E+03	5.30E+01	5.30E+01	1.03E+01	4.19E-01	3.31E+03	1.70E+02
24	1.00E+00	1.59E+02	9.19E+01	1.92E+01	-2.98E+00	1.77E+03	5.22E+01	5.21E+01	1.01E+01	5.28E-01	1.65E+03	1.13E+02
25	2.00E+00	1.09E+02	6.33E+01	1.42E+01	-4.03E+00	8.99E+02	5.12E+01	5.11E+01	9.95E+00	7.00E-01	8.27E+02	7.22E+01
26	3.00E+00	9.07E+01	5.26E+01	1.15E+01	-4.96E+00	6.06E+02	5.05E+01	5.03E+01	9.81E+00	8.51E-01	5.51E+02	5.51E+01
27	5.00E+00	7.46E+01	4.33E+01	8.55E+00	-6.67E+00	3.70E+02	4.96E+01	4.92E+01	9.59E+00	1.12E+00	3.31E+02	3.91E+01
28	1.00E+01	6.06E+01	3.51E+01	5.42E+00	-1.05E+01	1.90E+02	4.76E+01	4.69E+01	9.13E+00	1.68E+00	1.65E+02	2.51E+01
29	2.00E+01	5.12E+01	2.97E+01	3.35E+00	-1.66E+01	9.96E+01	4.44E+01	4.26E+01	8.29E+00	2.47E+00	8.27E+01	1.69E+01
30	3.00E+01	4.75E+01	2.75E+01	2.51E+00	-2.18E+01	6.90E+01	4.14E+01	3.85E+01	7.49E+00	2.99E+00	5.51E+01	1.38E+01
31	4.00E+01	4.51E+01	2.61E+01	2.06E+00	-2.59E+01	5.37E+01	3.86E+01	3.47E+01	6.77E+00	3.29E+00	4.13E+01	1.24E+01

	AA	AB	AC	AD	AE
1	Avra - 25% Sample				
2	Low Frequency - Coaxial				
3					
4					
5					
6					
7					
8					
9					
10	$100/\omega^{0.5}$	f, MHz	σ'', S/m	$1/\omega^{0.5}$	C(s), pF
11					
12	1.26E+00	1.00E-03	4.64E-04	1.26E-02	3.00E+07
13	8.92E-01	2.00E-03	3.75E-05	8.92E-03	2.23E+07
14	7.28E-01	3.00E-03	2.22E-05	7.28E-03	1.82E+07
15	5.64E-01	5.00E-03	1.61E-04	5.64E-03	1.36E+07
16	3.99E-01	1.00E-02	5.40E-04	3.99E-03	8.31E+06
17	2.82E-01	2.00E-02	9.83E-04	2.82E-03	4.64E+06
18	2.30E-01	3.00E-02	1.28E-03	2.30E-03	3.16E+06
19	1.78E-01	5.00E-02	1.67E-03	1.78E-03	1.88E+06
20	1.26E-01	1.00E-01	2.23E-03	1.26E-03	8.88E+05
21	8.92E-02	2.00E-01	2.86E-03	8.92E-04	4.05E+05
22	7.28E-02	3.00E-01	3.29E-03	7.28E-04	2.52E+05
23	5.64E-02	5.00E-01	3.93E-03	5.64E-04	1.36E+05
24	3.99E-02	1.00E+00	5.11E-03	3.99E-04	5.71E+04
25	2.82E-02	2.00E+00	7.04E-03	2.82E-04	2.25E+04
26	2.30E-02	3.00E+00	8.78E-03	2.30E-04	1.26E+04
27	1.78E-02	5.00E+00	1.20E-02	1.78E-04	5.88E+03
28	1.26E-02	1.00E+01	1.95E-02	1.26E-04	2.03E+03
29	8.92E-03	2.00E+01	3.30E-02	8.92E-05	7.24E+02
30	7.28E-03	3.00E+01	4.59E-02	7.28E-05	4.16E+02
31	6.31E-03	4.00E+01	5.82E-02	6.31E-05	2.98E+02

Avra (45%)

Avra - 45% Sample													
High Frequency - Coaxial				Wet sample (30%)		Teflon spacers, h=2.0675mm(pattern)							
Subtracting the Impedance of Electrode Polarization after Lopatin.						Not using Rms for coaxial							

Length,cm	R, cm	r, cm	Weight	Dens.(wet)	Drying, weight in g.		Water cont.		Bulk dens.	σc S/m	σc' S/m		
2.5865	0.7144	0.3102	M, g	γ, g/cm³	Wet	Dry	W, %	Wv, %	g/cm³	0.0903	0.09434		
L, nH	C$_s$, pF	V, cm³	6.61	1.96421	28.64	22.01	30.123	45.470	1.51	From PL	From Arg-2		
1.59	0.55642	3.365217	Cs is taken from Air meas. 6.3.98 with same Tefl. thickness (closest data)							Method	diagr:1/ρ_0		

f, MHz	Z, Ohm	ϕ, deg	tanδ	Res.meas.	Xmeas.	$X_L=\omega L$	Rm-Rms	$Xm-\omega L$	G, S	σ', S/m	B, S	Zcorr	Rcorr
				R(s), Ohm	Zsinϕ	Ohm	R'=Rm	X'				Ohm	Ohm
1.00E+00	5.20E+01	-3.25E+00	1.76E+01	5.19E+01	-2.95E+00	9.99E-03	5.19E+01	-2.96E+00	1.92E-02	9.85E-02	-1.09E-03	5.20E+01	5.19E+01
1.00E+01	4.73E+01	-1.26E+01	4.47E+00	4.61E+01	-1.03E+01	9.99E-02	4.61E+01	-1.04E+01	2.06E-02	1.06E-01	-4.62E-03	4.73E+01	4.62E+01
2.00E+01	4.34E+01	-2.11E+01	2.59E+00	4.05E+01	-1.56E+01	2.00E-01	4.05E+01	-1.58E+01	2.14E-02	1.10E-01	-8.32E-03	4.35E+01	4.05E+01
3.00E+01	3.97E+01	-2.77E+01	1.90E+00	3.51E+01	-1.85E+01	3.00E-01	3.51E+01	-1.88E+01	2.22E-02	1.14E-01	-1.17E-02	3.99E+01	3.53E+01
4.00E+01	3.63E+01	-3.30E+01	1.54E+00	3.04E+01	-1.97E+01	4.00E-01	3.04E+01	-2.01E+01	2.29E-02	1.17E-01	-1.50E-02	3.66E+01	3.06E+01
5.00E+01	3.32E+01	-3.73E+01	1.31E+00	2.64E+01	-2.01E+01	5.00E-01	2.64E+01	-2.06E+01	2.35E-02	1.21E-01	-1.82E-02	3.36E+01	2.66E+01
6.00E+01	3.05E+01	-4.08E+01	1.16E+00	2.31E+01	-1.99E+01	5.99E-01	2.31E+01	-2.05E+01	2.42E-02	1.24E-01	-2.13E-02	3.10E+01	2.33E+01
7.00E+01	2.82E+01	-4.37E+01	1.05E+00	2.04E+01	-1.95E+01	6.99E-01	2.04E+01	-2.02E+01	2.48E-02	1.27E-01	-2.43E-02	2.88E+01	2.06E+01
8.00E+01	2.61E+01	-4.61E+01	9.63E-01	1.81E+01	-1.88E+01	7.99E-01	1.81E+01	-1.96E+01	2.54E-02	1.31E-01	-2.73E-02	2.68E+01	1.83E+01
9.00E+01	2.43E+01	-4.81E+01	8.96E-01	1.62E+01	-1.81E+01	8.99E-01	1.62E+01	-1.90E+01	2.60E-02	1.34E-01	-3.02E-02	2.51E+01	1.64E+01
1.00E+02	2.27E+01	-4.99E+01	8.43E-01	1.46E+01	-1.73E+01	9.99E-01	1.46E+01	-1.83E+01	2.66E-02	1.37E-01	-3.30E-02	2.36E+01	1.48E+01
2.00E+02	1.29E+01	-5.82E+01	6.20E-01	6.77E+00	-1.09E+01	2.00E+00	6.77E+00	-1.29E+01	3.18E-02	1.63E-01	-6.00E-02	1.47E+01	6.90E+00
3.00E+02	7.98E+00	-5.85E+01	6.13E-01	4.17E+00	-6.80E+00	3.00E+00	4.17E+00	-9.80E+00	3.68E-02	1.89E-01	-8.53E-02	1.08E+01	4.26E+00
4.00E+02	4.84E+00	-5.14E+01	7.98E-01	3.02E+00	-3.78E+00	4.00E+00	3.02E+00	-7.78E+00	4.34E-02	2.23E-01	-1.10E-01	8.44E+00	3.09E+00
5.00E+02	2.75E+00	-2.68E+01	1.98E+00	2.45E+00	-1.24E+00	5.00E+00	2.45E+00	-6.23E+00	5.47E-02	2.81E-01	-1.37E-01	6.77E+00	2.51E+00
6.00E+02	2.53E+00	2.79E+01	-1.89E+00	2.24E+00	1.18E+00	5.99E+00	2.24E+00	-4.81E+00	7.94E-02	4.08E-01	-1.69E-01	5.36E+00	2.28E+00
7.00E+02	4.46E+00	5.85E+01	-6.13E-01	2.33E+00	3.80E+00	6.99E+00	2.33E+00	-3.19E+00	1.49E-01	7.67E-01	-2.02E-01	3.98E+00	2.37E+00
8.00E+02	7.58E+00	6.74E+01	-4.17E-01	2.92E+00	7.00E+00	7.99E+00	2.92E+00	-9.93E-01	3.07E-01	1.58E+00	-1.02E-01	3.09E+00	2.93E+00
9.00E+02	1.23E+01	6.75E+01	-4.14E-01	4.72E+00	1.14E+01	8.99E+00	4.72E+00	2.41E+00	1.68E-01	8.63E-01	8.90E-02	5.26E+00	4.65E+00
1.00E+03	2.06E+01	5.89E+01	-6.04E-01	1.07E+01	1.77E+01	9.99E+00	1.07E+01	7.68E+00	6.18E-01	3.17E-01	4.79E-02	1.28E+01	1.01E+01

	O	P	Q	R	S	T	U	V	W	X	Y	Z	AA	AB
1	Avra - 45% Sample													
2	High Frequency - Coaxial													
3														
4														
5														
6									With sc = 0.09434 (used)					
7									(From Argand 2)					
8									Conductive	Dielectric		Conductive	Dielectric	
9	f, MHz	C, pF	ε'	tanδcorr	φcorr	ε''	ρ', Ohm-m	ρ''	losses,ε''c	losses	$100/\omega{^\wedge}0.5$	losses,ε''c	losses	σ'', S/m
10		B/w		G/B		ε'tand	(Rcorr)		sc/we0	ε''-ε''c		sc/we0	ε''-ε''c	
11	1.00E+00	1.74E+02	1.01E+02	1.76E+01	-3.25E+00	1.77E+03	1.01E+01	5.75E-01	1.62E+03	1.48E+02	3.99E-02	1.70E+03	7.54E+01	5.60E-03
12	1.00E+01	7.36E+01	4.27E+01	4.46E+00	-1.26E+01	1.90E+02	8.99E+00	2.02E+00	1.62E+02	2.80E+01	1.26E-02	1.70E+02	2.08E+01	2.37E-02
13	2.00E+01	6.62E+01	3.84E+01	2.58E+00	-2.12E+01	9.89E+01	7.90E+00	3.07E+00	8.12E+01	1.77E+01	8.92E-03	8.48E+01	1.41E+01	4.27E-02
14	3.00E+01	6.22E+01	3.61E+01	1.89E+00	-2.79E+01	6.81E+01	6.87E+00	3.64E+00	5.41E+01	1.40E+01	7.28E-03	5.65E+01	1.16E+01	6.02E-02
15	4.00E+01	5.97E+01	3.46E+01	1.52E+00	-3.33E+01	5.27E+01	5.96E+00	3.91E+00	4.06E+01	1.22E+01	6.31E-03	4.24E+01	1.03E+01	7.70E-02
16	5.00E+01	5.79E+01	3.36E+01	1.29E+00	-3.77E+01	4.34E+01	5.18E+00	4.01E+00	3.25E+01	1.10E+01	5.64E-03	3.39E+01	9.51E+00	9.33E-02
17	6.00E+01	5.64E+01	3.27E+01	1.14E+00	-4.13E+01	3.72E+01	4.54E+00	4.00E+00	2.71E+01	1.01E+01	5.15E-03	2.83E+01	8.92E+00	1.09E-01
18	7.00E+01	5.52E+01	3.20E+01	1.02E+00	-4.44E+01	3.27E+01	4.01E+00	3.92E+00	2.32E+01	9.52E+00	4.77E-03	2.42E+01	8.48E+00	1.25E-01
19	8.00E+01	5.42E+01	3.14E+01	9.33E-01	-4.70E+01	2.93E+01	3.57E+00	3.82E+00	2.03E+01	9.04E+00	4.46E-03	2.12E+01	8.13E+00	1.40E-01
20	9.00E+01	5.34E+01	3.09E+01	8.63E-01	-4.92E+01	2.67E+01	3.19E+00	3.70E+00	1.80E+01	8.65E+00	4.21E-03	1.88E+01	7.84E+00	1.55E-01
21	1.00E+02	5.26E+01	3.05E+01	8.05E-01	-5.12E+01	2.46E+01	2.88E+00	3.58E+00	1.62E+01	8.32E+00	3.99E-03	1.70E+01	7.59E+00	1.70E-01
22	2.00E+02	4.77E+01	2.77E+01	5.30E-01	-6.21E+01	1.47E+01	1.34E+00	2.54E+00	8.12E+00	6.55E+00	2.82E-03	8.48E+00	6.19E+00	3.08E-01
23	3.00E+02	4.53E+01	2.62E+01	4.31E-01	-6.67E+01	1.13E+01	8.30E-01	1.93E+00	5.41E+00	5.90E+00	2.30E-03	5.65E+00	5.66E+00	4.38E-01
24	4.00E+02	4.39E+01	2.54E+01	3.93E-01	-6.85E+01	1.00E+01	6.01E-01	1.53E+00	4.06E+00	5.95E+00	1.99E-03	4.24E+00	5.76E+00	5.66E-01
25	5.00E+02	4.37E+01	2.53E+01	3.98E-01	-6.83E+01	1.01E+01	4.88E-01	1.23E+00	3.25E+00	6.84E+00	1.78E-03	3.39E+00	6.69E+00	7.04E-01
26	6.00E+02	4.48E+01	2.60E+01	4.70E-01	-6.48E+01	1.22E+01	4.44E-01	9.45E-01	2.71E+00	9.51E+00	1.63E-03	2.83E+00	9.39E+00	8.67E-01
27	7.00E+02	4.59E+01	2.66E+01	7.40E-01	-5.35E+01	1.97E+01	4.61E-01	6.24E-01	2.32E+00	1.74E+01	1.51E-03	2.42E+00	1.73E+01	1.04E+00
28	8.00E+02	2.02E+01	1.17E+01	3.02E+00	-1.83E+01	3.54E+01	5.71E-01	1.89E-01	2.03E+00	3.34E+01	1.41E-03	2.12E+00	3.33E+01	5.22E-01
29	9.00E+02	-1.57E+01	-9.13E+00	-1.89E+00	2.79E+01	1.72E+01	9.05E-01	-4.80E-01	1.80E+00	1.54E+01	1.33E-03	1.88E+00	1.53E+01	-4.57E-01
30	1.00E+03	-7.62E+00	-4.42E+00	-1.29E+00	3.78E+01	5.70E+00	1.97E+00	-1.53E+00	1.62E+00	4.07E+00	1.26E-03	1.70E+00	4.00E+00	-2.46E-01

	A	B	C	D	E	F	G	H	I	J	K	L	M
1	Avra - 45% Sample												
2	Low Frequency - Coaxial					Wet sample (30%)	Teflon spacers, h=2.0675mm(pattern)			Not using Rms for coaxial			
3		Subtracting the Impedance of Electrode Polarization after Lopatin.											
4													
5	Length,cm	R, cm	r, cm	Weight	Dens.(wet)	Drying, weight in g.		Water cont.		Bulk dens.	σ'_{co} S/m		
6	2.5865	0.7144	0.3102	M, g	γ, g/cm^3	Dry	22.01	W, %	30.123	g/cm^3	0.0903		
7	L, nH	Cs, pF	V, cm^3	6.61	1.96421	Wet	28.64	Wv, %	45.470	1.51 from FL			
8	0	0.4876	3.365217	Cs from Air (AVG, with same Tefl. thickness-"CoaxStray-07Reverse").									
9										R'	X'		
10	f, MHz	Z, Ohm	ϕ, deg	tanδ	Res.meas.	Xmeas.	$X_L = \omega L$	R_{el}	X_{el}	Rm-Rms-	Xm-ωL-	G, S	σ', S/m
11					R(s), Ohm	Zsinϕ	Ohm	$y=8.225 \times R_{el}$	$y=-1.232 \times R_{el}$	R_{el}	X_{el}		
12	1.00E-02	6.01E+01	-1.96E+00	2.92E+01	6.00E+01	-2.05E+00	0.00E+00	3.28E+00	-4.92E-01	5.67E+01	-1.56E+00	1.76E-02	9.04E-02
13	2.00E-02	5.93E+01	-1.95E+00	2.94E+01	5.92E+01	-2.01E+00	0.00E+00	2.32E+00	-3.48E-01	5.69E+01	-1.66E+00	1.76E-02	9.01E-02
14	3.00E-02	5.88E+01	-1.91E+00	3.00E+01	5.88E+01	-1.96E+00	0.00E+00	1.89E+00	-2.84E-01	5.69E+01	-1.68E+00	1.76E-02	9.02E-02
15	5.00E-02	5.82E+01	-1.85E+00	3.09E+01	5.82E+01	-1.88E+00	0.00E+00	1.47E+00	-2.20E-01	5.67E+01	-1.66E+00	1.76E-02	9.04E-02
16	1.00E-01	5.75E+01	-1.97E+00	2.90E+01	5.74E+01	-1.98E+00	0.00E+00	1.04E+00	-1.55E-01	5.64E+01	-1.82E+00	1.77E-02	9.09E-02
17	2.00E-01	5.67E+01	-2.16E+00	2.65E+01	5.66E+01	-2.13E+00	0.00E+00	7.34E-01	-1.10E-01	5.59E+01	-2.02E+00	1.79E-02	9.17E-02
18	3.00E-01	5.62E+01	-2.34E+00	2.44E+01	5.61E+01	-2.30E+00	0.00E+00	5.99E-01	-8.98E-02	5.55E+01	-2.21E+00	1.80E-02	9.23E-02
19	5.00E-01	5.55E+01	-2.64E+00	2.17E+01	5.55E+01	-2.55E+00	0.00E+00	4.64E-01	-6.95E-02	5.50E+01	-2.48E+00	1.81E-02	9.31E-02
20	1.00E+00	5.46E+01	-3.36E+00	1.70E+01	5.45E+01	-3.20E+00	0.00E+00	3.28E-01	-4.92E-02	5.42E+01	-3.15E+00	1.84E-02	9.44E-02
21	2.00E+00	5.36E+01	-4.67E+00	1.22E+01	5.34E+01	-4.37E+00	0.00E+00	2.32E-01	-3.48E-02	5.32E+01	-4.33E+00	1.87E-02	9.59E-02
22	3.00E+00	5.29E+01	-5.94E+00	9.61E+00	5.26E+01	-5.48E+00	0.00E+00	1.89E-01	-2.84E-02	5.24E+01	-5.45E+00	1.89E-02	9.68E-02
23	5.00E+00	5.18E+01	-8.28E+00	6.87E+00	5.13E+01	-7.46E+00	0.00E+00	1.47E-01	-2.20E-02	5.11E+01	-7.44E+00	1.91E-02	9.83E-02
24	1.00E+01	4.95E+01	-1.33E+01	4.21E+00	4.82E+01	-1.14E+01	0.00E+00	1.04E-01	-1.55E-02	4.81E+01	-1.14E+01	1.97E-02	1.01E-01
25	2.00E+01	4.51E+01	-2.22E+01	2.45E+00	4.18E+01	-1.71E+01	0.00E+00	7.34E-02	-1.10E-02	4.17E+01	-1.71E+01	2.05E-02	1.05E-01
26	3.00E+01	4.10E+01	-2.91E+01	1.80E+00	3.58E+01	-1.99E+01	0.00E+00	5.99E-02	-8.98E-03	3.57E+01	-1.99E+01	2.14E-02	1.10E-01
27	4.00E+01	3.72E+01	-3.45E+01	1.45E+00	3.06E+01	-2.11E+01	0.00E+00	5.19E-02	-7.77E-03	3.06E+01	-2.11E+01	2.22E-02	1.14E-01
28	5.00E+01	3.39E+01	-3.89E+01	1.24E+00	2.64E+01	-2.13E+01	0.00E+00	4.64E-02	-6.95E-03	2.63E+01	-2.13E+01	2.30E-02	1.18E-01
29	6.00E+01	3.10E+01	-4.25E+01	1.09E+00	2.28E+01	-2.09E+01	0.00E+00	4.24E-02	-6.35E-03	2.28E+01	-2.09E+01	2.38E-02	1.22E-01
30	7.00E+01	2.85E+01	-4.55E+01	9.83E-01	2.00E+01	-2.03E+01	0.00E+00	3.92E-02	-5.88E-03	1.99E+01	-2.03E+01	2.46E-02	1.27E-01
31	8.00E+01	2.62E+01	-4.79E+01	9.02E-01	1.76E+01	-1.95E+01	0.00E+00	3.67E-02	-5.50E-03	1.75E+01	-1.95E+01	2.55E-02	1.31E-01
32	9.00E+01	2.43E+01	-5.00E+01	8.38E-01	1.56E+01	-1.86E+01	0.00E+00	3.46E-02	-5.18E-03	1.56E+01	-1.86E+01	2.64E-02	1.36E-01
33	1.00E+02	2.26E+01	-5.18E+01	7.86E-01	1.39E+01	-1.77E+01	0.00E+00	3.28E-02	-4.92E-03	1.39E+01	-1.77E+01	2.74E-02	1.41E-01

Avra - 45% Sample

Low Frequency - Coaxial

B, S	f, MHz	C, pF	ε'	tanδcorr	φcorr	ε''	Zcorr	Rcorr	ρ', Ohm-m	ρ''	Conductive losses,ε''c	Dielectric losses
		B/w		G/B		ε'tand	Ohm	Ohm	(Rcorr)		sc/we0	ε''-ε''c
-4.85E-04	1.00E-02	7.71E+03	4.47E+03	3.63E+01	-1.58E+00	1.63E+05	5.68E+01	5.67E+01	1.11E+01	3.04E-01	1.62E+05	1.95E+02
-5.13E-04	2.00E-02	4.08E+03	2.37E+03	3.42E+01	-1.67E+00	8.10E+04	5.70E+01	5.69E+01	1.11E+01	3.24E-01	8.12E+04	-1.85E+02
-5.18E-04	3.00E-02	2.75E+03	1.59E+03	3.39E+01	-1.69E+00	5.40E+04	5.69E+01	5.69E+01	1.11E+01	3.27E-01	5.41E+04	-5.79E+01
-5.16E-04	5.00E-02	1.64E+03	9.53E+02	3.41E+01	-1.68E+00	3.25E+04	5.68E+01	5.67E+01	1.11E+01	3.24E-01	3.25E+04	3.12E+01
-5.72E-04	1.00E-01	9.11E+02	5.28E+02	3.10E+01	-1.85E+00	1.63E+04	5.64E+01	5.64E+01	1.10E+01	3.55E-01	1.62E+04	1.16E+02
-6.47E-04	2.00E-01	5.14E+02	2.98E+02	2.76E+01	-2.07E+00	8.25E+03	5.59E+01	5.59E+01	1.09E+01	3.94E-01	8.12E+03	1.29E+02
-7.14E-04	3.00E-01	3.79E+02	2.20E+02	2.52E+01	-2.27E+00	5.53E+03	5.56E+01	5.55E+01	1.08E+01	4.29E-01	5.41E+03	1.22E+02
-8.18E-04	5.00E-01	2.60E+02	1.51E+02	2.22E+01	-2.58E+00	3.35E+03	5.51E+01	5.50E+01	1.07E+01	4.83E-01	3.25E+03	1.02E+02
-1.07E-03	1.00E+00	1.70E+02	9.84E+01	1.72E+01	-3.32E+00	1.70E+03	5.43E+01	5.42E+01	1.06E+01	6.12E-01	1.62E+03	7.34E+01
-1.51E-03	2.00E+00	1.20E+02	6.98E+01	1.23E+01	-4.64E+00	8.61E+02	5.34E+01	5.32E+01	1.04E+01	8.40E-01	8.12E+02	4.99E+01
-1.95E-03	3.00E+00	1.03E+02	6.00E+01	9.67E+00	-5.90E+00	5.80E+02	5.27E+01	5.25E+01	1.02E+01	1.06E+00	5.41E+02	3.91E+01
-2.77E-03	5.00E+00	8.82E+01	5.11E+01	6.91E+00	-8.23E+00	3.53E+02	5.17E+01	5.12E+01	9.97E+00	1.44E+00	3.25E+02	2.87E+01
-4.65E-03	1.00E+01	7.40E+01	4.29E+01	4.24E+00	-1.33E+01	1.82E+02	4.94E+01	4.81E+01	9.37E+00	2.21E+00	1.62E+02	1.95E+01
-8.35E-03	2.00E+01	6.64E+01	3.85E+01	2.46E+00	-2.21E+01	9.48E+01	4.51E+01	4.18E+01	8.14E+00	3.31E+00	8.12E+01	1.36E+01
-1.18E-02	3.00E+01	6.27E+01	3.63E+01	1.81E+00	-2.89E+01	6.57E+01	4.10E+01	3.59E+01	6.99E+00	3.86E+00	5.41E+01	1.16E+01
-1.52E-02	4.00E+01	6.03E+01	3.50E+01	1.46E+00	-3.44E+01	5.11E+01	3.72E+01	3.07E+01	5.99E+00	4.09E+00	4.06E+01	1.06E+01
-1.84E-02	5.00E+01	5.87E+01	3.40E+01	1.25E+00	-3.87E+01	4.24E+01	3.39E+01	2.65E+01	5.16E+00	4.14E+00	3.25E+01	9.95E+00
-2.17E-02	6.00E+01	5.75E+01	3.33E+01	1.10E+00	-4.23E+01	3.66E+01	3.11E+01	2.30E+01	4.48E+00	4.08E+00	2.71E+01	9.54E+00
-2.49E-02	7.00E+01	5.66E+01	3.28E+01	9.90E-01	-4.53E+01	3.25E+01	2.85E+01	2.01E+01	3.91E+00	3.95E+00	2.32E+01	9.30E+00
-2.81E-02	8.00E+01	5.59E+01	3.24E+01	9.09E-01	-4.77E+01	2.94E+01	2.63E+01	1.77E+01	3.45E+00	3.80E+00	2.03E+01	9.16E+00
-3.13E-02	9.00E+01	5.54E+01	3.21E+01	8.44E-01	-4.98E+01	2.71E+01	2.44E+01	1.57E+01	3.06E+00	3.63E+00	1.80E+01	9.08E+00
-3.46E-02	1.00E+02	5.51E+01	3.19E+01	7.91E-01	-5.17E+01	2.53E+01	2.27E+01	1.41E+01	2.74E+00	3.46E+00	1.62E+01	9.03E+00

	AA	AB	AC	AD	AE
1			Avra - 45% Sample		
2			Low Frequency - Coaxial		
3					
4					
5					
6					
7					
8					
9					
10	$100/\omega^{0.5}$	C(s), pF	f, MHz	σ'', S/m	ρ''
11					
12	3.99E-01	7.75E+06	1.00E-02	2.49E-03	3.04E-01
13	2.82E-01	3.96E+06	2.00E-02	2.63E-03	3.24E-01
14	2.30E-01	2.71E+06	3.00E-02	2.66E-03	3.27E-01
15	1.78E-01	1.69E+06	5.00E-02	2.65E-03	3.24E-01
16	1.26E-01	8.05E+05	1.00E-01	2.94E-03	3.55E-01
17	8.92E-02	3.73E+05	2.00E-01	3.32E-03	3.94E-01
18	7.28E-02	2.31E+05	3.00E-01	3.66E-03	4.29E-01
19	5.64E-02	1.25E+05	5.00E-01	4.20E-03	4.83E-01
20	3.99E-02	4.97E+04	1.00E+00	5.47E-03	6.12E-01
21	2.82E-02	1.82E+04	2.00E+00	7.77E-03	8.40E-01
22	2.30E-02	9.69E+03	3.00E+00	1.00E-02	1.06E+00
23	1.78E-02	4.27E+03	5.00E+00	1.42E-02	1.44E+00
24	1.26E-02	1.39E+03	1.00E+01	2.39E-02	2.21E+00
25	8.92E-03	4.66E+02	2.00E+01	4.28E-02	3.31E+00
26	7.28E-03	2.66E+02	3.00E+01	6.06E-02	3.86E+00
27	6.31E-03	1.89E+02	4.00E+01	7.78E-02	4.09E+00
28	5.64E-03	1.50E+02	5.00E+01	9.46E-02	4.14E+00
29	5.15E-03	1.27E+02	6.00E+01	1.11E-01	4.08E+00
30	4.77E-03	1.12E+02	7.00E+01	1.28E-01	3.95E+00
31	4.46E-03	1.02E+02	8.00E+01	1.44E-01	3.80E+00
32	4.21E-03	9.50E+01	9.00E+01	1.61E-01	3.63E+00
33	3.99E-03	8.97E+01	1.00E+02	1.78E-01	3.46E+00

Brookhaven (8.11%)

	A	B	C	D	E	F	G	H	I	J	K	L	M
1	Brookhaven - 8.11%												
2	High Frequencey												
3													
4	HP 4191A										From PL		
5	Length,cm	R, cm	r, cm	Weight	Dens.(wet)	Drying, weight in g.		Water cont.		Bulk densi	method		
6	2.676	0.7144	0.3102	M, g	γ, g/cm^3	Wet	Dry	W, %	Wv, %	g/cm^3	σ_{st}, S/m		
7	L, nH	Cs, pF	V, cm^3	5.77	1.65725	28.42	27.03	5.142	8.106	1.576	0.00044		
8	1.736	0.74922	3.4816623	Stray capacitance Cs - from 3.28.97(Air w/same thickness of Tef.spacers)									
9	L-Calc(from Air), Avg. for 1996-1997 years												
10	f, MHz	Z, Ohm	ϕ, deg	tanδ	Res.meas.	Xmeas.	$X_L=\omega L$	Rm-Rms	Xm-ωL	G, S	σ', S/m	B, S	C, pF
11					R(s), Ohm	Zsinϕ	Ohm	R'=Rm	X'				B/w
12	1.00E+00	7.40E+03	-3.70E+01	1.33E+00	5.91E+03	-4.45E+03	1.09E-02	5.91E+03	-4.45E+03	1.1E-04	5.35E-04	-7.7E-05	1.22E+01
13	1.00E+01	1.49E+03	-7.73E+01	2.25E-01	3.28E+02	-1.46E+03	1.09E-01	3.28E+02	-1.46E+03	1.5E-04	7.31E-04	-6.1E-04	9.66E+00
14	2.00E+01	7.18E+02	-8.42E+01	1.02E-01	7.26E+01	-7.14E+02	2.18E-01	7.26E+01	-7.15E+02	1.4E-04	6.98E-04	-1.3E-03	1.03E+01
15	3.00E+01	4.80E+02	-8.61E+01	6.82E-02	3.26E+01	-4.78E+02	3.27E-01	3.26E+01	-4.79E+02	1.4E-04	7.03E-04	-1.9E-03	1.03E+01
16	4.00E+01	3.62E+02	-8.68E+01	5.63E-02	2.03E+01	-3.61E+02	4.36E-01	2.03E+01	-3.61E+02	1.5E-04	7.69E-04	-2.6E-03	1.02E+01
17	5.00E+01	2.90E+02	-8.73E+01	4.65E-02	1.34E+01	-2.89E+02	5.45E-01	1.34E+01	-2.90E+02	1.6E-04	7.92E-04	-3.2E-03	1.02E+01
18	1.00E+02	1.45E+02	-8.80E+01	3.46E-02	5.01E+00	-1.45E+02	1.09E+00	5.01E+00	-1.46E+02	2.3E-04	1.17E-03	-6.4E-03	1.01E+01
19	1.50E+02	9.61E+01	-8.84E+01	2.86E-02	2.75E+00	-9.61E+01	1.64E+00	2.75E+00	-9.77E+01	2.9E-04	1.43E-03	-9.5E-03	1.01E+01
20	2.00E+02	7.14E+01	-8.81E+01	3.30E-02	2.36E+00	-7.14E+01	2.18E+00	2.36E+00	-7.36E+01	4.3E-04	2.16E-03	-1.3E-02	1.01E+01
21	2.50E+02	5.65E+01	-8.82E+01	3.09E-02	1.74E+00	-5.64E+01	2.73E+00	1.74E+00	-5.92E+01	5.0E-04	2.47E-03	-1.6E-02	1.00E+01
22	3.00E+02	4.63E+01	-8.84E+01	2.74E-02	1.27E+00	-4.62E+01	3.27E+00	1.27E+00	-4.95E+01	5.2E-04	2.56E-03	-1.9E-02	9.96E+00
23	3.50E+02	3.89E+01	-8.81E+01	3.25E-02	1.26E+00	-3.89E+01	3.82E+00	1.26E+00	-4.27E+01	6.9E-04	3.43E-03	-2.2E-02	9.89E+00
24	4.00E+02	3.33E+01	-8.82E+01	3.21E-02	1.07E+00	-3.33E+01	4.36E+00	1.07E+00	-3.76E+01	7.5E-04	3.74E-03	-2.5E-02	9.81E+00
25	4.50E+02	2.88E+01	-8.80E+01	3.44E-02	9.91E-01	-2.88E+01	4.91E+00	9.91E-01	-3.37E+01	8.7E-04	4.32E-03	-2.8E-02	9.73E+00
26	5.00E+02	2.52E+01	-8.79E+01	3.63E-02	9.15E-01	-2.52E+01	5.45E+00	9.15E-01	-3.06E+01	9.7E-04	4.83E-03	-3.0E-02	9.63E+00
27	6.00E+02	1.95E+01	-8.78E+01	3.86E-02	7.51E-01	-1.95E+01	6.54E+00	7.51E-01	-2.60E+01	1.1E-03	5.50E-03	-3.6E-02	9.44E+00
28	7.00E+02	1.51E+01	-8.75E+01	4.31E-02	6.49E-01	-1.51E+01	7.64E+00	6.49E-01	-2.27E+01	1.3E-03	6.25E-03	-4.1E-02	9.26E+00
29	8.00E+02	1.15E+01	-8.75E+01	4.45E-02	5.13E-01	-1.15E+01	8.73E+00	5.13E-01	-2.02E+01	1.3E-03	6.21E-03	-4.6E-02	9.08E+00
30	9.00E+02	8.36E+00	-8.74E+01	4.59E-02	3.84E-01	-8.35E+00	9.82E+00	3.84E-01	-1.82E+01	1.2E-03	5.76E-03	-5.1E-02	8.98E+00
31	1.00E+03	5.46E+00	-8.58E+01	7.31E-02	3.98E-01	-5.45E+00	1.09E+01	3.98E-01	-1.64E+01	1.5E-03	7.38E-03	-5.6E-02	8.98E+00
32													

	N	O	P	Q	R	S	T	U	V	W	X	Y
1	Brookhaven - 8.11 %											
2	High Frequencey											
3												
4												
5												
6												
7												
8												
9												
10	f, MHz	ε'	$\tan\delta$corr	ϕcorr	ε''	Zcorr	Rcorr	ρ', Ohm-m	ρ''	losses,ε''c	ε''diel	σ'', S/m
11			G/B		$\varepsilon'\tan\delta$	Ohm	Ohm	(Rcorr)		$\sigma_{st}/\omega\varepsilon_0$	$\varepsilon''-\varepsilon''$c	
12	2.00E+00	6.83E+00	1.41E+00	-3.54E+01	9.63E+00	7.56E+03	6.16E+03	1.24E+03	8.82E+02	3.95E+00	5.67E+00	3.80E-04
13	1.10E+01	5.41E+00	2.43E-01	-7.64E+01	1.31E+00	1.60E+03	3.78E+02	7.62E+01	3.14E+02	7.19E-01	5.95E-01	3.01E-03
14	2.10E+01	5.76E+00	1.09E-01	-8.38E+01	6.27E-01	7.70E+02	8.34E+01	1.68E+01	1.54E+02	3.77E-01	2.51E-01	6.41E-03
15	3.10E+01	5.76E+00	7.31E-02	-8.58E+01	4.21E-01	5.15E+02	3.75E+01	7.56E+00	1.03E+02	2.55E-01	1.66E-01	9.62E-03
16	4.10E+01	5.73E+00	6.03E-02	-8.65E+01	3.46E-01	3.88E+02	2.34E+01	4.71E+00	7.82E+01	1.93E-01	1.53E-01	1.27E-02
17	5.10E+01	5.72E+00	4.98E-02	-8.72E+01	2.85E-01	3.12E+02	1.55E+01	3.12E+00	6.27E+01	1.55E-01	1.30E-01	1.59E-02
18	1.01E+02	5.69E+00	3.68E-02	-8.79E+01	2.10E-01	1.57E+02	5.77E+00	1.16E+00	3.16E+01	7.83E-02	1.31E-01	3.16E-02
19	1.51E+02	5.66E+00	3.02E-02	-8.83E+01	1.71E-01	1.05E+02	3.17E+00	6.40E-01	2.12E+01	5.24E-02	1.19E-01	4.72E-02
20	2.01E+02	5.64E+00	3.44E-02	-8.80E+01	1.94E-01	7.91E+01	2.72E+00	5.48E-01	1.59E+01	3.93E-02	1.55E-01	6.27E-02
21	2.51E+02	5.61E+00	3.17E-02	-8.82E+01	1.78E-01	6.36E+01	2.01E+00	4.06E-01	1.28E+01	3.15E-02	1.46E-01	7.80E-02
22	3.01E+02	5.58E+00	2.75E-02	-8.84E+01	1.54E-01	5.33E+01	1.47E+00	2.95E-01	1.07E+01	2.63E-02	1.27E-01	9.31E-02
23	3.51E+02	5.54E+00	3.18E-02	-8.82E+01	1.76E-01	4.60E+01	1.46E+00	2.95E-01	9.26E+00	2.25E-02	1.54E-01	1.08E-01
24	4.01E+02	5.50E+00	3.06E-02	-8.82E+01	1.68E-01	4.05E+01	1.24E+00	2.50E-01	8.16E+00	1.97E-02	1.48E-01	1.22E-01
25	4.51E+02	5.45E+00	3.17E-02	-8.82E+01	1.73E-01	3.63E+01	1.15E+00	2.32E-01	7.32E+00	1.75E-02	1.55E-01	1.37E-01
26	5.01E+02	5.40E+00	3.22E-02	-8.82E+01	1.74E-01	3.30E+01	1.06E+00	2.14E-01	6.66E+00	1.58E-02	1.58E-01	1.50E-01
27	6.01E+02	5.29E+00	3.12E-02	-8.82E+01	1.65E-01	2.81E+01	8.75E-01	1.76E-01	5.66E+00	1.32E-02	1.52E-01	1.77E-01
28	7.01E+02	5.19E+00	3.09E-02	-8.82E+01	1.61E-01	2.45E+01	7.59E-01	1.53E-01	4.94E+00	1.13E-02	1.49E-01	2.02E-01
29	8.01E+02	5.09E+00	2.74E-02	-8.84E+01	1.39E-01	2.19E+01	6.01E-01	1.21E-01	4.41E+00	9.87E-03	1.30E-01	2.26E-01
30	9.01E+02	5.03E+00	2.29E-02	-8.87E+01	1.15E-01	1.97E+01	4.50E-01	9.08E-02	3.97E+00	8.78E-03	1.06E-01	2.52E-01
31	1.00E+03	5.03E+00	2.64E-02	-8.85E+01	1.33E-01	1.77E+01	4.67E-01	9.42E-02	3.57E+00	7.90E-03	1.25E-01	2.80E-01
32												

	A	B	C	D	E	F	G	H	I	J
1	Brookhaven - 8.11 %									
2	Low Frequencey									
3										
4				Coaxial # 154, spacers 1.5 mm. (2 flat spacers, avg: 0.324 cm)						
5	HP 4194A, Fixture			Brookhaven, I, 4-6' (Data from Coaxian1, sht.3)		From PL method	σ_{st}, S/m	0.00044		Bulk density
6	Length,cm	R, cm	r, cm	Weight M, g	Dens.(wet) ‌ing, weight in g. g/cm^3	Wet	Dry	Water cont. W, %	Wv, %	g/cm^3
7	2.676	0.7144	0.3102	5.77	1.65725	28.42	27.03	5.142	8.110	1.576
8	L, nH	Cs, pF	V, cm^3							
9	14.7	0.55700	3.4816623	8.97(Air w/same thickness of Tef.spacers)						
10	‌orted coax, brass disk)		Not using Rms for coaxial			Subtracting the Electrode Polarization				
11	f, MHz	Z, Ohm	φ, deg	$\tan\delta$	Res.meas.	Xmeas.	$X_L=wL$	R_{el}	X_{el}	R'
12					R(s), Ohm	Zsinf	Ohm	y=135.86*x		R'=Rm-Rel
13	1.00E-03	1.16E+04	-6.56E-01	8.73E+01	1.16E+04	-1.32E+02	9.24E-05	1.71E+02		1.14E+04
14	2.00E-03	1.15E+04	-6.52E-01	8.79E+01	1.15E+04	-1.31E+02	1.85E-04	1.21E+02		1.14E+04
15	3.00E-03	1.15E+04	-6.90E-01	8.31E+01	1.15E+04	-1.38E+02	2.77E-04	9.90E+01		1.14E+04
16	5.00E-03	1.15E+04	-8.19E-01	7.00E+01	1.15E+04	-1.64E+02	4.62E-04	7.67E+01		1.14E+04
17	1.00E-02	1.14E+04	-1.12E+00	5.11E+01	1.14E+04	-2.23E+02	9.24E-04	5.42E+01		1.14E+04
18	2.00E-02	1.14E+04	-1.73E+00	3.31E+01	1.14E+04	-3.43E+02	1.85E-03	3.83E+01		1.13E+04
19	3.00E-02	1.13E+04	-2.33E+00	2.46E+01	1.13E+04	-4.61E+02	2.77E-03	3.13E+01		1.13E+04
20	5.00E-02	1.13E+04	-3.49E+00	1.64E+01	1.12E+04	-6.85E+02	4.62E-03	2.42E+01		1.12E+04
21	1.00E-01	1.11E+04	-6.24E+00	9.15E+00	1.11E+04	-1.21E+03	9.24E-03	1.71E+01		1.10E+04
22	2.00E-01	1.08E+04	-1.13E+01	5.00E+00	1.06E+04	-2.13E+03	1.85E-02	1.21E+01		1.06E+04
23	3.00E-01	1.05E+04	-1.59E+01	3.50E+00	1.01E+04	-2.88E+03	2.77E-02	9.90E+00		1.01E+04
24	5.00E-01	9.75E+03	-2.39E+01	2.25E+00	8.91E+03	-3.96E+03	4.62E-02	7.67E+00		8.91E+03
25	1.00E+00	7.96E+03	-3.84E+01	1.26E+00	6.24E+03	-4.95E+03	9.24E-02	5.42E+00		6.23E+03
26	2.00E+00	5.54E+03	-5.43E+01	7.18E-01	3.24E+03	-4.50E+03	1.85E-01	3.83E+00		3.23E+03
27	3.00E+00	4.17E+03	-6.28E+01	5.14E-01	1.91E+03	-3.71E+03	2.77E-01	3.13E+00		1.90E+03
28	5.00E+00	2.75E+03	-7.16E+01	3.33E-01	8.68E+02	-2.61E+03	4.62E-01	2.42E+00		8.65E+02
29	1.00E+01	1.46E+03	-7.97E+01	1.82E-01	2.61E+02	-1.43E+03	9.24E-01	1.71E+00		2.59E+02
30	2.00E+01	7.43E+02	-8.45E+01	9.69E-02	7.17E+01	-7.40E+02	1.85E+00	1.21E+00		7.05E+01
31	3.00E+01	4.97E+02	-8.61E+01	6.84E-02	3.39E+01	-4.96E+02	2.77E+00	9.90E-01		3.29E+01
32	4.00E+01	3.72E+02	-8.69E+01	5.50E-02	2.04E+01	-3.71E+02	3.69E+00	8.57E-01		1.96E+01
33	Note: There is no linear dependence of X_{meas} vs. $w^{0.5}$; no corrections for X_{el} are made.									

Brookhaven - 8.11 %
Low Frequencey

	f, MHz	X'	G, S	σ', S/m	B, S	C, pF	ε'	tanδcorr	φcorr	ε''
		Xm-wL				B/w		G/B		e'tanδ
13	1.00E-03	-1.32E+02	8.78E-05	4.36E-04	-1.02E-06	1.62E+02	9.08E+01	8.63E+01	-6.64E-01	7.83E+03
14	2.00E-03	-1.31E+02	8.78E-05	4.36E-04	-1.00E-06	7.98E+01	4.47E+01	8.76E+01	-6.54E-01	3.91E+03
15	3.00E-03	-1.38E+02	8.78E-05	4.36E-04	-1.06E-06	5.60E+01	3.14E+01	8.32E+01	-6.89E-01	2.61E+03
16	5.00E-03	-1.64E+02	8.79E-05	4.36E-04	-1.25E-06	3.97E+01	2.22E+01	7.05E+01	-8.13E-01	1.57E+03
17	1.00E-02	-2.23E+02	8.80E-05	4.37E-04	-1.70E-06	2.70E+01	1.51E+01	5.19E+01	-1.10E+00	7.85E+02
18	2.00E-02	-3.43E+02	8.83E-05	4.38E-04	-2.61E-06	2.07E+01	1.16E+01	3.39E+01	-1.69E+00	3.94E+02
19	3.00E-02	-4.61E+02	8.85E-05	4.39E-04	-3.51E-06	1.86E+01	1.04E+01	2.52E+01	-2.27E+00	2.63E+02
20	5.00E-02	-6.85E+02	8.88E-05	4.41E-04	-5.25E-06	1.67E+01	9.36E+00	1.69E+01	-3.38E+00	1.58E+02
21	1.00E-01	-1.21E+03	8.95E-05	4.44E-04	-9.44E-06	1.50E+01	8.42E+00	9.48E+00	-6.02E+00	7.98E+01
22	2.00E-01	-2.13E+03	9.06E-05	4.50E-04	-1.74E-05	1.39E+01	7.78E+00	5.19E+00	-1.09E+01	4.04E+01
23	3.00E-01	-2.88E+03	9.17E-05	4.55E-04	-2.52E-05	1.33E+01	7.48E+00	3.64E+00	-1.53E+01	2.73E+01
24	5.00E-01	-3.96E+03	9.38E-05	4.65E-04	-3.99E-05	1.27E+01	7.12E+00	2.35E+00	-2.31E+01	1.67E+01
25	1.00E+00	-4.95E+03	9.84E-05	4.88E-04	-7.46E-05	1.19E+01	6.65E+00	1.32E+00	-3.72E+01	8.78E+00
26	2.00E+00	-4.50E+03	1.05E-04	5.22E-04	-1.40E-04	1.11E+01	6.22E+00	7.54E-01	-5.30E+01	4.69E+00
27	3.00E+00	-3.71E+03	1.09E-04	5.43E-04	-2.03E-04	1.08E+01	6.03E+00	5.40E-01	-6.16E+01	3.25E+00
28	5.00E+00	-2.61E+03	1.15E-04	5.69E-04	-3.28E-04	1.04E+01	5.85E+00	3.50E-01	-7.07E+01	2.05E+00
29	1.00E+01	-1.43E+03	1.22E-04	6.06E-04	-6.41E-04	1.02E+01	5.71E+00	1.91E-01	-7.92E+01	1.09E+00
30	2.00E+01	-7.42E+02	1.27E-04	6.30E-04	-1.27E-03	1.01E+01	5.65E+00	1.00E-01	-8.43E+01	5.66E-01
31	3.00E+01	-4.99E+02	1.32E-04	6.54E-04	-1.89E-03	1.00E+01	5.62E+00	6.97E-02	-8.60E+01	3.92E-01
32	4.00E+01	-3.75E+02	1.39E-04	6.89E-04	-2.52E-03	1.00E+01	5.62E+00	5.51E-02	-8.68E+01	3.09E-01

					Conduct				
f, MHz	Zcorr	Rcorr	ρ', Ohm-m	ρ''	losses,$\varepsilon''c$	ε''diel	100/w^0.5	C(s), pF	σ'', S/m
	Ohm	Ohm	(Rcorr)		σ_{st}/we_0	e"-e"c			
1.00E-03	1.14E+04	1.14E+04	2.29E+03	2.66E+01	7.91E+03	-7.44E+01	1.26E+00	1.20E+06	5.05E-06
2.00E-03	1.14E+04	1.14E+04	2.30E+03	2.62E+01	3.95E+03	-3.97E+01	8.92E-01	6.08E+05	4.97E-06
3.00E-03	1.14E+04	1.14E+04	2.30E+03	2.76E+01	2.64E+03	-2.63E+01	7.28E-01	3.84E+05	5.24E-06
5.00E-03	1.14E+04	1.14E+04	2.29E+03	3.25E+01	1.58E+03	-1.47E+01	5.64E-01	1.94E+05	6.18E-06
1.00E-02	1.14E+04	1.14E+04	2.29E+03	4.41E+01	7.91E+02	-5.81E+00	3.99E-01	7.12E+04	8.42E-06
2.00E-02	1.13E+04	1.13E+04	2.28E+03	6.74E+01	3.95E+02	-1.83E+00	2.82E-01	2.32E+04	1.29E-05
3.00E-02	1.13E+04	1.13E+04	2.27E+03	9.02E+01	2.64E+02	-6.20E-01	2.30E-01	1.15E+04	1.74E-05
5.00E-02	1.12E+04	1.12E+04	2.26E+03	1.34E+02	1.58E+02	2.11E-01	1.78E-01	4.65E+03	2.60E-05
1.00E-01	1.11E+04	1.11E+04	2.23E+03	2.35E+02	7.91E+01	7.00E-01	1.26E-01	1.32E+03	4.68E-05
2.00E-01	1.08E+04	1.06E+04	2.15E+03	4.13E+02	3.95E+01	8.55E-01	8.92E-02	3.74E+02	8.66E-05
3.00E-01	1.05E+04	1.01E+04	2.04E+03	5.61E+02	2.64E+01	8.87E-01	7.28E-02	1.84E+02	1.25E-04
5.00E-01	9.81E+03	9.03E+03	1.82E+03	7.75E+02	1.58E+01	9.05E-01	5.64E-02	8.04E+01	1.98E-04
1.00E+00	8.10E+03	6.45E+03	1.30E+03	9.86E+02	7.91E+00	8.70E-01	3.99E-02	3.22E+01	3.70E-04
2.00E+00	5.72E+03	3.44E+03	6.94E+02	9.21E+02	3.95E+00	7.36E-01	2.82E-02	1.77E+01	6.93E-04
3.00E+00	4.34E+03	2.06E+03	4.15E+02	7.70E+02	2.64E+00	6.17E-01	2.30E-02	1.43E+01	1.01E-03
5.00E+00	2.88E+03	9.50E+02	1.91E+02	5.47E+02	1.58E+00	4.65E-01	1.78E-02	1.22E+01	1.63E-03
1.00E+01	1.53E+03	2.87E+02	5.79E+01	3.04E+02	7.91E-01	2.99E-01	1.26E-02	1.11E+01	3.18E-03
2.00E+01	7.86E+02	7.84E+01	1.58E+01	1.58E+02	3.95E-01	1.71E-01	8.92E-03	1.08E+01	6.28E-03
3.00E+01	5.27E+02	3.67E+01	7.39E+00	1.06E+02	2.64E-01	1.28E-01	7.28E-03	1.07E+01	9.39E-03
4.00E+01	3.96E+02	2.18E+01	4.39E+00	7.98E+01	1.98E-01	1.12E-01	6.31E-03	1.07E+01	1.25E-02

Brookhaven (8.69%)

	A	B	C	D	E	F	G	H	I	J	K	L	M
1	8.69 % - Brookhaven												
2	High Frequency												
3													
4	Brookhaven, I, 2-4". Some water is added.					Data from "Coaxian1", sheet 3						
5	Sample holder # 154, 3 cm Spacers are identical: 1.5 cm.(AVG h: 1.622 mm)								2h = 0.324 cm				
6	Length	R, cm	r, cm	Volume	Weight	Dens.(wet)	Drying, weight in g.		Water cont.		Bulk dens.	method	
7	cm			V, cm^3	M, g	γ, g/cm^3	Wet	Dry	W, %	Wv, %	g/cm^3	σst', S/m	
8	2.676	0.7144	0.3102	3.481662	5.85	1.68023	27.08	25.68	5.452	8.687	1.593	0.000504	
9	L, nH		Cs, pF	Stray capacitance Cs - from 3.28.97(Air w/same thickness of Tef.spacers)									
10	1.736	0.74922	L - Calc(Air), Avg.for 1996 - 1997 years										
11	f, MHz	Z, Ohm	φ, deg	tanδ	Res.meas.	Xmeas.	$X_L=\omega L$	Rm-Rms	Xm-wL	G, S	σ', S/m	B, S	C, pF
12					R(s), Ohm	Zsinφ	Ohm	R'=Rm	X'				B/w
13	1.00E+00	6.60E+03	-3.30E+01	1.54E+00	5.54E+03	-3.59E+03	1.09E-02	5.54E+03	-3.59E+03	1.27E-04	6.30E-04	-7.78E-05	1.24E+01
14	1.00E+01	1.50E+03	-7.60E+01	2.49E-01	3.63E+02	-1.46E+03	1.09E-01	3.63E+02	-1.46E+03	1.61E-04	8.00E-04	-5.99E-04	9.54E+00
15	2.00E+01	7.24E+02	-8.35E+01	1.14E-01	8.20E+01	-7.19E+02	2.18E-01	8.20E+01	-7.20E+02	1.56E-04	7.75E-04	-1.28E-03	1.02E+01
16	3.00E+01	4.84E+02	-8.56E+01	7.69E-02	3.71E+01	-4.82E+02	3.27E-01	3.71E+01	-4.83E+02	1.58E-04	7.86E-04	-1.92E-03	1.02E+01
17	4.00E+01	3.65E+02	-8.66E+01	6.01E-02	2.19E+01	-3.64E+02	4.36E-01	2.19E+01	-3.64E+02	1.64E-04	8.15E-04	-2.55E-03	1.01E+01
18	5.00E+01	2.92E+02	-8.71E+01	5.01E-02	1.46E+01	-2.92E+02	5.45E-01	1.46E+01	-2.92E+02	1.71E-04	8.47E-04	-3.18E-03	1.01E+01
19	1.00E+02	1.46E+02	-8.83E+01	3.04E-02	4.43E+00	-1.46E+02	1.09E+00	4.43E+00	-1.47E+02	2.05E-04	1.02E-03	-6.33E-03	1.01E+01
20	1.50E+02	9.65E+01	-8.85E+01	2.62E-02	2.53E+00	-9.65E+01	1.64E+00	2.53E+00	-9.81E+01	2.62E-04	1.30E-03	-9.48E-03	1.01E+01
21	2.00E+02	7.15E+01	-8.86E+01	2.48E-02	1.77E+00	-7.15E+01	2.18E+00	1.77E+00	-7.37E+01	3.26E-04	1.62E-03	-1.26E-02	1.00E+01
22	2.50E+02	5.63E+01	-8.86E+01	2.50E-02	1.41E+00	-5.63E+01	2.73E+00	1.41E+00	-5.90E+01	4.03E-04	2.00E-03	-1.58E-02	1.00E+01
23	3.00E+02	4.60E+01	-8.85E+01	2.71E-02	1.24E+00	-4.60E+01	3.27E+00	1.24E+00	-4.92E+01	5.13E-04	2.54E-03	-1.89E-02	1.00E+01
24	3.50E+02	3.85E+01	-8.84E+01	2.83E-02	1.09E+00	-3.85E+01	3.82E+00	1.09E+00	-4.23E+01	6.08E-04	3.02E-03	-2.20E-02	9.99E+00
25	4.00E+02	3.27E+01	-8.79E+01	3.60E-02	1.18E+00	-3.27E+01	4.36E+00	1.18E+00	-3.71E+01	8.55E-04	4.24E-03	-2.51E-02	9.97E+00
26	4.50E+02	2.81E+01	-8.79E+01	3.70E-02	1.04E+00	-2.81E+01	4.91E+00	1.04E+00	-3.30E+01	9.54E-04	4.73E-03	-2.81E-02	9.95E+00
27	5.00E+02	2.44E+01	-8.78E+01	3.93E-02	9.58E-01	-2.44E+01	5.45E+00	9.58E-01	-2.98E+01	1.08E-03	5.33E-03	-3.11E-02	9.91E+00
28	6.00E+02	1.84E+01	-8.54E+01	7.99E-02	1.47E+00	-1.84E+01	6.54E+00	1.47E+00	-2.49E+01	2.36E-03	1.17E-02	-3.72E-02	9.86E+00
29	7.00E+02	1.38E+01	-8.48E+01	9.07E-02	1.25E+00	-1.37E+01	7.64E+00	1.25E+00	-2.14E+01	2.72E-03	1.35E-02	-4.33E-02	9.85E+00
30	8.00E+02	9.98E+00	-8.48E+01	9.17E-02	9.12E-01	-9.94E+00	8.73E+00	9.12E-01	-1.87E+01	2.61E-03	1.30E-02	-4.97E-02	9.88E+00
31	9.00E+02	6.62E+00	-8.38E+01	1.09E-01	7.18E-01	-6.58E+00	9.82E+00	7.18E-01	-1.64E+01	2.67E-03	1.32E-02	-5.66E-02	1.00E+01
32	1.00E+03	3.51E+00	-8.15E+01	1.49E-01	5.19E-01	-3.47E+00	1.09E+01	5.19E-01	-1.44E+01	2.51E-03	1.24E-02	-6.47E-02	1.03E+01
33													

	N	O	P	Q	R	S	T	U	V	W	X	Y
1	8.69 % - Brookhaven											
2	High Frequency											
3												
4												
5												
6												
7												
8												
9												
10												
11	f, MHz	ε'	tanδcorr	φcorr	ε''	Zcorr	Rcorr	ρ', Ohm-m	ρ''	σ'', S/m	losses,ε''_c	ε''diel
12			G/B		ε'tanδ	Ohm	Ohm	(Rcorr)			$\sigma_c/\omega\varepsilon_0$	$\varepsilon''-\varepsilon''_c$
13	1.00E+00	6.94E+00	1.63E+00	-3.15E+01	1.13E+01	6.71E+03	5.72E+03	1.15E+03	7.06E+02	3.86E-04	9.06E+00	2.27E+00
14	1.00E+01	5.35E+00	2.69E-01	-7.49E+01	1.44E+00	1.61E+03	4.18E+02	8.43E+01	3.14E+02	2.97E-03	9.06E-01	5.31E-01
15	2.00E+01	5.70E+00	1.22E-01	-8.30E+01	6.97E-01	7.77E+02	9.43E+01	1.90E+01	1.55E+02	6.34E-03	4.53E-01	2.44E-01
16	3.00E+01	5.70E+00	8.26E-02	-8.53E+01	4.71E-01	5.20E+02	4.27E+01	8.62E+00	1.04E+02	9.52E-03	3.02E-01	1.69E-01
17	4.00E+01	5.68E+00	6.45E-02	-8.63E+01	3.66E-01	3.92E+02	2.52E+01	5.08E+00	7.88E+01	1.26E-02	2.26E-01	1.40E-01
18	5.00E+01	5.67E+00	5.37E-02	-8.69E+01	3.05E-01	3.14E+02	1.69E+01	3.40E+00	6.32E+01	1.58E-02	1.81E-01	1.23E-01
19	1.00E+02	5.65E+00	3.24E-02	-8.81E+01	1.83E-01	1.58E+02	5.11E+00	1.03E+00	3.18E+01	3.14E-02	9.06E-02	9.24E-02
20	1.50E+02	5.63E+00	2.77E-02	-8.84E+01	1.56E-01	1.05E+02	2.92E+00	5.88E-01	2.13E+01	4.70E-02	6.04E-02	9.55E-02
21	2.00E+02	5.63E+00	2.58E-02	-8.85E+01	1.46E-01	7.92E+01	2.05E+00	4.12E-01	1.60E+01	6.26E-02	4.53E-02	1.00E-01
22	2.50E+02	5.62E+00	2.56E-02	-8.85E+01	1.44E-01	6.34E+01	1.62E+00	3.27E-01	1.28E+01	7.82E-02	3.62E-02	1.08E-01
23	3.00E+02	5.61E+00	2.71E-02	-8.84E+01	1.52E-01	5.29E+01	1.44E+00	2.90E-01	1.07E+01	9.37E-02	3.02E-02	1.22E-01
24	3.50E+02	5.60E+00	2.77E-02	-8.84E+01	1.55E-01	4.55E+01	1.26E+00	2.53E-01	9.16E+00	1.09E-01	2.59E-02	1.29E-01
25	4.00E+02	5.59E+00	3.41E-02	-8.80E+01	1.91E-01	3.99E+01	1.36E+00	2.74E-01	8.03E+00	1.24E-01	2.26E-02	1.68E-01
26	4.50E+02	5.58E+00	3.39E-02	-8.81E+01	1.89E-01	3.55E+01	1.20E+00	2.42E-01	7.15E+00	1.40E-01	2.01E-02	1.69E-01
27	5.00E+02	5.55E+00	3.45E-02	-8.80E+01	1.92E-01	3.21E+01	1.11E+00	2.23E-01	6.47E+00	1.54E-01	1.81E-02	1.74E-01
28	6.00E+02	5.53E+00	6.34E-02	-8.64E+01	3.50E-01	2.68E+01	1.70E+00	3.42E-01	5.40E+00	1.84E-01	1.51E-02	3.35E-01
29	7.00E+02	5.52E+00	6.27E-02	-8.64E+01	3.46E-01	2.30E+01	1.44E+00	2.90E-01	4.63E+00	2.15E-01	1.29E-02	3.33E-01
30	8.00E+02	5.54E+00	5.25E-02	-8.70E+01	2.91E-01	2.01E+01	1.05E+00	2.13E-01	4.05E+00	2.46E-01	1.13E-02	2.80E-01
31	9.00E+02	5.61E+00	4.71E-02	-8.73E+01	2.64E-01	1.76E+01	8.29E-01	1.67E-01	3.55E+00	2.81E-01	1.01E-02	2.54E-01
32	1.00E+03	5.77E+00	3.87E-02	-8.78E+01	2.24E-01	1.54E+01	5.97E-01	1.20E-01	3.11E+00	3.21E-01	9.06E-03	2.14E-01
33												

Low Frequency

Brookhaven, l, 2-4". Some water is added. — From FL method — σ_{st}', S/m

HP 4194A — Sample holder # 154, 3 cm Spacers are identical: 1.5 cm.(AVG h: 1.622 mm) — 0.000504 — 2h = 0.324 cm

Length	R, cm	r, cm	Volume	Weight	Dens.(wet)	Drying, weight in g.		Water cont.		Bulk dens.
cm			V, cm^3	M, g	γ, g/cm^3	Wet	Dry	W, %	Wv, %	g/cm^3
2.676	0.7144	0.3102	3.48166228	5.85	1.68023	27.08	25.68	5.452	8.687	1.593

L, nH: 14.7 — C_s, pF: 0.55700 — L-from shorted coax, Brass disk — Rms not used for coaxial

Stray capacitance - from 3.28.97(Air w/same thickness of Tef.spacers)

Note: The data for additional samples are not fully updated, because we need from them only ρ_{st} values.

f, MHz	Z, Ohm	φ, deg	$\tan\delta$	Res.meas. R(s), Ohm	Xmeas. $Z\sin\varphi$	$X_L=\omega L$ Ohm	Rm-Rms- Rel	Xm-wL X'	G, S	σ', S/m
1.00E-03	9.96E+03	-5.83E-01	9.83E+01	9.96E+03	-1.01E+02	9.24E-05	9.84E+02	-1.01E+02	1.02E-04	5.04E-04
2.00E-03	9.93E+03	-5.47E-01	1.05E+02	9.93E+03	-9.49E+01	1.85E-04	9.84E+01	-9.49E+01	1.02E-04	5.04E-04
3.00E-03	9.91E+03	-5.70E-01	1.00E+02	9.91E+03	-9.86E+01	2.77E-04	9.84E+01	-9.86E+01	1.02E-04	5.04E-04
5.00E-03	9.89E+03	-6.53E-01	8.77E+01	9.89E+03	-1.13E+02	4.62E-04	9.84E+02	-1.13E+02	1.02E-04	5.04E-04
1.00E-02	9.86E+03	-9.03E-01	6.35E+01	9.86E+03	-1.55E+02	9.24E-04	9.82E+02	-1.55E+02	1.02E-04	5.05E-04
2.00E-02	9.83E+03	-1.42E+00	4.04E+01	9.83E+03	-2.43E+02	1.85E-03	9.80E+02	-2.43E+02	1.02E-04	5.06E-04
3.00E-02	9.81E+03	-1.94E+00	2.96E+01	9.80E+03	-3.32E+02	2.77E-03	9.78E+02	-3.32E+02	1.02E-04	5.07E-04
5.00E-02	9.77E+03	-2.94E+00	1.95E+01	9.76E+03	-5.01E+02	4.62E-03	9.74E+02	-5.01E+02	1.02E-04	5.08E-04
1.00E-01	9.68E+03	-5.35E+00	1.07E+01	9.64E+03	-9.02E+02	9.24E-03	9.63E+02	-9.02E+02	1.03E-04	5.11E-04
2.00E-01	9.48E+03	-9.85E+00	5.76E+00	9.34E+03	-1.62E+03	1.85E-02	9.33E+02	-1.62E+03	1.04E-04	5.16E-04
3.00E-01	9.24E+03	-1.40E+01	4.01E+00	8.97E+03	-2.24E+03	2.77E-02	8.96E+02	-2.24E+03	1.05E-04	5.21E-04
5.00E-01	8.70E+03	-2.13E+01	2.56E+00	8.10E+03	-3.16E+03	4.62E-02	8.10E+02	-3.16E+03	1.07E-04	5.32E-04
1.00E+00	7.31E+03	-3.50E+01	1.43E+00	5.99E+03	-4.20E+03	9.24E-02	5.99E+02	-4.20E+03	1.12E-04	5.56E-04
2.00E+00	5.28E+03	-5.10E+01	8.10E-01	3.33E+03	-4.11E+03	1.85E-01	3.32E+02	-4.11E+03	1.19E-04	5.91E-04
3.00E+00	4.05E+03	-5.99E+01	5.79E-01	2.03E+03	-3.51E+03	2.77E-01	2.03E+02	-3.51E+03	1.24E-04	6.13E-04
5.00E+00	2.71E+03	-6.94E+01	3.75E-01	9.51E+02	-2.53E+03	4.62E-01	9.50E+02	-2.54E+02	1.30E-04	6.43E-04
1.00E+01	1.45E+03	-7.85E+01	2.03E-01	2.89E+02	-1.42E+03	9.24E-01	2.87E+02	-1.42E+02	1.36E-04	6.76E-04
2.00E+01	7.43E+02	-8.39E+01	1.06E-01	7.85E+01	-7.39E+02	1.85E+00	7.77E+01	-7.40E+02	1.40E-04	6.95E-04
3.00E+01	4.97E+02	-8.58E+01	7.36E-02	3.65E+01	-4.96E+02	2.77E+00	3.58E+01	-4.98E+02	1.43E-04	7.11E-04
4.00E+01	3.72E+02	-8.66E+01	5.86E-02	2.18E+01	-3.72E+02	3.69E+00	2.12E+01	-3.75E+02	1.50E-04	7.44E-04

	L	M	N	O	P	Q	R	S	T	U	V
1	Brookhaven - 8.69 %										
2	Low Frequency										
3											
4											
5											
6											
7											
8											
9											
10											
11											
12	f, MHz	B, S	C, pF	ε'	tanδcorr	φcorr	ε''	Zcorr	Rcorr	ρ', Ohm-m	ρ"
13		B/w	B/w		G/B		ε'tanδ	Ohm	Ohm	(Rcorr)	
14	1.00E-03	-1.04E-06	1.66E+02	9.31E+01	9.74E+01	-5.88E-01	9.06E+03	9.84E+03	9.84E+03	1.98E+03	2.04E+01
15	2.00E-03	-9.72E-07	7.74E+01	4.33E+01	1.05E+02	-5.48E-01	4.53E+03	9.84E+03	9.84E+03	1.98E+03	1.90E+01
16	3.00E-03	-1.01E-06	5.35E+01	3.00E+01	1.01E+02	-5.68E-01	3.02E+03	9.84E+03	9.84E+03	1.98E+03	1.97E+01
17	5.00E-03	-1.15E-06	3.65E+01	2.05E+01	8.86E+01	-6.47E-01	1.81E+03	9.84E+03	9.84E+03	1.98E+03	2.24E+01
18	1.00E-02	-1.57E-06	2.51E+01	1.40E+01	6.46E+01	-8.87E-01	9.08E+02	9.82E+03	9.82E+03	1.98E+03	3.06E+01
19	2.00E-02	-2.46E-06	1.96E+01	1.10E+01	4.14E+01	-1.38E+00	4.55E+02	9.80E+03	9.80E+03	1.98E+03	4.77E+01
20	3.00E-02	-3.36E-06	1.78E+01	9.98E+00	3.04E+01	-1.88E+00	3.04E+02	9.79E+03	9.78E+03	1.97E+03	6.48E+01
21	5.00E-02	-5.09E-06	1.62E+01	9.08E+00	2.01E+01	-2.85E+00	1.83E+02	9.75E+03	9.74E+03	1.96E+03	9.76E+01
22	1.00E-01	-9.30E-06	1.48E+01	8.29E+00	1.11E+01	-5.16E+00	9.18E+01	9.67E+03	9.63E+03	1.94E+03	1.75E+02
23	2.00E-01	-1.74E-05	1.38E+01	7.75E+00	5.98E+00	-9.49E+00	4.64E+01	9.48E+03	9.35E+03	1.89E+03	3.15E+02
24	3.00E-01	-2.52E-05	1.34E+01	7.48E+00	4.17E+00	-1.35E+01	3.12E+01	9.26E+03	9.00E+03	1.81E+03	4.35E+02
25	5.00E-01	-4.01E-05	1.28E+01	7.15E+00	2.67E+00	-2.05E+01	1.91E+01	8.74E+03	8.19E+03	1.65E+03	6.18E+02
26	1.00E+00	-7.50E-05	1.19E+01	6.69E+00	1.49E+00	-3.38E+01	9.99E+00	7.42E+03	6.16E+03	1.24E+03	8.32E+02
27	2.00E+00	-1.40E-04	1.12E+01	6.25E+00	8.50E-01	-4.96E+01	5.31E+00	5.44E+03	3.52E+03	7.10E+02	8.35E+02
28	3.00E+00	-2.03E-04	1.08E+01	6.04E+00	6.08E-01	-5.87E+01	3.68E+00	4.20E+03	2.18E+03	4.40E+02	7.24E+02
29	5.00E+00	-3.28E-04	1.05E+01	5.86E+00	3.94E-01	-6.85E+01	2.31E+00	2.83E+03	1.04E+03	2.10E+02	5.31E+02
30	1.00E+01	-6.40E-04	1.02E+01	5.71E+00	2.13E-01	-7.80E+01	1.22E+00	1.53E+03	3.18E+02	6.41E+01	3.01E+02
31	2.00E+01	-1.27E-03	1.01E+01	5.65E+00	1.11E-01	-8.37E+01	6.25E-01	7.85E+02	8.64E+01	1.74E+01	1.57E+02
32	3.00E+01	-1.89E-03	1.00E+01	5.62E+00	7.58E-02	-8.57E+01	4.26E-01	5.27E+02	3.99E+01	8.03E+00	1.06E+02
33	4.00E+01	-2.52E-03	1.00E+01	5.61E+00	5.96E-02	-8.66E+01	3.35E-01	3.97E+02	2.36E+01	4.76E+00	7.98E+01

	W	X	Y	Z	AA	AB	AC
1	Brookhaven - 8.69 %						
2	Low Frequency						
3							
4							
5							
6							
7							
8							
9							
10							
11							
12	f, MHz	$100/\omega^{0.5}$	R_{el}	σ'', S/m	losses,$\varepsilon''c$	losses,$\varepsilon''c$	ε''diel
13			y=94.3663*x		C(s), pF	$\sigma_{st}/\omega\varepsilon0$	$\varepsilon''-\varepsilon''c$
14	1.00E-03	1.26E+00	1.19E+02	5.18E-06	1.57E+06	9.06E+03	5.07E+00
15	2.00E-03	8.92E-01	8.42E+01	4.82E-06	8.39E+05	4.53E+03	4.37E-01
16	3.00E-03	7.28E-01	6.87E+01	5.00E-06	5.38E+05	3.02E+03	7.20E-01
17	5.00E-03	5.64E-01	5.32E+01	5.69E-06	2.82E+05	1.81E+03	1.14E+00
18	1.00E-02	3.99E-01	3.76E+01	7.81E-06	1.02E+05	9.06E+02	1.71E+00
19	2.00E-02	2.82E-01	2.66E+01	1.22E-05	3.27E+04	4.53E+02	1.80E+00
20	3.00E-02	2.30E-01	2.17E+01	1.67E-05	1.60E+04	3.02E+02	1.63E+00
21	5.00E-02	1.78E-01	1.68E+01	2.53E-05	6.35E+03	1.81E+02	1.47E+00
22	1.00E-01	1.26E-01	1.19E+01	4.61E-05	1.76E+03	9.06E+01	1.24E+00
23	2.00E-01	8.92E-02	8.42E+00	8.62E-05	4.90E+02	4.53E+01	1.08E+00
24	3.00E-01	7.28E-02	6.87E+00	1.25E-04	2.37E+02	3.02E+01	1.03E+00
25	5.00E-01	5.64E-02	5.32E+00	1.99E-04	1.01E+02	1.81E+01	9.91E-01
26	1.00E+00	3.99E-02	3.76E+00	3.72E-04	3.79E+01	9.06E+00	9.28E-01
27	2.00E+00	2.82E-02	2.66E+00	6.95E-04	1.94E+01	4.53E+00	7.81E-01
28	3.00E+00	2.30E-02	2.17E+00	1.01E-03	1.51E+01	3.02E+00	6.56E-01
29	5.00E+00	1.78E-02	1.68E+00	1.63E-03	1.26E+01	1.81E+00	4.99E-01
30	1.00E+01	1.26E-02	1.19E+00	3.18E-03	1.12E+01	9.06E-01	3.10E-01
31	2.00E+01	8.92E-03	8.42E-01	6.28E-03	1.08E+01	4.53E-01	1.72E-01
32	3.00E+01	7.28E-03	6.87E-01	9.38E-03	1.07E+01	3.02E-01	1.24E-01
33	4.00E+01	6.31E-03	5.95E-01	1.25E-02	1.07E+01	2.26E-01	1.08E-01

Brookhaven (9.75%)

	A	B	C	D	E	F	G	H	I	J	K	L	M
1	Brookhaven - 9.75 %												
2	High Frequency												
3													
4	Wet sample: Brookhaven, #1, 6-8'	PL method: σc', S/m :			0.002193		Bulk density, g/cm^3 :		1.50				
5	HP 4191A	Coaxial sample holder #154, new teflon spacers (without upper pin); Avg thickness: 0.432 cm											
6	Length,cm	R, cm	r, cm	Weight	Dens.(wet)	Drying, weight in g.		Water cont.					
7	2.568	0.7144	0.3102	M, g	γ, g/cm^3	Wet	Dry	W, %	Wv, %				
8	L, nH	Cs, pF	V, cm^3	5.34	1.59825	20.66	19.4	6.495	9.747				
9	1.566	0.55	3.341147										
10	L-Calc(Air, AVG: 1998-1999 years)												

11	f, MHz	Z, Ohm	φ, deg	tanδ	Res.meas.	Xmeas.	$X_L=\omega L$	Rm-Rms	Xm-wL	G, S	σ', S/m	B, S	C, pF
12					R(s), Ohm	Zsinφ	Ohm	R'=Rm	X'				B/w
13	1.00E+00	2.19E+03	-1.03E+01	5.50E+00	2.15E+03	-3.92E+02	9.84E-03	2.15E+03	-3.92E+02	4.49E-04	2.32E-03	-7.82E-05	1.24E+01
14	1.00E+01	1.16E+03	-5.23E+01	7.73E-01	7.12E+02	-9.21E+02	9.84E-02	7.12E+02	-9.21E+02	5.25E-04	2.72E-03	-6.45E-04	1.03E+01
15	2.00E+01	7.11E+02	-6.63E+01	4.39E-01	2.86E+02	-6.51E+02	1.97E-01	2.86E+02	-6.51E+02	5.65E-04	2.92E-03	-1.22E-03	9.70E+00
16	3.00E+01	5.05E+02	-7.28E+01	3.10E-01	1.49E+02	-4.82E+02	2.95E-01	1.49E+02	-4.83E+02	5.85E-04	3.02E-03	-1.79E-03	9.48E+00
17	4.00E+01	3.89E+02	-7.65E+01	2.40E-01	9.10E+01	-3.79E+02	3.94E-01	9.10E+01	-3.79E+02	5.99E-04	3.10E-03	-2.36E-03	9.38E+00
18	5.00E+01	3.16E+02	-7.88E+01	1.97E-01	6.12E+01	-3.10E+02	4.92E-01	6.12E+01	-3.11E+02	6.10E-04	3.16E-03	-2.93E-03	9.31E+00
19	1.07E+02	1.51E+02	-8.44E+01	9.86E-02	1.48E+01	-1.50E+02	1.05E+00	1.48E+01	-1.51E+02	6.41E-04	3.31E-03	-6.18E-03	9.19E+00
20	1.45E+02	1.11E+02	-8.56E+01	7.78E-02	8.64E+00	-1.11E+02	1.43E+00	8.64E+00	-1.12E+02	6.79E-04	3.51E-03	-8.34E-03	9.15E+00
21	1.83E+02	8.79E+01	-8.62E+01	6.62E-02	5.81E+00	-8.77E+01	1.80E+00	5.81E+00	-8.95E+01	7.23E-04	3.74E-03	-1.05E-02	9.13E+00
22	2.78E+02	5.61E+01	-8.70E+01	5.29E-02	2.96E+00	-5.60E+01	2.74E+00	2.96E+00	-5.87E+01	8.57E-04	4.43E-03	-1.60E-02	9.18E+00
23	2.97E+02	5.27E+01	-8.71E+01	5.14E-02	2.70E+00	-5.26E+01	2.92E+00	2.70E+00	-5.55E+01	8.74E-04	4.52E-03	-1.69E-02	9.08E+00
24	3.73E+02	4.08E+01	-8.72E+01	4.89E-02	1.99E+00	-4.08E+01	3.67E+00	1.99E+00	-4.44E+01	1.01E-03	5.21E-03	-2.12E-02	9.03E+00
25	4.11E+02	3.64E+01	-8.72E+01	4.82E-02	1.75E+00	-3.64E+01	4.04E+00	1.75E+00	-4.04E+01	1.07E-03	5.54E-03	-2.33E-02	9.01E+00
26	4.49E+02	3.27E+01	-8.73E+01	4.80E-02	1.57E+00	-3.27E+01	4.42E+00	1.57E+00	-3.71E+01	1.14E-03	5.88E-03	-2.53E-02	8.98E+00
27	5.82E+02	2.33E+01	-8.72E+01	4.98E-02	1.16E+00	-2.32E+01	5.73E+00	1.16E+00	-2.89E+01	1.38E-03	7.12E-03	-3.25E-02	8.88E+00
28	6.20E+02	2.12E+01	-8.71E+01	5.12E-02	1.08E+00	-2.12E+01	6.10E+00	1.08E+00	-2.73E+01	1.46E-03	7.52E-03	-3.45E-02	8.85E+00
29	7.72E+02	1.45E+01	-8.65E+01	6.08E-02	8.83E-01	-1.45E+01	7.60E+00	8.83E-01	-2.21E+01	1.80E-03	9.32E-03	-4.25E-02	8.76E+00
30	8.48E+02	1.19E+01	-8.61E+01	6.87E-02	8.14E-01	-1.18E+01	8.34E+00	8.14E-01	-2.02E+01	1.99E-03	1.03E-02	-4.65E-02	8.73E+00
31	9.62E+02	8.32E+00	-8.50E+01	8.82E-02	7.31E-01	-8.28E+00	9.47E+00	7.31E-01	-1.78E+01	2.32E-03	1.20E-02	-5.29E-02	8.75E+00
32	1.00E+03	7.22E+00	-8.44E+01	9.79E-02	7.03E-01	-7.18E+00	9.84E+00	7.03E-01	-1.70E+01	2.42E-03	1.25E-02	-5.52E-02	8.79E+00

	N	O	P	Q	R	S	T	U	V	W	X	Y
1	Brookhaven - 9.75 %											
2	High Frequency											
3												
4												
5												
6												
7												
8												
9												
10												
11	f, MHz	ε'	tanδcorr	φcorr	ε''	Zcorr	Rcorr	ρ', Ohm-m	ρ''	losses,e''c|e''diel	$\varepsilon''-\varepsilon''c$	σ'', S/m
12			G/B		ε'tanδ	Ohm	Ohm	(Rcorr)		$\sigma_c/\omega\varepsilon_0$		
13	1.00E+00	7.27E+00	5.75E+00	-9.87E+00	4.18E+01	2.19E+03	2.16E+03	4.18E+02	7.27E+02	3.94E+01	2.33E+00	4.04E-04
14	1.00E+01	6.00E+00	8.14E-01	-5.08E+01	4.88E+00	1.20E+03	7.59E+02	1.47E+02	1.80E+02	3.94E+00	9.40E-01	3.34E-03
15	2.00E+01	5.66E+00	4.64E-01	-6.51E+01	2.63E+00	7.45E+02	3.13E+02	6.06E+01	1.31E+02	1.97E+00	6.55E-01	6.30E-03
16	3.00E+01	5.54E+00	3.27E-01	-7.19E+01	1.81E+00	5.32E+02	1.65E+02	3.20E+01	9.78E+01	1.31E+00	4.98E-01	9.24E-03
17	4.00E+01	5.47E+00	2.54E-01	-7.57E+01	1.39E+00	4.11E+02	1.01E+02	1.96E+01	7.71E+01	9.86E-01	4.07E-01	1.22E-02
18	5.00E+01	5.44E+00	2.09E-01	-7.82E+01	1.13E+00	3.35E+02	6.83E+01	1.32E+01	6.34E+01	7.88E-01	3.46E-01	1.51E-02
19	1.07E+02	5.36E+00	1.04E-01	-8.41E+01	5.57E-01	1.61E+02	1.66E+01	3.21E+00	3.10E+01	3.68E-01	1.88E-01	3.19E-02
20	1.45E+02	5.35E+00	8.15E-02	-8.53E+01	4.35E-01	1.20E+02	9.70E+00	1.88E+00	2.30E+01	2.72E-01	1.64E-01	4.31E-02
21	1.83E+02	5.33E+00	6.88E-02	-8.61E+01	3.67E-01	9.50E+01	6.52E+00	1.26E+00	1.83E+01	2.15E-01	1.52E-01	5.43E-02
22	2.78E+02	5.36E+00	5.35E-02	-8.69E+01	2.87E-01	6.23E+01	3.33E+00	6.44E-01	1.20E+01	1.42E-01	1.45E-01	8.29E-02
23	2.97E+02	5.30E+00	5.16E-02	-8.70E+01	2.74E-01	5.89E+01	3.04E+00	5.88E-01	1.14E+01	1.33E-01	1.41E-01	8.76E-02
24	3.73E+02	5.28E+00	4.76E-02	-8.73E+01	2.51E-01	4.72E+01	2.24E+00	4.34E-01	9.11E+00	1.06E-01	1.45E-01	1.09E-01
25	4.11E+02	5.26E+00	4.60E-02	-8.74E+01	2.42E-01	4.29E+01	1.97E+00	3.82E-01	8.29E+00	9.59E-02	1.46E-01	1.20E-01
26	4.49E+02	5.24E+00	4.49E-02	-8.74E+01	2.36E-01	3.94E+01	1.77E+00	3.42E-01	7.62E+00	8.78E-02	1.48E-01	1.31E-01
27	5.82E+02	5.19E+00	4.24E-02	-8.76E+01	2.20E-01	3.08E+01	1.30E+00	2.52E-01	5.94E+00	6.77E-02	1.52E-01	1.68E-01
28	6.20E+02	5.17E+00	4.22E-02	-8.76E+01	2.18E-01	2.90E+01	1.22E+00	2.36E-01	5.60E+00	6.36E-02	1.55E-01	1.78E-01
29	7.72E+02	5.12E+00	4.24E-02	-8.76E+01	2.17E-01	2.35E+01	9.97E-01	1.93E-01	4.54E+00	5.11E-02	1.66E-01	2.20E-01
30	8.48E+02	5.10E+00	4.28E-02	-8.75E+01	2.18E-01	2.15E+01	9.19E-01	1.78E-01	4.15E+00	4.65E-02	1.72E-01	2.41E-01
31	9.62E+02	5.11E+00	4.37E-02	-8.75E+01	2.24E-01	1.89E+01	8.25E-01	1.60E-01	3.65E+00	4.10E-02	1.83E-01	2.74E-01
32	1.00E+03	5.13E+00	4.39E-02	-8.75E+01	2.25E-01	1.81E+01	7.93E-01	1.53E-01	3.50E+00	3.94E-02	1.86E-01	2.85E-01

Brookhaven - 9.75 %

Low Frequency

Wet sample: Brookhaven, # I, 6-8'

HP 4194A, Fixture — Coaxial sample holder #154, new teflon spacers (without upper pin); Avg. thickness: 0.432 cm

Argand or PL method: σ_c', S/m = 0.002193

	Length, cm	R, cm	r, cm	Weight	Dens. (wet)	Drying, weight in g.		Water cont.		Bulk dens.
	2.568	0.7144	0.3102	M, g	γ, g/cm^3	Wet	Dry	W, %	Wv, %	g/cm^3
	L, nH	Cs, pF	V, cm^3	5.34	1.59825	20.66	19.4	6.495	9.747	1.50
	14.7	0.43982	3.34114676					R'	X'	

f, MHz	Z, Ohm	ϕ, deg	tanδ	Res.meas.	Xmeas.	$X_L = \omega L$	Rm-Rms-	Xm-wL-	G, S	σ', S/m
				R(s), Ohm	Zsinϕ	Ohm	Rel	Xel		
1.00E-03	2.39E+03	-9.18E-01	6.24E+01	2.39E+03	-3.83E+01	9.24E-05	2.36E+03	-3.94E+00	4.24E-04	2.19E-03
2.00E-03	2.38E+03	-6.60E-01	8.68E+01	2.38E+03	-2.74E+01	1.85E-04	2.36E+03	-3.13E+00	4.24E-04	2.19E-03
3.00E-03	2.38E+03	-5.58E-01	1.03E+02	2.38E+03	-2.32E+01	2.77E-04	2.36E+03	-3.33E+00	4.24E-04	2.19E-03
5.00E-03	2.37E+03	-4.71E-01	1.22E+02	2.37E+03	-1.95E+01	4.62E-04	2.36E+03	-4.13E+00	4.24E-04	2.19E-03
1.00E-02	2.37E+03	-4.34E-01	1.32E+02	2.37E+03	-1.79E+01	9.24E-04	2.36E+03	-7.09E+00	4.24E-04	2.19E-03
2.00E-02	2.36E+03	-4.95E-01	1.16E+02	2.36E+03	-2.04E+01	1.85E-03	2.36E+03	-1.27E+01	4.24E-04	2.19E-03
3.00E-02	2.36E+03	-5.84E-01	9.81E+01	2.36E+03	-2.41E+01	2.77E-03	2.36E+03	-1.78E+01	4.25E-04	2.19E-03
5.00E-02	2.36E+03	-7.95E-01	7.21E+01	2.36E+03	-3.27E+01	4.62E-03	2.35E+03	-2.79E+01	4.25E-04	2.20E-03
1.00E-01	2.35E+03	-1.35E+00	4.23E+01	2.35E+03	-5.56E+01	9.24E-03	2.35E+03	-5.22E+01	4.25E-04	2.20E-03
2.00E-01	2.35E+03	-2.46E+00	2.33E+01	2.34E+03	-1.01E+02	1.85E-02	2.34E+03	-9.82E+01	4.26E-04	2.20E-03
3.00E-01	2.34E+03	-3.54E+00	1.62E+01	2.34E+03	-1.45E+02	2.77E-02	2.33E+03	-1.43E+02	4.27E-04	2.21E-03
5.00E-01	2.32E+03	-5.66E+00	1.01E+01	2.31E+03	-2.29E+02	4.62E-02	2.31E+03	-2.28E+02	4.28E-04	2.21E-03
1.00E+00	2.27E+03	-1.06E+01	5.32E+00	2.24E+03	-4.20E+02	9.24E-02	2.23E+03	-4.19E+02	4.32E-04	2.24E-03
2.00E+00	2.14E+03	-1.94E+01	2.84E+00	2.02E+03	-7.10E+02	1.85E-01	2.02E+03	-7.09E+02	4.41E-04	2.28E-03
3.00E+00	1.98E+03	-2.66E+01	2.00E+00	1.77E+03	-8.88E+02	2.77E-01	1.77E+03	-8.88E+02	4.51E-04	2.33E-03
5.00E+00	1.69E+03	-3.74E+01	1.31E+00	1.34E+03	-1.03E+03	4.62E-01	1.34E+03	-1.03E+03	4.70E-04	2.43E-03
1.00E+01	1.19E+03	-5.30E+01	7.54E-01	7.16E+02	-9.50E+02	9.24E-01	7.16E+02	-9.50E+02	5.06E-04	2.61E-03
2.00E+01	7.21E+02	-6.69E+01	4.27E-01	2.83E+02	-6.63E+02	1.85E+00	2.83E+02	-6.65E+02	5.42E-04	2.80E-03
3.00E+01	5.09E+02	-7.31E+01	3.03E-01	1.48E+02	-4.87E+02	2.77E+00	1.48E+02	-4.90E+02	5.64E-04	2.92E-03
4.00E+01	3.91E+02	-7.65E+01	2.40E-01	9.12E+01	-3.80E+02	3.69E+00	9.10E+01	-3.83E+02	5.86E-04	3.03E-03

f, MHz	B, S	C, pF	ε'	tanδcorr	φcorr	ε"	Zcorr	Rcorr	ρ', Ohm-m	ρ"
		B/w		G/B		ε'tanδ	Ohm	Ohm	(Rcorr)	
1.00E-03	-7.06E-07	1.12E+02	6.56E+01	6.01E+02	-9.53E-02	3.94E+04	2.36E+03	2.36E+03	4.56E+02	7.58E-01
2.00E-03	-5.57E-07	4.44E+01	2.59E+01	7.61E+02	-7.53E-02	1.97E+04	2.36E+03	2.36E+03	4.56E+02	5.99E-01
3.00E-03	-5.90E-07	3.13E+01	1.83E+01	7.18E+02	-7.98E-02	1.31E+04	2.36E+03	2.36E+03	4.56E+02	6.35E-01
5.00E-03	-7.28E-07	2.32E+01	1.35E+01	5.82E+02	-9.84E-02	7.88E+03	2.36E+03	2.36E+03	4.56E+02	7.83E-01
1.00E-02	-1.25E-06	1.99E+01	1.16E+01	3.40E+02	-1.69E-01	3.94E+03	2.36E+03	2.36E+03	4.56E+02	1.34E+00
2.00E-02	-2.24E-06	1.78E+01	1.04E+01	1.90E+02	-3.02E-01	1.97E+03	2.36E+03	2.36E+03	4.56E+02	2.40E+00
3.00E-02	-3.13E-06	1.66E+01	9.69E+00	1.36E+02	-4.22E-01	1.32E+03	2.36E+03	2.36E+03	4.56E+02	3.36E+00
5.00E-02	-4.89E-06	1.56E+01	9.09E+00	8.69E+01	-6.60E-01	7.90E+02	2.35E+03	2.35E+03	4.55E+02	5.24E+00
1.00E-01	-9.17E-06	1.46E+01	8.52E+00	4.64E+01	-1.23E+00	3.95E+02	2.35E+03	2.35E+03	4.55E+02	9.80E+00
2.00E-01	-1.73E-05	1.38E+01	8.05E+00	2.46E+01	-2.33E+00	1.98E+02	2.34E+03	2.34E+03	4.53E+02	1.84E+01
3.00E-01	-2.53E-05	1.34E+01	7.83E+00	1.69E+01	-3.39E+00	1.32E+02	2.34E+03	2.33E+03	4.51E+02	2.67E+01
5.00E-01	-4.08E-05	1.30E+01	7.58E+00	1.05E+01	-5.44E+00	7.96E+01	2.32E+03	2.31E+03	4.47E+02	4.26E+01
1.00E+00	-7.83E-05	1.25E+01	7.28E+00	5.52E+00	-1.03E+01	4.02E+01	2.28E+03	2.24E+03	4.33E+02	7.85E+01
2.00E+00	-1.50E-04	1.19E+01	6.96E+00	2.95E+00	-1.87E+01	2.05E+01	2.15E+03	2.03E+03	3.93E+02	1.33E+02
3.00E+00	-2.18E-04	1.15E+01	6.74E+00	2.07E+00	-2.58E+01	1.40E+01	2.00E+03	1.80E+03	3.48E+02	1.68E+02
5.00E+00	-3.46E-04	1.10E+01	6.43E+00	1.36E+00	-3.63E+01	8.74E+00	1.71E+03	1.38E+03	2.67E+02	1.96E+02
1.00E+01	-6.44E-04	1.02E+01	5.98E+00	7.86E-01	-5.18E+01	4.70E+00	1.22E+03	7.55E+02	1.46E+02	1.86E+02
2.00E+01	-1.22E-03	9.69E+00	5.66E+00	4.45E-01	-6.60E+01	2.52E+00	7.50E+02	3.05E+02	5.90E+01	1.33E+02
3.00E+01	-1.79E-03	9.49E+00	5.54E+00	3.15E-01	-7.25E+01	1.75E+00	5.33E+02	1.60E+02	3.10E+01	9.84E+01
4.00E+01	-2.36E-03	9.38E+00	5.48E+00	2.49E-01	-7.60E+01	1.36E+00	4.11E+02	9.93E+01	1.92E+01	7.72E+01

	W	X	Y	Z	AA	AB	AC	AD
1	Brookhaven - 9.75 %							
2	Low Frequency							
3								
4								
5								
6								
7								
8								
9								
10								
11	f, MHz	$100/\omega^{0.5}$	R_{el}	Xel	C, pF	σ'', S/m	losses,ε''c	ε''diel
12			y=25.66*x	y=-27.225*x	B/w		C(s), pF	$\varepsilon''-\varepsilon''$c
13	1.00E-03	1.26E+00	3.24E+01	-3.43E+01	4.16E+06	3.65E-06	3.94E+04	4.21E+00
14	2.00E-03	8.92E-01	2.29E+01	-2.43E+01	2.90E+06	2.88E-06	1.97E+04	-3.40E+00
15	3.00E-03	7.28E-01	1.87E+01	-1.98E+01	2.29E+06	3.05E-06	1.31E+04	-5.81E+00
16	5.00E-03	5.64E-01	1.45E+01	-1.54E+01	1.63E+06	3.77E-06	7.88E+03	-1.54E+00
17	1.00E-02	3.99E-01	1.02E+01	-1.09E+01	8.87E+05	6.45E-06	3.94E+03	9.26E-01
18	2.00E-02	2.82E-01	7.24E+00	-7.68E+00	3.90E+05	1.16E-05	1.97E+03	1.29E+00
19	3.00E-02	2.30E-01	5.91E+00	-6.27E+00	2.20E+05	1.62E-05	1.31E+03	1.11E+00
20	5.00E-02	1.78E-01	4.58E+00	-4.86E+00	9.73E+04	2.53E-05	7.88E+02	1.16E+00
21	1.00E-01	1.26E-01	3.24E+00	-3.43E+00	2.86E+04	4.74E-05	3.94E+02	1.08E+00
22	2.00E-01	8.92E-02	2.29E+00	-2.43E+00	7.91E+03	8.95E-05	1.97E+02	9.47E-01
23	3.00E-01	7.28E-02	1.87E+00	-1.98E+00	3.67E+03	1.31E-04	1.31E+02	8.73E-01
24	5.00E-01	5.64E-02	1.45E+00	-1.54E+00	1.39E+03	2.11E-04	7.88E+01	7.91E-01
25	1.00E+00	3.99E-02	1.02E+00	-1.09E+00	3.79E+02	4.05E-04	3.94E+01	7.66E-01
26	2.00E+00	2.82E-02	7.24E-01	-7.68E-01	1.12E+02	7.75E-04	1.97E+01	8.02E-01
27	3.00E+00	2.30E-02	5.91E-01	-6.27E-01	5.97E+01	1.13E-03	1.31E+01	8.34E-01
28	5.00E+00	1.78E-02	4.58E-01	-4.86E-01	3.10E+01	1.79E-03	7.88E+00	8.54E-01
29	1.00E+01	1.26E-02	3.24E-01	-3.43E-01	1.68E+01	3.33E-03	3.94E+00	7.58E-01
30	2.00E+01	8.92E-03	2.29E-01	-2.43E-01	1.20E+01	6.30E-03	1.97E+00	5.50E-01
31	3.00E+01	7.28E-03	1.87E-01	-1.98E-01	1.09E+01	9.25E-03	1.31E+00	4.33E-01
32	4.00E+01	6.31E-03	1.62E-01	-1.72E-01	1.05E+01	1.22E-02	9.86E-01	3.77E-01

Brookhaven (10.85%)

Brookhaven - 10.85 %

High Frequency

Brookhaven, II, 2-4". Some water is added. (Data from Coaxian1, sht.3) Air spase: 0.5 mm From PL

Sample holder # 154, 3 cm Spacers are identical: 1.5 cm. (2 spacer, avg:0.324) Water cont. Bulk dens. method

Length,cm	R, cm	r, cm	Weight M, g	Dens.(wet) γ, g/cm^3	Drying, weight in g. Wet	Dry	W, %	Wv, %	g/cm^3	σc', S/m
2.676	0.7144	0.3102	5.86	1.68310	5.83	5.46	6.777	10.682	1.58	0.000858

L, nH 1.736   Cs, pF 0.74922   V, cm^3 3.4816623

Stray capacitance Cs - from 3.28.97 (Air w/same thickness of Tef.spacers)

L - Calc(Air). Avg.for 1996 - 1997 years   Not using Rms for coaxial

f, MHz	Z, Ohm	φ, deg	tanδ	Res.meas. R(s), Ohm	Xmeas. Zsinφ	X_L=ωL Ohm	Rm-Rms R'=Rm	Xm-wL X'	G, S	σ', S/m	B, S	C, pF B/w	ε'
1.00E+00	3.58E+03	-1.97E+01	2.79E+00	3.37E+03	-1.21E+03	1.09E-02	3.37E+03	-1.21E+03	2.63E-04	1.30E-03	-8.95E-05	1.42E+01	7.98E+00
1.00E+01	1.22E+03	-6.67E+01	4.31E-01	4.83E+02	-1.12E+03	1.09E-01	4.83E+02	-1.12E+03	3.24E-04	1.61E-03	-7.04E-04	1.12E+01	6.28E+00
2.00E+01	6.27E+02	-7.84E+01	2.05E-01	1.26E+02	-6.14E+02	2.18E-01	1.26E+02	-6.14E+02	3.20E-04	1.59E-03	-1.47E-03	1.17E+01	6.54E+00
3.00E+01	4.24E+02	-8.21E+01	1.39E-01	5.83E+01	-4.20E+02	3.27E-01	5.83E+01	-4.20E+02	3.24E-04	1.61E-03	-2.19E-03	1.16E+01	6.52E+00
4.00E+01	3.21E+02	-8.39E+01	1.08E-01	3.43E+01	-3.19E+02	4.36E-01	3.43E+01	-3.19E+02	3.33E-04	1.65E-03	-2.91E-03	1.16E+01	6.49E+00
5.00E+01	2.57E+02	-8.50E+01	8.84E-02	2.26E+01	-2.56E+02	5.45E-01	2.26E+01	-2.57E+02	3.41E-04	1.69E-03	-3.63E-03	1.15E+01	6.47E+00
1.00E+02	1.29E+02	-8.71E+01	5.05E-02	6.49E+00	-1.29E+02	1.09E+00	6.49E+00	-1.30E+02	3.85E-04	1.91E-03	-7.22E-03	1.15E+01	6.44E+00
1.50E+02	8.52E+01	-8.77E+01	4.00E-02	3.41E+00	-8.52E+01	1.64E+00	3.41E+00	-8.68E+01	4.51E-04	2.24E-03	-1.08E-02	1.15E+01	6.42E+00
2.00E+02	6.31E+01	-8.80E+01	3.58E-02	2.26E+00	-6.30E+01	2.18E+00	2.26E+00	-6.52E+01	5.30E-04	2.63E-03	-1.44E-02	1.14E+01	6.41E+00
2.50E+02	4.96E+01	-8.80E+01	3.49E-02	1.73E+00	-4.96E+01	2.73E+00	1.73E+00	-5.23E+01	6.32E-04	3.14E-03	-1.79E-02	1.14E+01	6.39E+00
3.00E+02	4.04E+01	-8.80E+01	3.46E-02	1.40E+00	-4.04E+01	3.27E+00	1.40E+00	-4.37E+01	7.32E-04	3.63E-03	-2.15E-02	1.14E+01	6.38E+00
3.50E+02	3.37E+01	-8.80E+01	3.53E-02	1.19E+00	-3.37E+01	3.82E+00	1.19E+00	-3.75E+01	8.44E-04	4.19E-03	-2.50E-02	1.14E+01	6.37E+00
4.00E+02	2.86E+01	-8.79E+01	3.74E-02	1.07E+00	-2.86E+01	4.36E+00	1.07E+00	-3.29E+01	9.84E-04	4.88E-03	-2.85E-02	1.13E+01	6.35E+00
4.50E+02	2.45E+01	-8.77E+01	3.96E-02	9.69E-01	-2.44E+01	4.91E+00	9.69E-01	-2.93E+01	1.12E-03	5.57E-03	-3.19E-02	1.13E+01	6.33E+00
5.00E+02	2.11E+01	-8.75E+01	4.40E-02	9.27E-01	-2.11E+01	5.45E+00	9.27E-01	-2.65E+01	1.32E-03	6.53E-03	-3.53E-02	1.13E+01	6.30E+00
6.00E+02	1.57E+01	-8.72E+01	4.98E-02	7.81E-01	-1.57E+01	6.54E+00	7.81E-01	-2.22E+01	1.58E-03	7.83E-03	-4.21E-02	1.12E+01	6.26E+00
7.00E+02	1.15E+01	-8.65E+01	6.17E-02	7.07E-01	-1.15E+01	7.64E+00	7.07E-01	-1.91E+01	1.94E-03	9.61E-03	-4.90E-02	1.11E+01	6.24E+00
8.00E+02	7.94E+00	-8.55E+01	7.87E-02	6.23E-01	-7.91E+00	8.73E+00	6.23E-01	-1.66E+01	2.25E-03	1.11E-02	-5.63E-02	1.11E+01	6.27E+00
9.00E+02	4.78E+00	-8.32E+01	1.19E-01	5.66E-01	-4.75E+00	9.82E+00	5.66E-01	-1.46E+01	2.66E-03	1.32E-02	-6.43E-02	1.12E+01	6.37E+00
1.00E+03	1.86E+00	-7.39E+01	2.89E-01	5.16E-01	-1.79E+00	1.09E+01	5.16E-01	-1.27E+01	3.20E-03	1.59E-02	-7.39E-02	1.18E+01	6.59E+00

f, MHz	tanδcorr	φcorr	ε''	Zcorr	Rcorr	ρ', Ohm-m	ρ''	losses,ε''c	ε''diel	σ'', S/m
	G/B		ε'tanδ	Ohm	Ohm	(Rcorr)		$\sigma_{st}/\omega\varepsilon_0$	ε''-ε''c	
1.00E+00	2.94E+00	-1.88E+01	2.35E+01	3.60E+03	3.41E+03	6.87E+02	2.34E+02	1.54E+02	8.03E+00	4.44E-04
1.00E+01	4.59E-01	-6.53E+01	2.89E+00	1.29E+03	5.38E+02	1.09E+02	2.36E+02	1.54E+02	1.34E+00	3.50E-03
2.00E+01	2.18E-01	-7.77E+01	1.43E+00	6.66E+02	1.42E+02	2.86E+01	1.31E+02	7.71E-01	6.58E-01	7.28E-03
3.00E+01	1.48E-01	-8.16E+01	9.64E-01	4.51E+02	6.59E+01	1.33E+01	8.99E+01	5.14E-01	4.50E-01	1.09E-02
4.00E+01	1.15E-01	-8.35E+01	7.43E-01	3.42E+02	3.89E+01	7.84E+00	6.84E+01	3.86E-01	3.57E-01	1.44E-02
5.00E+01	9.39E-02	-8.46E+01	6.08E-01	2.74E+02	2.57E+01	5.17E+00	5.51E+01	3.08E-01	2.99E-01	1.80E-02
1.00E+02	5.33E-02	-8.69E+01	3.43E-01	1.38E+02	7.36E+00	1.48E+00	2.78E+01	1.54E-01	1.89E-01	3.58E-02
1.50E+02	4.18E-02	-8.76E+01	2.68E-01	9.25E+01	3.86E+00	7.79E-01	1.86E+01	1.03E-01	1.66E-01	5.36E-02
2.00E+02	3.69E-02	-8.79E+01	2.36E-01	6.95E+01	2.56E+00	5.16E-01	1.40E+01	7.71E-02	1.59E-01	7.13E-02
2.50E+02	3.53E-02	-8.80E+01	2.26E-01	5.58E+01	1.97E+00	3.96E-01	1.12E+01	6.17E-02	1.64E-01	8.89E-02
3.00E+02	3.41E-02	-8.80E+01	2.18E-01	4.66E+01	1.59E+00	3.20E-01	9.38E+00	5.14E-02	1.66E-01	1.07E-01
3.50E+02	3.38E-02	-8.81E+01	2.15E-01	4.00E+01	1.35E+00	2.72E-01	8.06E+00	4.41E-02	1.71E-01	1.24E-01
4.00E+02	3.46E-02	-8.80E+01	2.19E-01	3.51E+01	1.21E+00	2.44E-01	7.07E+00	3.86E-02	1.81E-01	1.41E-01
4.50E+02	3.52E-02	-8.80E+01	2.23E-01	3.13E+01	1.10E+00	2.22E-01	6.31E+00	3.43E-02	1.88E-01	1.58E-01
5.00E+02	3.73E-02	-8.79E+01	2.35E-01	2.83E+01	1.05E+00	2.13E-01	5.70E+00	3.08E-02	2.04E-01	1.75E-01
6.00E+02	3.75E-02	-8.79E+01	2.35E-01	2.37E+01	8.89E-01	1.79E-01	4.78E+00	2.57E-02	2.09E-01	2.09E-01
7.00E+02	3.95E-02	-8.77E+01	2.47E-01	2.04E+01	8.05E-01	1.62E-01	4.11E+00	2.20E-02	2.25E-01	2.43E-01
8.00E+02	3.99E-02	-8.77E+01	2.50E-01	1.78E+01	7.09E-01	1.43E-01	3.58E+00	1.93E-02	2.31E-01	2.79E-01
9.00E+02	4.14E-02	-8.76E+01	2.64E-01	1.55E+01	6.43E-01	1.30E-01	3.13E+00	1.71E-02	2.47E-01	3.19E-01
1.00E+03	4.32E-02	-8.75E+01	2.85E-01	1.35E+01	5.83E-01	1.18E-01	2.72E+00	1.54E-02	2.70E-01	3.67E-01

Brookhaven - 10.85 %

Low Frequency

Brookhaven, II, 2-4' (Data from Avrarep1, sht.2)

Subtracting the Electrode Polarization.

	Diameter	Thickness	Volume	Weight	Dens.(wet)	Drying, weight in g.		Water cont.		Bulk density	
	D, cm	h, cm	V, cm^3	M, g	γ, g/cm^3	Wet	Dry	W, %	Wv, %	g/cm^3	From PL
	5	0.920317	18.070382	31.52	1.74429	31.51	29.55	6.633	10.850	1.64	method
	L, nH	Cs, pF	Inductance L - AVG from Air & Teflon meas.File "MethodsP"								σc', S/m
	50	2.1839362	Cs - calc. (Regression eqn. For Air meas. File "MethodsP", Sheet 1: Cs-Ct vs.h)								0.000858

f, MHz	Z, Ohm	ϕ, deg	tanδ	Res.meas. R(s), Ohm	Xmeas. Zsinϕ	$X_L=\omega L$ Ohm	Rms # 1 Ohm	Rm-Rms Rel	Xm-wL X'	G, S	σ', S/m
1.00E-03	5.51E+03	-4.01E-01	1.43E+02	5.51E+03	-3.86E+01	3.14E-04	1.51E-02	5.45E+03	-3.86E+01	1.83E-04	8.59E-04
2.00E-03	5.50E+03	-3.85E-01	1.49E+02	5.50E+03	-3.69E+01	6.28E-04	1.51E-02	5.46E+03	-3.69E+01	1.83E-04	8.59E-04
3.00E-03	5.49E+03	-3.96E-01	1.45E+02	5.49E+03	-3.79E+01	9.42E-04	1.51E-02	5.46E+03	-3.79E+01	1.83E-04	8.58E-04
5.00E-03	5.48E+03	-4.35E-01	1.32E+02	5.48E+03	-4.16E+01	1.57E-03	1.51E-02	5.46E+03	-4.16E+01	1.83E-04	8.59E-04
1.00E-02	5.47E+03	-6.60E-01	8.68E+01	5.47E+03	-6.30E+01	3.14E-03	1.51E-02	5.45E+03	-6.30E+01	1.83E-04	8.59E-04
2.00E-02	5.46E+03	-1.05E+00	5.46E+01	5.46E+03	-9.99E+01	6.28E-03	1.51E-02	5.45E+03	-9.99E+01	1.84E-04	8.60E-04
3.00E-02	5.45E+03	-1.42E+00	4.03E+01	5.45E+03	-1.35E+02	9.42E-03	1.51E-02	5.44E+03	-1.35E+02	1.84E-04	8.61E-04
5.00E-02	5.44E+03	-2.15E+00	2.66E+01	5.43E+03	-2.04E+02	1.57E-02	1.51E-02	5.42E+03	-2.04E+02	1.84E-04	8.63E-04
1.00E-01	5.41E+03	-3.92E+00	1.46E+01	5.39E+03	-3.69E+02	3.14E-02	1.51E-02	5.39E+03	-3.69E+02	1.85E-04	8.66E-04
2.00E-01	5.34E+03	-7.29E+00	7.81E+00	5.30E+03	-6.78E+02	6.28E-02	1.51E-02	5.29E+03	-6.78E+02	1.86E-04	8.71E-04
3.00E-01	5.26E+03	-1.05E+01	5.39E+00	5.18E+03	-9.60E+02	9.42E-02	2.12E-02	5.17E+03	-9.60E+02	1.87E-04	8.76E-04
5.00E-01	5.08E+03	-1.65E+01	3.38E+00	4.87E+03	-1.44E+03	1.57E-01	2.42E-02	4.86E+03	-1.44E+03	1.89E-04	8.86E-04
1.00E+00	4.51E+03	-2.88E+01	1.82E+00	3.95E+03	-2.17E+03	3.14E-01	2.89E-02	3.95E+03	-2.17E+03	1.94E-04	9.11E-04
2.00E+00	3.48E+03	-4.48E+01	1.01E+00	2.46E+03	-2.45E+03	6.28E-01	3.45E-02	2.46E+03	-2.45E+03	2.04E-04	9.56E-04
3.00E+00	2.75E+03	-5.44E+01	7.16E-01	1.60E+03	-2.24E+03	9.42E-01	3.83E-02	1.60E+03	-2.24E+03	2.11E-04	9.90E-04
5.00E+00	1.90E+03	-6.51E+01	4.65E-01	8.02E+02	-1.72E+03	1.57E+00	4.37E-02	8.01E+02	-1.72E+03	2.21E-04	1.04E-03
1.00E+01	1.04E+03	-7.57E+01	2.54E-01	2.57E+02	-1.01E+03	3.14E+00	5.22E-02	2.56E+02	-1.01E+03	2.34E-04	1.10E-03
2.00E+01	5.37E+02	-8.23E+01	1.35E-01	7.21E+01	-5.33E+02	6.28E+00	6.23E-02	7.16E+01	-5.39E+02	2.42E-04	1.14E-03
3.00E+01	3.57E+02	-8.46E+01	9.54E-02	3.39E+01	-3.55E+02	9.42E+00	6.91E-02	3.35E+01	-3.65E+02	2.50E-04	1.17E-03
4.00E+01	2.64E+02	-8.56E+01	7.61E-02	2.01E+01	-2.63E+02	1.26E+01	7.44E-02	1.97E+01	-2.76E+02	2.57E-04	1.21E-03

Brookhaven - 10.85 %
Low Frequency

f,MHz	B, S	C, pF	ε'	tanδcorr	φcorr	ε"tanδ	Zcorr	Rcorr	ρ', Ohm-m	ρ"	100/ω^0.5
		B/w		G/B			Ohm	Ohm	(Rcorr)		
1.00E-03	-1.28E-06	2.04E+02	1.08E+02	1.43E+02	-4.01E-01	1.54E+04	5.45E+03	5.45E+03	1.16E+03	8.15E+00	1.26E+00
2.00E-03	-1.21E-06	9.64E+01	5.10E+01	1.51E+02	-3.79E-01	7.72E+03	5.46E+03	5.46E+03	1.16E+03	7.71E+00	8.92E-01
3.00E-03	-1.23E-06	6.53E+01	3.46E+01	1.49E+02	-3.85E-01	5.14E+03	5.46E+03	5.46E+03	1.16E+03	7.83E+00	7.28E-01
5.00E-03	-1.33E-06	4.23E+01	2.24E+01	1.38E+02	-4.16E-01	3.09E+03	5.46E+03	5.46E+03	1.16E+03	8.45E+00	5.64E-01
1.00E-02	-1.98E-06	3.15E+01	1.67E+01	9.25E+01	-6.19E-01	1.54E+03	5.45E+03	5.45E+03	1.16E+03	1.26E+01	3.99E-01
2.00E-02	-3.09E-06	2.46E+01	1.30E+01	5.93E+01	-9.66E-01	7.73E+02	5.45E+03	5.45E+03	1.16E+03	1.96E+01	2.82E-01
3.00E-02	-4.16E-06	2.21E+01	1.17E+01	4.42E+01	-1.30E+00	5.16E+02	5.44E+03	5.44E+03	1.16E+03	2.63E+01	2.30E-01
5.00E-02	-6.24E-06	1.99E+01	1.05E+01	2.95E+01	-1.94E+00	3.10E+02	5.43E+03	5.43E+03	1.16E+03	3.92E+01	1.78E-01
1.00E-01	-1.13E-05	1.80E+01	9.51E+00	1.64E+01	-3.50E+00	1.56E+02	5.40E+03	5.39E+03	1.15E+03	7.03E+01	1.26E-01
2.00E-01	-2.11E-05	1.68E+01	8.87E+00	8.83E+00	-6.46E+00	7.83E+01	5.35E+03	5.31E+03	1.13E+03	1.28E+02	8.92E-02
3.00E-01	-3.06E-05	1.62E+01	8.59E+00	6.11E+00	-9.29E+00	5.25E+01	5.28E+03	5.21E+03	1.11E+03	1.82E+02	7.28E-02
5.00E-01	-4.91E-05	1.56E+01	8.28E+00	3.85E+00	-1.46E+01	3.18E+01	5.12E+03	4.96E+03	1.06E+03	2.75E+02	5.64E-02
1.00E+00	-9.32E-05	1.48E+01	7.85E+00	2.09E+00	-2.56E+01	1.64E+01	4.64E+03	4.18E+03	8.92E+02	4.28E+02	3.99E-02
2.00E+00	-1.76E-04	1.40E+01	7.40E+00	1.16E+00	-4.07E+01	8.59E+00	3.72E+03	2.82E+03	6.01E+02	5.17E+02	2.82E-02
3.00E+00	-2.54E-04	1.35E+01	7.14E+00	8.31E-01	-5.03E+01	5.93E+00	3.02E+03	1.93E+03	4.12E+02	4.96E+02	2.30E-02
5.00E+00	-4.08E-04	1.30E+01	6.88E+00	5.42E-01	-6.15E+01	3.73E+00	2.15E+03	1.03E+03	2.19E+02	4.04E+02	1.78E-02
1.00E+01	-7.89E-04	1.26E+01	6.65E+00	2.97E-01	-7.35E+01	1.97E+00	1.22E+03	3.46E+02	7.37E+01	2.49E+02	1.26E-02
2.00E+01	-1.55E-03	1.23E+01	6.53E+00	1.56E-01	-8.11E+01	1.02E+00	6.38E+02	9.86E+01	2.10E+01	1.34E+02	8.92E-03
3.00E+01	-2.31E-03	1.22E+01	6.48E+00	1.08E-01	-8.38E+01	7.01E-01	4.31E+02	4.64E+01	9.90E+00	9.14E+01	7.28E-03
4.00E+01	-3.06E-03	1.22E+01	6.44E+00	8.42E-02	-8.52E+01	5.42E-01	3.26E+02	2.74E+01	5.84E+00	6.93E+01	6.31E-03

	Y	Z	AA	AB	AC
1	Brookhaven - 10.85 %				
2	Low Frequency				
3					
4					
5					
6					
7					
8					
9					
10					
11	f, MHz	R_{el}	losses,ε"c	ε"diel	σ", S/m
12		y=44.758*x	$\sigma c/we_0$	ε"-ε"c	
13	1.00E-03	5.65E+01	1.54E+04	2.19E+01	6.02E-06
14	2.00E-03	3.99E+01	7.71E+03	4.65E+00	5.68E-06
15	3.00E-03	3.26E+01	5.14E+03	2.09E+00	5.77E-06
16	5.00E-03	2.53E+01	3.08E+03	1.82E+00	6.23E-06
17	1.00E-02	1.79E+01	1.54E+03	2.23E+00	9.29E-06
18	2.00E-02	1.26E+01	7.71E+02	2.01E+00	1.45E-05
19	3.00E-02	1.03E+01	5.14E+02	1.91E+00	1.95E-05
20	5.00E-02	7.99E+00	3.08E+02	1.73E+00	2.92E-05
21	1.00E-01	5.65E+00	1.54E+02	1.45E+00	5.29E-05
22	2.00E-01	3.99E+00	7.71E+01	1.19E+00	9.87E-05
23	3.00E-01	3.26E+00	5.14E+01	1.08E+00	1.43E-04
24	5.00E-01	2.53E+00	3.08E+01	1.00E+00	2.30E-04
25	1.00E+00	1.79E+00	1.54E+01	9.56E-01	4.37E-04
26	2.00E+00	1.26E+00	7.71E+00	8.82E-01	8.23E-04
27	3.00E+00	1.03E+00	5.14E+00	7.92E-01	1.19E-03
28	5.00E+00	7.99E-01	3.08E+00	6.47E-01	1.91E-03
29	1.00E+01	5.65E-01	1.54E+00	4.30E-01	3.70E-03
30	2.00E+01	3.99E-01	7.71E-01	2.50E-01	7.26E-03
31	3.00E+01	3.26E-01	5.14E-01	1.87E-01	1.08E-02
32	4.00E+01	2.82E-01	3.86E-01	1.56E-01	1.43E-02

Brookhaven (14.67%)

	A	B	C	D	E	F	G	H	I	J	K	L	M
1		Brookhaven - 14.67 %											
2		High Frequency											
3													
4					Measuring a sample of Brookhaven, l, 6-8' with 10% water (by weight)								
5	HP 4191A				Coaxial sample holder #154, patterned teflon spacers (Data from Coaxstan, sht. 2)						Argand or		
6	Length,cm	R, cm	r, cm	Weight	Dens.(wet)	Drying, weight in g.					PL		
7	2.5865	0.7144	0.3102	M, g	γ, g/cm^3	Wet	Dry	Wv, %			method:		
8	L, nH		Cs, pF	5.44	V, cm^3	18.95	17.23	W, %	9.983		σ_c', S/m		
9	1.612	0.55642	3.365217	1.61654	Stray capacitance Cs - from Air meas., 6.3.98 (with same Tefl. spacers)			14.673			0.0022		
10	L-Calc(Air), meas 6.3.98							(Data from Coaxstan, Sheet2)					
11	f, MHz	Z, Ohm	φ, deg	$\tan\delta$	Res.meas.	Xmeas.	$X_L=\omega L$	Rm-Rms	Xm-wL	G, S	σ', S/m	B, S	C, pF
12					R(s), Ohm	$Z\sin\varphi$	Ohm	R'=Rm	X'				B/w
13	5.00E+00	1.53E+03	-3.28E+01	1.55E+00	1.28E+03	-8.27E+02	5.06E-02	1.28E+03	-8.27E+02	5.50E-04	2.83E-03	-3.37E-04	1.07E+01
14	1.00E+01	1.03E+03	-5.33E+01	7.45E-01	6.18E+02	-8.29E+02	1.01E-01	6.18E+02	-8.29E+02	5.78E-04	2.97E-03	-7.40E-04	1.18E+01
15	2.00E+01	5.93E+02	-6.92E+01	3.80E-01	2.11E+02	-5.54E+02	2.03E-01	2.11E+02	-5.55E+02	5.98E-04	3.07E-03	-1.51E-03	1.20E+01
16	3.00E+01	4.13E+02	-7.53E+01	2.62E-01	1.05E+02	-3.99E+02	3.04E-01	1.05E+02	-3.99E+02	6.14E-04	3.15E-03	-2.24E-03	1.19E+01
17	4.00E+01	3.15E+02	-7.85E+01	2.04E-01	6.29E+01	-3.09E+02	4.05E-01	6.29E+01	-3.09E+02	6.33E-04	3.25E-03	-2.97E-03	1.18E+01
18	5.00E+01	2.55E+02	-8.03E+01	1.71E-01	4.30E+01	-2.51E+02	5.06E-01	4.30E+01	-2.52E+02	6.61E-04	3.39E-03	-3.69E-03	1.17E+01
19	6.00E+01	2.14E+02	-8.17E+01	1.46E-01	3.08E+01	-2.12E+02	6.08E-01	3.08E+01	-2.12E+02	6.69E-04	3.44E-03	-4.40E-03	1.17E+01
20	7.00E+01	1.84E+02	-8.27E+01	1.28E-01	2.34E+01	-1.83E+02	7.09E-01	2.34E+01	-1.83E+02	6.85E-04	3.51E-03	-5.12E-03	1.16E+01
21	8.00E+01	1.62E+02	-8.37E+01	1.10E-01	1.77E+01	-1.61E+02	8.10E-01	1.77E+01	-1.62E+02	6.69E-04	3.44E-03	-5.83E-03	1.16E+01
22	9.00E+01	1.44E+02	-8.39E+01	1.07E-01	1.53E+01	-1.44E+02	9.12E-01	1.53E+01	-1.44E+02	7.26E-04	3.73E-03	-6.53E-03	1.15E+01
23	1.00E+02	1.30E+02	-8.47E+01	9.28E-02	1.20E+01	-1.30E+02	1.01E+00	1.20E+01	-1.31E+02	6.98E-04	3.59E-03	-7.24E-03	1.15E+01
24	2.00E+02	6.49E+01	-8.73E+01	4.79E-02	3.10E+00	-6.48E+01	2.03E+00	3.10E+00	-6.68E+01	6.93E-04	3.56E-03	-1.42E-02	1.13E+01
25	3.00E+02	4.23E+01	-8.74E+01	4.56E-02	1.93E+00	-4.23E+01	3.04E+00	1.93E+00	-4.53E+01	9.37E-04	4.81E-03	-2.10E-02	1.11E+01
26	4.00E+02	3.04E+01	-8.86E+01	2.46E-02	7.48E-01	-3.04E+01	4.05E+00	7.48E-01	-3.44E+01	6.30E-04	3.24E-03	-2.76E-02	1.10E+01
27	5.00E+02	2.29E+01	-8.87E+01	2.34E-02	5.34E-01	-2.28E+01	5.06E+00	5.34E-01	-2.79E+01	6.86E-04	3.52E-03	-3.41E-02	1.08E+01
28	6.00E+02	1.75E+01	-8.84E+01	2.76E-02	4.83E-01	-1.75E+01	6.08E+00	4.83E-01	-2.36E+01	8.68E-04	4.45E-03	-4.03E-02	1.07E+01
29	7.00E+02	1.34E+01	-8.81E+01	3.35E-02	4.49E-01	-1.34E+01	7.09E+00	4.49E-01	-2.05E+01	1.07E-03	5.49E-03	-4.63E-02	1.05E+01
30	8.00E+02	1.00E+01	-8.72E+01	4.93E-02	4.92E-01	-9.98E+00	8.10E+00	4.92E-01	-1.81E+01	1.50E-03	7.71E-03	-5.25E-02	1.04E+01
31	9.00E+02	7.02E+00	-8.42E+01	1.01E-01	7.07E-01	-6.99E+00	9.12E+00	7.07E-01	-1.61E+01	2.72E-03	1.40E-02	-5.88E-02	1.04E+01
32	1.00E+03	4.29E+00	-8.23E+01	1.35E-01	5.74E-01	-4.25E+00	1.01E+01	5.74E-01	-1.44E+01	2.77E-03	1.42E-02	-6.60E-02	1.05E+01

	N	O	P	Q	R	S	T	U	V	W	X	Y
1	Brookhaven - 14.67 %											
2	High Frequency											
3												
4												
5												
6												
7												
8												
9												
10												
11	f, MHz	ε'	tanδcorr	φcorr	ε"	Zcorr	Rcorr	ρ', Ohm-m	ρ"	losses,ε"c	ε"diel	σ", S/m
12			G/B		e'tanδ	Ohm	Ohm	(Rcorr)		$\sigma_c/\omega\varepsilon_0$	ε"-ε"c	
13	5.00E+00	6.22E+00	1.63E+00	-3.15E+01	1.02E+01	1.55E+03	1.32E+03	2.57E+02	1.58E+02	7.91E+02	2.25E+00	1.73E-03
14	1.00E+01	6.83E+00	7.80E-01	-5.20E+01	5.33E+00	1.06E+03	6.55E+02	1.28E+02	1.64E+02	3.95E+02	1.38E+00	3.80E-03
15	2.00E+01	6.95E+00	3.97E-01	-6.83E+01	2.76E+00	6.17E+02	2.28E+02	4.44E+01	1.12E+02	1.98E+02	7.84E-01	7.73E-03
16	3.00E+01	6.88E+00	2.74E-01	-7.47E+01	1.89E+00	4.31E+02	1.14E+02	2.22E+01	8.09E+01	1.32E+02	5.71E-01	1.15E-02
17	4.00E+01	6.85E+00	2.13E-01	-7.80E+01	1.46E+00	3.30E+02	6.87E+01	1.34E+01	6.28E+01	9.89E+01	4.72E-01	1.52E-02
18	5.00E+01	6.81E+00	1.79E-01	-7.98E+01	1.22E+00	2.67E+02	4.71E+01	9.17E+00	5.12E+01	7.91E+01	4.29E-01	1.89E-02
19	6.00E+01	6.77E+00	1.52E-01	-8.14E+01	1.03E+00	2.24E+02	3.37E+01	6.57E+00	4.32E+01	6.59E+01	3.70E-01	2.26E-02
20	7.00E+01	6.75E+00	1.34E-01	-8.24E+01	9.02E-01	1.94E+02	2.57E+01	5.00E+00	3.74E+01	5.65E+01	3.37E-01	2.63E-02
21	8.00E+01	6.73E+00	1.15E-01	-8.35E+01	7.72E-01	1.70E+02	1.94E+01	3.78E+00	3.30E+01	4.94E+01	2.78E-01	2.99E-02
22	9.00E+01	6.70E+00	1.11E-01	-8.37E+01	7.44E-01	1.52E+02	1.68E+01	3.27E+00	2.95E+01	4.39E+01	3.05E-01	3.35E-02
23	1.00E+02	6.68E+00	9.65E-02	-8.45E+01	6.44E-01	1.38E+02	1.32E+01	2.57E+00	2.67E+01	3.95E+01	2.49E-01	3.72E-02
24	2.00E+02	6.57E+00	4.87E-02	-8.72E+01	3.20E-01	7.02E+01	3.41E+00	6.65E-01	1.37E+01	1.98E+01	1.22E-01	7.31E-02
25	3.00E+02	6.46E+00	4.47E-02	-8.74E+01	2.88E-01	4.76E+01	2.12E+00	4.14E-01	9.26E+00	1.32E+01	1.56E-01	1.08E-01
26	4.00E+02	6.37E+00	2.28E-02	-8.87E+01	1.45E-01	3.62E+01	8.25E-01	1.61E-01	7.05E+00	9.89E+00	4.66E-02	1.42E-01
27	5.00E+02	6.29E+00	2.01E-02	-8.88E+01	1.27E-01	2.93E+01	5.91E-01	1.15E-01	5.72E+00	7.91E+00	4.75E-02	1.75E-01
28	6.00E+02	6.19E+00	2.15E-02	-8.88E+01	1.33E-01	2.48E+01	5.35E-01	1.04E-01	4.83E+00	6.59E+00	6.75E-02	2.07E-01
29	7.00E+02	6.11E+00	2.31E-02	-8.87E+01	1.41E-01	2.16E+01	4.98E-01	9.70E-02	4.20E+00	5.65E+00	8.45E-02	2.38E-01
30	8.00E+02	6.05E+00	2.86E-02	-8.84E+01	1.73E-01	1.91E+01	5.46E-01	1.06E-01	3.71E+00	4.94E+00	1.24E-01	2.69E-01
31	9.00E+02	6.03E+00	4.63E-02	-8.74E+01	2.79E-01	1.70E+01	7.85E-01	1.53E-01	3.30E+00	4.39E+00	2.35E-01	3.02E-01
32	1.00E+03	6.09E+00	4.21E-02	-8.76E+01	2.56E-01	1.51E+01	6.36E-01	1.24E-01	2.95E+00	3.95E+00	2.16E-01	3.39E-01

	A	B	C	D	E	F	G	H	I	J	K	L
1	Brookhaven - 14.67 %											
2	Low Frequency											
3												
4	Measuring a sample of Brookhaven, l. 6-8' with 10% water (by weight)											
5	HP 4194A, Fixture				Coaxial sample holder #154, patterned teflon spacers		Average thickness for 2 spacers: 0.4135 cm					Argand or PL Method:
6	Length,cm	R, cm	r, cm	Weight	Dens.(wet)	Drying, weight in g.		Water cont.			sc', S/m	
7	2.5865	0.7144	0.3102	M, g	g, g/cm³	Wet	Dry	W, %	Wv, %			
8	L, nH	Cs, pF	V, cm³	5.44	1.61654	18.95	17.23	9.983	14.673		0.0022	
9	14.7	4.1757	3.3652166	Cs - from air meas. 2.20.99 (with calibration on the Fixture).								
10	f, MHz	Z, Ohm	φ, deg	$\tan\delta$	Res.meas.	Xmeas.	$X_L=wL$	Rm-Rms-	Xm-wL-	G, S	σ', S/m	B, S
11					R(s), Ohm	$Z\sin\varphi$	Ohm	Rel	Xel			
12	1.00E-03	2.36E+03	-7.08E-01	8.09E+01	2.36E+03	-2.91E+01	9.24E-05	2.33E+03	-4.35E+00	4.29E-04	2.20E-03	-7.76E-07
13	2.00E-03	2.35E+03	-5.24E-01	1.09E+02	2.35E+03	-2.15E+01	1.85E-04	2.33E+03	-3.98E+00	4.29E-04	2.20E-03	-6.79E-07
14	3.00E-03	2.35E+03	-4.43E-01	1.29E+02	2.35E+03	-1.81E+01	2.77E-04	2.33E+03	-3.84E+00	4.29E-04	2.20E-03	-6.29E-07
15	5.00E-03	2.34E+03	-3.82E-01	1.50E+02	2.34E+03	-1.56E+01	4.62E-04	2.33E+03	-4.54E+00	4.29E-04	2.20E-03	-7.05E-07
16	1.00E-02	2.34E+03	-4.17E-01	1.37E+02	2.34E+03	-1.70E+01	9.24E-04	2.33E+03	-9.19E+00	4.29E-04	2.20E-03	-1.43E-06
17	2.00E-02	2.34E+03	-5.55E-01	1.03E+02	2.33E+03	-2.26E+01	1.85E-03	2.33E+03	-1.71E+01	4.29E-04	2.20E-03	-2.62E-06
18	3.00E-02	2.33E+03	-7.11E-01	8.06E+01	2.33E+03	-2.90E+01	2.77E-03	2.33E+03	-2.44E+01	4.29E-04	2.20E-03	-3.72E-06
19	5.00E-02	2.33E+03	-1.04E+00	5.49E+01	2.33E+03	-4.25E+01	4.62E-03	2.33E+03	-3.90E+01	4.30E-04	2.21E-03	-5.88E-06
20	1.00E-01	2.33E+03	-1.89E+00	3.04E+01	2.32E+03	-7.65E+01	9.24E-03	2.32E+03	-7.40E+01	4.30E-04	2.21E-03	-1.11E-05
21	2.00E-01	2.32E+03	-3.54E+00	1.62E+01	2.31E+03	-1.43E+02	1.85E-02	2.31E+03	-1.41E+02	4.31E-04	2.21E-03	-2.11E-05
22	3.00E-01	2.31E+03	-5.15E+00	1.11E+01	2.30E+03	-2.07E+02	2.77E-02	2.30E+03	-2.06E+02	4.32E-04	2.22E-03	-3.09E-05
23	5.00E-01	2.28E+03	-8.29E+00	6.86E+00	2.26E+03	-3.29E+02	4.62E-02	2.26E+03	-3.28E+02	4.34E-04	2.23E-03	-4.99E-05
24	1.00E+00	2.19E+03	-1.55E+01	3.59E+00	2.11E+03	-5.88E+02	9.24E-02	2.11E+03	-5.87E+02	4.40E-04	2.26E-03	-9.59E-05
25	2.00E+00	1.97E+03	-2.74E+01	1.93E+00	1.74E+03	-9.05E+02	1.85E-01	1.74E+03	-9.04E+02	4.52E-04	2.32E-03	-1.82E-04
26	3.00E+00	1.74E+03	-3.63E+01	1.36E+00	1.40E+03	-1.03E+03	2.77E-01	1.40E+03	-1.03E+03	4.64E-04	2.38E-03	-2.62E-04
27	5.00E+00	1.37E+03	-4.83E+01	8.91E-01	9.11E+02	-1.02E+03	4.62E-01	9.11E+02	-1.02E+03	4.86E-04	2.49E-03	-4.14E-04
28	1.00E+01	8.62E+02	-6.32E+01	5.05E-01	3.88E+02	-7.70E+02	9.24E-01	3.88E+02	-7.70E+02	5.22E-04	2.68E-03	-7.73E-04
29	2.00E+01	4.79E+02	-7.45E+01	2.77E-01	1.28E+02	-4.62E+02	1.85E+00	1.28E+02	-4.64E+02	5.53E-04	2.84E-03	-1.48E-03
30	3.00E+01	3.27E+02	-7.90E+01	1.95E-01	6.25E+01	-3.21E+02	2.77E+00	6.23E+01	-3.24E+02	5.74E-04	2.95E-03	-2.19E-03
31	4.00E+01	2.46E+02	-8.13E+01	1.53E-01	3.73E+01	-2.43E+02	3.69E+00	3.71E+01	-2.47E+02	5.96E-04	3.06E-03	-2.91E-03

	M	N	O	P	Q	R	S	T	U	V	W	X
1	Brookhaven - 14.67 %											
2	Low Frequency											
3												
4												
5												
6												
7												
8												
9												
10	f, MHz	C, pF	ε'	tanδcorr	φcorr	ε''	Zcorr	Rcorr	ρ', Ohm-m	ρ''	100/w^0.5	R_{el}
11		B/w		G/B		ε'tanδ	Ohm	Ohm	(Rcorr)			y=20.557*x
12	1.00E-03	1.23E+02	7.16E+01	5.53E+02	-1.04E-01	3.96E+04	2.33E+03	2.33E+03	4.54E+02	8.20E-01	1.26E+00	2.59E+01
13	2.00E-03	5.41E+01	3.13E+01	6.32E+02	-9.07E-02	1.98E+04	2.33E+03	2.33E+03	4.54E+02	7.19E-01	8.92E-01	1.83E+01
14	3.00E-03	3.34E+01	1.93E+01	6.82E+02	-8.40E-02	1.32E+04	2.33E+03	2.33E+03	4.54E+02	6.66E-01	7.28E-01	1.50E+01
15	5.00E-03	2.24E+01	1.30E+01	6.09E+02	-9.42E-02	7.92E+03	2.33E+03	2.33E+03	4.54E+02	7.46E-01	5.64E-01	1.16E+01
16	1.00E-02	2.28E+01	1.32E+01	3.00E+02	-1.91E-01	3.96E+03	2.33E+03	2.33E+03	4.54E+02	1.51E+00	3.99E-01	8.20E+00
17	2.00E-02	2.09E+01	1.21E+01	1.64E+02	-3.50E-01	1.98E+03	2.33E+03	2.33E+03	4.54E+02	2.77E+00	2.82E-01	5.80E+00
18	3.00E-02	1.97E+01	1.14E+01	1.15E+02	-4.96E-01	1.32E+03	2.33E+03	2.33E+03	4.54E+02	3.93E+00	2.30E-01	4.73E+00
19	5.00E-02	1.87E+01	1.09E+01	7.30E+01	-7.85E-01	7.93E+02	2.33E+03	2.33E+03	4.53E+02	6.21E+00	1.78E-01	3.67E+00
20	1.00E-01	1.77E+01	1.02E+01	3.88E+01	-1.48E+00	3.97E+02	2.32E+03	2.32E+03	4.53E+02	1.17E+01	1.26E-01	2.59E+00
21	2.00E-01	1.68E+01	9.74E+00	2.04E+01	-2.80E+00	1.99E+02	2.32E+03	2.31E+03	4.51E+02	2.21E+01	8.92E-02	1.83E+00
22	3.00E-01	1.64E+01	9.49E+00	1.40E+01	-4.08E+00	1.33E+02	2.31E+03	2.30E+03	4.49E+02	3.20E+01	7.28E-02	1.50E+00
23	5.00E-01	1.59E+01	9.21E+00	8.69E+00	-6.56E+00	8.01E+01	2.29E+03	2.27E+03	4.43E+02	5.10E+01	5.64E-02	1.16E+00
24	1.00E+00	1.53E+01	8.85E+00	4.58E+00	-1.23E+01	4.06E+01	2.22E+03	2.17E+03	4.23E+02	9.23E+01	3.99E-02	8.20E-01
25	2.00E+00	1.45E+01	8.39E+00	2.48E+00	-2.19E+01	2.08E+01	2.05E+03	1.90E+03	3.71E+02	1.49E+02	2.82E-02	5.80E-01
26	3.00E+00	1.39E+01	8.07E+00	1.77E+00	-2.95E+01	1.43E+01	1.87E+03	1.63E+03	3.18E+02	1.80E+02	2.30E-02	4.73E-01
27	5.00E+00	1.32E+01	7.65E+00	1.17E+00	-4.04E+01	8.97E+00	1.57E+03	1.19E+03	2.32E+02	1.98E+02	1.78E-02	3.67E-01
28	1.00E+01	1.23E+01	7.13E+00	6.75E-01	-5.60E+01	4.81E+00	1.07E+03	6.00E+02	1.17E+02	1.73E+02	1.26E-02	2.59E-01
29	2.00E+01	1.18E+01	6.83E+00	3.74E-01	-6.95E+01	2.55E+00	6.33E+02	2.22E+02	4.32E+01	1.16E+02	8.92E-03	1.83E-01
30	3.00E+01	1.16E+01	6.74E+00	2.62E-01	-7.53E+01	1.76E+00	4.41E+02	1.12E+02	2.18E+01	8.32E+01	7.28E-03	1.50E-01
31	4.00E+01	1.16E+01	6.72E+00	2.05E-01	-7.84E+01	1.38E+00	3.36E+02	6.75E+01	1.31E+01	6.42E+01	6.31E-03	1.30E-01

	Y	Z	AA	AB	AC
1	Brookhaven - 14.67 %				
2	Low Frequency				
3					
4					
5					
6					
7					
8					
9					
10	f, MHz	X_{el}	losses,$\varepsilon"c$	$\varepsilon"$diel	$\sigma"$, S/m
11		$y = -19.626*x$	$\sigma_c/\omega\varepsilon_0$	$\varepsilon"-\varepsilon"c$	
12	1.00E-03	-2.48E+01	3.95E+04	5.76E+01	3.98E-06
13	2.00E-03	-1.75E+01	1.98E+04	2.31E+01	3.49E-06
14	3.00E-03	-1.43E+01	1.32E+04	1.32E+01	3.23E-06
15	5.00E-03	-1.11E+01	7.91E+03	9.95E+00	3.62E-06
16	1.00E-02	-7.83E+00	3.95E+03	5.97E+00	7.34E-06
17	2.00E-02	-5.54E+00	1.98E+03	3.46E+00	1.35E-05
18	3.00E-02	-4.52E+00	1.32E+03	2.23E+00	1.91E-05
19	5.00E-02	-3.50E+00	7.91E+02	1.93E+00	3.02E-05
20	1.00E-01	-2.48E+00	3.95E+02	1.49E+00	5.69E-05
21	2.00E-01	-1.75E+00	1.98E+02	1.19E+00	1.08E-04
22	3.00E-01	-1.43E+00	1.32E+02	1.06E+00	1.58E-04
23	5.00E-01	-1.11E+00	7.91E+01	9.83E-01	2.56E-04
24	1.00E+00	-7.83E-01	3.95E+01	1.01E+00	4.92E-04
25	2.00E+00	-5.54E-01	1.98E+01	1.08E+00	9.34E-04
26	3.00E+00	-4.52E-01	1.32E+01	1.10E+00	1.35E-03
27	5.00E+00	-3.50E-01	7.91E+00	1.06E+00	2.13E-03
28	1.00E+01	-2.48E-01	3.95E+00	8.59E-01	3.97E-03
29	2.00E+01	-1.75E-01	1.98E+00	5.74E-01	7.59E-03
30	3.00E+01	-1.43E-01	1.32E+00	4.47E-01	1.13E-02
31	4.00E+01	-1.24E-01	9.89E-01	3.87E-01	1.49E-02

Columbia (9.18%)

	A	B	C	D	E	F	G	H	I	J	K	L	M
1	Columbia - 9.18 %												
2	High Frequency												
3													
4	Sands from Columbia University. Seaved, 0.6 mm.					Distilled water is added.							
5	HP 4191A	Measured								Updated 4.27.09	From		
6	26-07-1998	Combination of a new teflon spacer (0.22025) and old one (0.20675). Sample: 2.573 cm											
7	Length,cm	R, cm	Without upper pin.	Weight	Dens.(wet)	Drying, weight in g.		Water cont.		Bulk dens.	PL-method		
8	2.573	0.7144	0.3102	M, g	γ, g/cm^3	Wet	Dry	W, %	Wv, %	g/cm^3	σc, S/m		
9	L, nH	Cs, pF	V, cm^3	5.26	1.57125	24.83	23.38	6.202	9.176	1.47949	0.001683		
10	1.56	0.5514	3.3476521	L & Cs- Avg. from air meas., 8.2.98; 9.9.98 (with same combination of Teflon spacers)									

f, MHz	Z, Ohm	φ, deg	tanδ	Res.meas. R(s), Ohm	Xmeas. Zsinφ	$X_L=\omega L$ Ohm	Rm-Rms R'=Rm	Xm-ωL X'	G, S	σ', S/m	B, S	C, pF B/w
1.00E+00	2.67E+03	-1.27E+01	4.44E+00	2.60E+03	-5.87E+02	9.80E-03	2.60E+03	-5.87E+02	3.65E-04	1.89E-03	-7.89E-05	1.26E+01
1.00E+01	1.32E+03	-5.19E+00	7.84E-01	8.16E+02	-1.04E+03	9.80E-02	8.16E+02	-1.04E+03	4.67E-04	2.41E-03	-5.61E-04	8.92E+00
2.00E+01	7.70E+02	-6.75E+01	4.14E-01	2.95E+02	-7.11E+02	1.96E-01	2.95E+02	-7.12E+02	4.97E-04	2.56E-03	-1.13E-03	8.99E+00
3.00E+01	5.45E+02	-7.37E+01	2.92E-01	1.53E+02	-5.23E+02	2.94E-01	1.53E+02	-5.23E+02	5.14E-04	2.65E-03	-1.66E-03	8.79E+00
4.00E+01	4.21E+02	-7.70E+01	2.31E-01	9.48E+01	-4.11E+02	3.92E-01	9.48E+01	-4.11E+02	5.33E-04	2.75E-03	-2.17E-03	8.64E+00
5.00E+01	3.42E+02	-7.92E+01	1.91E-01	6.41E+01	-3.36E+02	4.90E-01	6.41E+01	-3.37E+02	5.45E-04	2.81E-03	-2.69E-03	8.57E+00
6.00E+01	2.88E+02	-8.08E+01	1.63E-01	4.62E+01	-2.84E+02	5.88E-01	4.62E+01	-2.85E+02	5.55E-04	2.86E-03	-3.22E-03	8.53E+00
7.00E+01	2.48E+02	-8.19E+01	1.42E-01	3.48E+01	-2.46E+02	6.86E-01	3.48E+01	-2.46E+02	5.62E-04	2.90E-03	-3.74E-03	8.49E+00
8.00E+01	2.18E+02	-8.28E+01	1.27E-01	2.73E+01	-2.16E+02	7.84E-01	2.73E+01	-2.17E+02	5.72E-04	2.95E-03	-4.26E-03	8.47E+00
9.00E+01	1.94E+02	-8.35E+01	1.14E-01	2.20E+01	-1.93E+02	8.82E-01	2.20E+01	-1.94E+02	5.79E-04	2.99E-03	-4.78E-03	8.46E+00
1.00E+02	1.75E+02	-8.40E+01	1.05E-01	1.82E+01	-1.74E+02	9.80E-01	1.82E+01	-1.75E+02	5.88E-04	3.03E-03	-5.31E-03	8.44E+00
1.20E+02	1.46E+02	-8.48E+01	9.05E-02	1.32E+01	-1.46E+02	1.18E+00	1.32E+01	-1.47E+02	6.07E-04	3.13E-03	-6.35E-03	8.42E+00
1.50E+02	1.17E+02	-8.56E+01	7.62E-02	8.88E+00	-1.17E+02	1.47E+00	8.88E+00	-1.18E+02	6.35E-04	3.27E-03	-7.91E-03	8.39E+00
2.00E+02	8.71E+01	-8.64E+01	6.26E-02	5.44E+00	-8.69E+01	1.96E+00	5.44E+00	-8.89E+01	6.86E-04	3.54E-03	-1.05E-02	8.37E+00
2.50E+02	6.91E+01	-8.68E+01	5.59E-02	3.86E+00	-6.90E+01	2.45E+00	3.86E+00	-7.14E+01	7.54E-04	3.89E-03	-1.31E-02	8.33E+00
3.00E+02	5.68E+01	-8.71E+01	5.08E-02	2.88E+00	-5.67E+01	2.94E+00	2.88E+00	-5.97E+01	8.08E-04	4.17E-03	-1.57E-02	8.32E+00
3.50E+02	4.80E+01	-8.72E+01	4.87E-02	2.33E+00	-4.79E+01	3.43E+00	2.33E+00	-5.13E+01	8.84E-04	4.56E-03	-1.82E-02	8.29E+00
4.00E+02	4.12E+01	-8.73E+01	4.77E-02	1.96E+00	-4.12E+01	3.92E+00	1.96E+00	-4.51E+01	9.63E-04	4.97E-03	-2.07E-02	8.25E+00
4.50E+02	3.59E+01	-8.73E+01	4.68E-02	1.68E+00	-3.59E+01	4.41E+00	1.68E+00	-4.03E+01	1.03E-03	5.33E-03	-2.32E-02	8.21E+00
5.00E+02	3.16E+01	-8.73E+01	4.77E-02	1.50E+00	-3.15E+01	4.90E+00	1.50E+00	-3.64E+01	1.13E-03	5.84E-03	-2.57E-02	8.17E+00
6.00E+02	2.48E+01	-8.73E+01	4.77E-02	1.18E+00	-2.48E+01	5.88E+00	1.18E+00	-3.07E+01	1.25E-03	6.48E-03	-3.05E-02	8.09E+00
7.00E+02	1.97E+01	-8.70E+01	5.17E-02	1.02E+00	-1.97E+01	6.86E+00	1.02E+00	-2.66E+01	1.44E-03	7.44E-03	-3.52E-02	7.99E+00
8.00E+02	1.57E+01	-8.68E+01	5.64E-02	8.82E-01	-1.56E+01	7.84E+00	8.82E-01	-2.35E+01	1.60E-03	8.25E-03	-3.98E-02	7.91E+00
9.00E+02	1.22E+01	-8.63E+01	6.52E-02	7.94E-01	-1.22E+01	8.82E+00	7.94E-01	-2.10E+01	1.80E-03	9.28E-03	-4.44E-02	7.86E+00
9.50E+02	1.06E+01	-8.60E+01	7.08E-02	7.51E-01	-1.06E+01	9.31E+00	7.51E-01	-1.99E+01	1.89E-03	9.75E-03	-4.68E-02	7.84E+00
1.00E+03	9.15E+00	-8.56E+01	7.78E-02	7.10E-01	-9.13E+00	9.80E+00	7.10E-01	-1.89E+01	1.98E-03	1.02E-02	-4.93E-02	7.85E+00

	N	O	P	Q	R	S	T	U	V	W	X	Y
11	f, MHz	ε'	tandcorr	ϕcorr	ε''	Zcorr	Rcorr	ρ', Ohm-m	ρ''	losses,e"c	losses	σ'', S/m
12			G/B		e'tand	Ohm	Ohm	(Rcorr)	ρ''	sc/we0	e"-e"c	
13	1.00E+00	7.32E+00	4.63E+00	-1.22E+01	3.39E+01	2.68E+03	2.62E+03	5.07E+02	1.09E+02	3.02E+01	3.64E+00	4.07E-04
14	1.00E+01	5.20E+00	8.32E-01	-5.02E+01	4.33E+00	1.37E+03	8.77E+02	1.70E+02	2.04E+02	3.02E+00	1.30E+00	2.89E-03
15	2.00E+01	5.24E+00	4.39E-01	-6.63E+01	2.30E+00	8.10E+02	3.26E+02	6.32E+01	1.44E+02	1.51E+00	7.92E-01	5.83E-03
16	3.00E+01	5.12E+00	3.11E-01	-7.27E+01	1.59E+00	5.77E+02	1.71E+02	3.31E+01	1.07E+02	1.01E+00	5.82E-01	8.55E-03
17	4.00E+01	5.04E+00	2.45E-01	-7.62E+01	1.24E+00	4.47E+02	1.07E+02	2.07E+01	8.42E+01	7.56E-01	4.79E-01	1.12E-02
18	5.00E+01	5.00E+00	2.03E-01	-7.85E+01	1.01E+00	3.64E+02	7.23E+01	1.40E+01	6.91E+01	6.05E-01	4.07E-01	1.39E-02
19	6.00E+01	4.97E+00	1.73E-01	-8.02E+01	8.58E-01	3.07E+02	5.21E+01	1.01E+01	5.85E+01	5.04E-01	3.54E-01	1.66E-02
20	7.00E+01	4.95E+00	1.50E-01	-8.14E+01	7.44E-01	2.65E+02	3.94E+01	7.63E+00	5.07E+01	4.32E-01	3.12E-01	1.93E-02
21	8.00E+01	4.94E+00	1.34E-01	-8.24E+01	6.63E-01	2.33E+02	3.10E+01	6.00E+00	4.47E+01	3.78E-01	2.85E-01	2.20E-02
22	9.00E+01	4.93E+00	1.21E-01	-8.31E+01	5.97E-01	2.08E+02	2.49E+01	4.83E+00	3.99E+01	3.36E-01	2.61E-01	2.47E-02
23	1.00E+02	4.92E+00	1.11E-01	-8.37E+01	5.45E-01	1.87E+02	2.06E+01	4.00E+00	3.61E+01	3.02E-01	2.43E-01	2.74E-02
24	1.20E+02	4.91E+00	9.56E-02	-8.45E+01	4.69E-01	1.57E+02	1.49E+01	2.89E+00	3.03E+01	2.52E-01	2.17E-01	3.28E-02
25	1.50E+02	4.89E+00	8.02E-02	-8.54E+01	3.92E-01	1.26E+02	1.01E+01	1.95E+00	2.43E+01	2.02E-01	1.91E-01	4.08E-02
26	2.00E+02	4.88E+00	6.52E-02	-8.63E+01	3.18E-01	9.49E+01	6.18E+00	1.20E+00	1.84E+01	1.51E-01	1.67E-01	5.43E-02
27	2.50E+02	4.86E+00	5.76E-02	-8.67E+01	2.80E-01	7.63E+01	4.38E+00	8.49E-01	1.48E+01	1.21E-01	1.59E-01	6.76E-02
28	3.00E+02	4.85E+00	5.15E-02	-8.71E+01	2.50E-01	6.37E+01	3.28E+00	6.35E-01	1.23E+01	1.01E-01	1.49E-01	8.09E-02
29	3.50E+02	4.83E+00	4.85E-02	-8.72E+01	2.34E-01	5.48E+01	2.66E+00	5.15E-01	1.06E+01	8.64E-02	1.48E-01	9.40E-02
30	4.00E+02	4.81E+00	4.64E-02	-8.73E+01	2.23E-01	4.82E+01	2.23E+00	4.33E-01	9.32E+00	7.56E-02	1.48E-01	1.07E-01
31	4.50E+02	4.79E+00	4.45E-02	-8.75E+01	2.13E-01	4.30E+01	1.91E+00	3.70E-01	8.33E+00	6.72E-02	1.46E-01	1.20E-01
32	5.00E+02	4.76E+00	4.40E-02	-8.75E+01	2.10E-01	3.89E+01	1.71E+00	3.32E-01	7.53E+00	6.05E-02	1.49E-01	1.33E-01
33	6.00E+02	4.71E+00	4.12E-02	-8.76E+01	1.94E-01	3.28E+01	1.35E+00	2.61E-01	6.35E+00	5.04E-02	1.44E-01	1.57E-01
34	7.00E+02	4.66E+00	4.10E-02	-8.77E+01	1.91E-01	2.84E+01	1.16E+00	2.26E-01	5.50E+00	4.32E-02	1.48E-01	1.81E-01
35	8.00E+02	4.61E+00	4.02E-02	-8.77E+01	1.85E-01	2.51E+01	1.01E+00	1.95E-01	4.86E+00	3.78E-02	1.48E-01	2.05E-01
36	9.00E+02	4.58E+00	4.05E-02	-8.77E+01	1.85E-01	2.25E+01	9.09E-01	1.76E-01	4.35E+00	3.36E-02	1.52E-01	2.29E-01
37	9.50E+02	4.57E+00	4.04E-02	-8.77E+01	1.85E-01	2.13E+01	8.61E-01	1.67E-01	4.13E+00	3.18E-02	1.53E-01	2.42E-01
38	1.00E+03	4.57E+00	4.02E-02	-8.77E+01	1.84E-01	2.03E+01	8.13E-01	1.58E-01	3.93E+00	3.02E-02	1.53E-01	2.54E-01

#	A	B	C	D	E	F	G	H	I	J	K	L
1	Columbia - 9.18 %											
2	Low Frequency											
3												
4	Sands from Columbia University. Seaved, 0.6 mm.				Distilled water is added.							
5	HP 4194A	Z-Probe	Combination of a new Teflon spacer (0.22025) and old one (0.20675). Sample: 2.573 cm									
6	26-07-1998	(measured)		Without upper pin.					From			
7	Length,cm	R, cm	r, cm	Weight	Dens.(wet)	Drying, weight in g.		Water cont.		Bulk dens.	PL method	
8	2.573	0.7144	0.3102	M, g	γ, g/cm^3	Wet	Dry	W, %	Wv, %	g/cm^3	σ_c, S/m	
9	Subtracting the Electrode Polar.			5.26	1.57125	24.83	23.38	6.202	9.176	1.47949	0.001679	
10	L, nH	Cs, pF	V, cm^3		Cs=Air meas. with combination of Tef.spacers (8.2.98)							
11	4.74	0.45600	3.3476521		L - from coaxial shorted (with brass disk).							
12	f, MHz	Z, Ohm	ϕ, deg	tanδ	Res.meas.	Xmeas.	Xc=ωL	Rm-Rms-	Xm-wL-Xel	G, S	σ', S/m	B, S
13					R(s), Ohm	Zsinϕ	Ohm	Rel	X'			
14	1.00E-02	3.17E+03	-1.28E+00	4.48E+01	3.17E+03	-7.07E+03	2.98E-04	3.06E+03	-2.52E+00	3.27E-04	1.69E-03	-2.41E-07
15	2.00E-02	3.16E+03	-1.71E+00	3.36E+01	3.16E+03	-9.42E+01	5.96E-04	3.09E+03	-4.60E+01	3.24E-04	1.67E-03	-4.77E-06
16	3.00E-02	3.13E+03	-1.71E+00	3.36E+01	3.13E+03	-9.32E+01	8.93E-04	3.07E+03	-5.38E+01	3.26E-04	1.68E-03	-5.63E-06
17	5.00E-02	3.11E+03	-1.41E+00	4.07E+01	3.11E+03	-7.65E+01	1.49E-03	3.06E+03	-4.60E+01	3.26E-04	1.68E-03	-4.76E-06
18	1.00E-01	3.11E+03	-1.86E+00	3.09E+01	3.11E+03	-1.01E+02	2.98E-03	3.07E+03	-7.90E+01	3.25E-04	1.68E-03	-8.08E-06
19	2.00E-01	3.09E+03	-3.26E+00	1.75E+01	3.09E+03	-1.76E+02	5.96E-03	3.06E+03	-1.61E+02	3.25E-04	1.68E-03	-1.65E-05
20	3.00E-01	3.08E+03	-4.69E+00	1.22E+01	3.07E+03	-2.52E+02	8.93E-03	3.05E+03	-2.39E+02	3.26E-04	1.68E-03	-2.47E-05
21	5.00E-01	3.05E+03	-7.47E+00	7.63E+00	3.02E+03	-3.96E+02	1.49E-02	3.01E+03	-3.86E+02	3.27E-04	1.69E-03	-4.06E-05
22	1.00E+00	2.94E+03	-1.38E+01	4.07E+00	2.85E+03	-7.01E+02	2.98E-02	2.84E+03	-6.94E+02	3.32E-04	1.71E-03	-7.83E-05
23	2.00E+00	2.66E+03	-2.41E+01	2.24E+00	2.43E+03	-1.09E+03	5.96E-02	2.43E+03	-1.08E+03	3.44E-04	1.77E-03	-1.48E-04
24	3.00E+00	2.40E+03	-3.18E+01	1.61E+00	2.04E+03	-1.26E+03	8.93E-02	2.03E+03	-1.26E+03	3.56E-04	1.84E-03	-2.12E-04
25	5.00E+00	1.96E+03	-4.24E+01	1.10E+00	1.45E+03	-1.32E+03	1.49E-01	1.44E+03	-1.32E+03	3.78E-04	1.95E-03	-3.31E-04
26	1.00E+01	1.32E+03	-5.66E+01	6.60E-01	7.28E+02	-1.10E+03	2.98E-01	7.25E+02	-1.10E+03	4.17E-04	2.15E-03	-6.05E-04
27	2.00E+01	7.88E+02	-6.89E+01	3.86E-01	2.84E+02	-7.35E+02	5.96E-01	2.82E+02	-7.34E+02	4.56E-04	2.35E-03	-1.13E-03
28	3.00E+01	5.56E+02	-7.46E+01	2.75E-01	1.47E+02	-5.36E+02	8.93E-01	1.45E+02	-5.36E+02	4.72E-04	2.43E-03	-1.65E-03
29	4.00E+01	4.27E+02	-7.79E+01	2.14E-01	8.94E+01	-4.18E+02	1.19E+00	8.77E+01	-4.18E+02	4.80E-04	2.48E-03	-2.18E-03
30	5.00E+01	3.46E+02	-8.01E+01	1.75E-01	5.97E+01	-3.41E+02	1.49E+00	5.82E+01	-3.42E+02	4.85E-04	2.50E-03	-2.70E-03
31	6.00E+01	2.91E+02	-8.16E+01	1.48E-01	4.26E+01	-2.87E+02	1.79E+00	4.12E+01	-2.88E+02	4.86E-04	2.51E-03	-3.23E-03
32	7.00E+01	2.50E+02	-8.27E+01	1.29E-01	3.19E+01	-2.48E+02	2.08E+00	3.07E+01	-2.49E+02	4.86E-04	2.51E-03	-3.75E-03
33	8.00E+01	2.19E+02	-8.35E+01	1.14E-01	2.48E+01	-2.18E+02	2.38E+00	2.36E+01	-2.19E+02	4.85E-04	2.50E-03	-4.28E-03
34	9.00E+01	1.95E+02	-8.42E+01	1.02E-01	1.98E+01	-1.94E+02	2.68E+00	1.87E+01	-1.96E+02	4.84E-04	2.50E-03	-4.80E-03
35	1.00E+02	1.75E+02	-8.47E+01	9.28E-02	1.62E+01	-1.75E+02	2.98E+00	1.51E+01	-1.77E+02	4.80E-04	2.48E-03	-5.33E-03

Columbia - 9.18 %
Low Frequency

f, MHz	C, pF	ε'	tanδcorr	φcorr	ε"	Zcorr	Rcorr	ρ', Ohm-m	ρ"	losses,e"c	losses
	B/w		G/B		$\varepsilon'\tan\delta$	Ohm	Ohm	(Rcorr)		$\sigma_c/\omega\varepsilon_0$	$\varepsilon"-\varepsilon"c$
1.00E-02	3.83E+00	2.23E+00	1.36E+03	-4.22E-02	3.03E+03	3.06E+03	3.06E+03	5.93E+02	4.37E-01	3.02E+03	1.38E+01
2.00E-02	3.80E+01	2.21E+01	6.79E+01	-8.44E-01	1.50E+03	3.09E+03	3.09E+03	5.98E+02	8.81E+00	1.51E+03	-6.57E+00
3.00E-02	2.99E+01	1.74E+01	5.79E+01	-9.90E-01	1.01E+03	3.07E+03	3.07E+03	5.95E+02	1.03E+01	1.01E+03	1.43E+00
5.00E-02	1.52E+01	8.83E+00	6.85E+01	-8.36E-01	6.06E+02	3.06E+03	3.06E+03	5.94E+02	8.66E+00	6.04E+02	1.89E+00
1.00E-01	1.29E+01	7.50E+00	4.02E+01	-1.42E+00	3.02E+02	3.07E+03	3.07E+03	5.95E+02	1.48E+01	3.02E+02	1.58E-02
2.00E-01	1.31E+01	7.65E+00	1.97E+01	-2.90E+00	1.51E+02	3.07E+03	3.07E+03	5.94E+02	3.01E+01	1.51E+02	-3.21E-03
3.00E-01	1.31E+01	7.63E+00	1.32E+01	-4.33E+00	1.01E+02	3.06E+03	3.05E+03	5.91E+02	4.48E+01	1.01E+02	1.53E-01
5.00E-01	1.29E+01	7.54E+00	8.06E+00	-7.07E+00	6.07E+01	3.03E+03	3.01E+03	5.83E+02	7.24E+01	6.04E+01	3.55E-01
1.00E+00	1.25E+01	7.26E+00	4.24E+00	-1.33E+01	3.08E+01	2.93E+03	2.85E+03	5.53E+02	1.30E+02	3.02E+01	6.34E-01
2.00E+00	1.18E+01	6.85E+00	2.33E+00	-2.32E+01	1.59E+01	2.67E+03	2.46E+03	4.76E+02	2.04E+02	1.51E+01	8.56E-01
3.00E+00	1.12E+01	6.55E+00	1.68E+00	-3.08E+01	1.10E+01	2.41E+03	2.07E+03	4.02E+02	2.39E+02	1.01E+01	9.45E-01
5.00E+00	1.05E+01	6.14E+00	1.14E+00	-4.12E+01	7.01E+00	1.99E+03	1.50E+03	2.90E+02	2.54E+02	6.04E+00	9.78E-01
1.00E+01	9.63E+00	5.61E+00	6.90E-01	-5.54E+01	3.87E+00	1.36E+03	7.72E+02	1.50E+02	2.17E+02	3.02E+00	8.52E-01
2.00E+01	8.99E+00	5.24E+00	4.03E-01	-6.80E+01	2.11E+00	8.21E+02	3.07E+02	5.95E+01	1.48E+02	1.51E+00	6.04E-01
3.00E+01	8.77E+00	5.11E+00	2.85E-01	-7.41E+01	1.46E+00	5.82E+02	1.60E+02	3.09E+01	1.08E+02	1.01E+00	4.52E-01
4.00E+01	8.66E+00	5.05E+00	2.21E-01	-7.76E+01	1.11E+00	4.49E+02	9.67E+01	1.87E+01	8.49E+01	7.55E-01	3.60E-01
5.00E+01	8.60E+00	5.01E+00	1.79E-01	-7.98E+01	8.99E-01	3.64E+02	6.43E+01	1.25E+01	6.95E+01	6.04E-01	2.95E-01
6.00E+01	8.56E+00	4.99E+00	1.51E-01	-8.14E+01	7.51E-01	3.06E+02	4.56E+01	8.84E+00	5.87E+01	5.03E-01	2.48E-01
7.00E+01	8.53E+00	4.97E+00	1.30E-01	-8.26E+01	6.44E-01	2.64E+02	3.40E+01	6.58E+00	5.08E+01	4.31E-01	2.13E-01
8.00E+01	8.51E+00	4.96E+00	1.13E-01	-8.35E+01	5.62E-01	2.32E+02	2.62E+01	5.07E+00	4.47E+01	3.77E-01	1.85E-01
9.00E+01	8.49E+00	4.95E+00	1.01E-01	-8.43E+01	4.98E-01	2.07E+02	2.08E+01	4.02E+00	3.99E+01	3.35E-01	1.63E-01
1.00E+02	8.48E+00	4.94E+00	9.01E-02	-8.48E+01	4.45E-01	1.87E+02	1.68E+01	3.25E+00	3.61E+01	3.02E-01	1.44E-01

#	Y	Z	AA	AB	AC	AD	AE
1	Columbia - 9.18 %						
2	Low Frequency						
3							
4							
5							
6							
7							
8							
9							
10							
11							
12	f, MHz	$100/\omega^{0.5}$	R_{el}	X_{el}	C(s), pF	σ''	Rms
13			y=269.16*x	y=-170.9*x		$\sigma/tand$	Ohm
14	1.00E-02	3.99E-01	1.07E+02	-6.82E+01	2.25E+05	1.24E-06	1.00E-03
15	2.00E-02	2.82E-01	7.59E+01	-4.82E+01	8.44E+04	2.46E-05	1.00E-03
16	3.00E-02	2.30E-01	6.20E+01	-3.94E+01	5.69E+04	2.90E-05	1.00E-03
17	5.00E-02	1.78E-01	4.80E+01	-3.05E+01	4.16E+04	2.46E-05	1.00E-03
18	1.00E-01	1.26E-01	3.40E+01	-2.16E+01	1.58E+04	4.17E-05	9.99E-04
19	2.00E-01	8.92E-02	2.40E+01	-1.52E+01	4.52E+03	8.51E-05	9.98E-04
20	3.00E-01	7.28E-02	1.96E+01	-1.24E+01	2.11E+03	1.27E-04	9.97E-04
21	5.00E-01	5.64E-02	1.52E+01	-9.64E+00	8.04E+02	2.10E-04	9.96E-04
22	1.00E+00	3.99E-02	1.07E+01	-6.82E+00	2.27E+02	4.04E-04	9.94E-04
23	2.00E+00	2.82E-02	7.59E+00	-4.82E+00	7.32E+01	7.62E-04	9.96E-04
24	3.00E+00	2.30E-02	6.20E+00	-3.94E+00	4.21E+01	1.09E-03	1.01E-03
25	5.00E+00	1.78E-02	4.80E+00	-3.05E+00	2.41E+01	1.71E-03	1.05E-03
26	1.00E+01	1.26E-02	3.40E+00	-2.16E+00	1.44E+01	3.12E-03	1.30E-03
27	2.00E+01	8.92E-03	2.40E+00	-1.52E+00	1.08E+01	5.83E-03	2.40E-03
28	3.00E+01	7.28E-03	1.96E+00	-1.24E+00	9.90E+00	8.53E-03	4.30E-03
29	4.00E+01	6.31E-03	1.70E+00	-1.08E+00	9.52E+00	1.12E-02	7.00E-03
30	5.00E+01	5.64E-03	1.52E+00	-9.64E-01	9.33E+00	1.39E-02	1.05E-02
31	6.00E+01	5.15E-03	1.39E+00	-8.80E-01	9.23E+00	1.66E-02	1.48E-02
32	7.00E+01	4.77E-03	1.28E+00	-8.15E-01	9.17E+00	1.94E-02	1.99E-02
33	8.00E+01	4.46E-03	1.20E+00	-7.62E-01	9.14E+00	2.21E-02	2.58E-02
34	9.00E+01	4.21E-03	1.13E+00	-7.19E-01	9.12E+00	2.48E-02	3.25E-02
35	1.00E+02	3.99E-03	1.07E+00	-6.82E-01	9.12E+00	2.75E-02	4.00E-02

Columbia (14.2%)

#	A	B	C	D	E	F	G	H	I	J	K	L	M
	A	B	C	D	E	F	G	H	I	J	K	L	M
1	Columbia - 14.2 %												
2	High Frequency												
3													
4	Sands from Columbia University. Seaved, 0.6 mm.					Distilled water is added.							
5	HP 4191A	Combination of a new teflon spacer (0.22025) and old one (0.20675). Sample: 2.573 cm											
6	26-07-1998 (measured)			Without upper pin.			Updated 2.8.08; 5.2.09			From			
7	Length,cm	R, cm		Weight	Dens.(wet)	Drying, weight in g.		Water cont.		Bulk dens.	LFR-meth.		
8	2.573	0.7144	R, cm	M, g	γ, g/cm^3	Wet	Dry	W, %	Wv, %	g/cm^3	sc, S/m		
9	L, nH	Cs, pF	V, cm^3	0.3102	5.24	1.56528	27.96	25.42	9.992	14.220	1.42308	0.002323	
10	1.564	0.55200	3.3476521	L-AVG from air meas., 1998year.	Cs from air meas. with same Teflon spacers (AVG).								
11	f. MHz	Z, Ohm	φ, deg	tanδ	Res.meas.	Xmeas.	$X_L = \omega L$	Rm-Rms	Xm-wL	G, S	σ', S/m	B, S	C, pF
12					R(s), Ohm	Zsinφ	Ohm	R'=Rm	X'				B/w
13	1.00E+00	2.07E+03	-1.31E+01	4.30E+00	2.02E+03	-4.69E+02	9.83E-03	2.02E+03	-4.69E+02	4.71E-04	2.43E-03	-1.06E-04	1.69E+01
14	1.00E+01	1.02E+03	-5.27E+01	7.62E-01	6.16E+02	-8.08E+02	9.83E-02	6.16E+02	-8.08E+02	5.96E-04	3.08E-03	-7.48E-04	1.19E+01
15	2.00E+01	5.93E+02	-6.84E+01	3.96E-01	2.18E+02	-5.51E+02	1.97E-01	2.18E+02	-5.52E+02	6.20E-04	3.20E-03	-1.50E-03	1.19E+01
16	3.00E+01	4.16E+02	-7.46E+01	2.75E-01	1.11E+02	-4.01E+02	2.95E-01	1.11E+02	-4.02E+02	6.37E-04	3.29E-03	-2.21E-03	1.17E+01
17	4.00E+01	3.20E+02	-7.79E+01	2.14E-01	6.70E+01	-3.13E+02	3.93E-01	6.70E+01	-3.13E+02	6.52E-04	3.37E-03	-2.91E-03	1.16E+01
18	5.00E+01	2.59E+02	-8.01E+01	1.75E-01	4.48E+01	-2.55E+02	4.91E-01	4.48E+01	-2.56E+02	6.65E-04	3.43E-03	-3.62E-03	1.15E+01
19	6.00E+01	2.17E+02	-8.15E+01	1.49E-01	3.20E+01	-2.15E+02	5.90E-01	3.20E+01	-2.16E+02	6.74E-04	3.48E-03	-4.33E-03	1.15E+01
20	7.00E+01	1.87E+02	-8.26E+01	1.30E-01	2.41E+01	-1.85E+02	6.88E-01	2.41E+01	-1.86E+02	6.84E-04	3.53E-03	-5.04E-03	1.15E+01
21	8.00E+01	1.64E+02	-8.34E+01	1.16E-01	1.88E+01	-1.63E+02	7.86E-01	1.88E+01	-1.64E+02	6.94E-04	3.58E-03	-5.75E-03	1.14E+01
22	9.00E+01	1.46E+02	-8.40E+01	1.05E-01	1.52E+01	-1.45E+02	8.84E-01	1.52E+01	-1.46E+02	7.04E-04	3.63E-03	-6.46E-03	1.14E+01
23	1.00E+02	1.31E+02	-8.45E+01	9.61E-02	1.26E+01	-1.31E+02	9.83E-01	1.26E+01	-1.32E+02	7.17E-04	3.70E-03	-7.17E-03	1.14E+01
24	1.20E+02	1.10E+02	-8.53E+01	8.31E-02	9.07E+00	-1.09E+02	1.18E+00	9.07E+00	-1.10E+02	7.40E-04	3.82E-03	-8.59E-03	1.14E+01
25	1.50E+02	8.74E+01	-8.60E+01	7.06E-02	6.16E+00	-8.72E+01	1.47E+00	6.16E+00	-8.86E+01	7.80E-04	4.02E-03	-1.07E-02	1.14E+01
26	2.00E+02	6.49E+01	-8.66E+01	5.89E-02	3.82E+00	-6.48E+01	1.97E+00	3.82E+00	-6.68E+01	8.53E-04	4.40E-03	-1.42E-02	1.13E+01
27	2.50E+02	5.12E+01	-8.70E+01	5.33E-02	2.72E+00	-5.11E+01	2.46E+00	2.72E+00	-5.36E+01	9.46E-04	4.88E-03	-1.77E-02	1.13E+01
28	3.00E+02	4.19E+01	-8.72E+01	4.91E-02	2.05E+00	-4.18E+01	2.95E+00	2.05E+00	-4.48E+01	1.02E-03	5.28E-03	-2.13E-02	1.13E+01
29	3.50E+02	3.51E+01	-8.73E+01	4.80E-02	1.68E+00	-3.51E+01	3.44E+00	1.68E+00	-3.85E+01	1.13E-03	5.85E-03	-2.47E-02	1.12E+01
30	4.00E+02	2.99E+01	-8.73E+01	4.79E-02	1.43E+00	-2.99E+01	3.93E+00	1.43E+00	-3.38E+01	1.25E-03	6.44E-03	-2.81E-02	1.12E+01
31	4.50E+02	2.58E+01	-8.73E+01	4.77E-02	1.23E+00	-2.58E+01	4.42E+00	1.23E+00	-3.02E+01	1.35E-03	6.95E-03	-3.15E-02	1.11E+01
32	5.00E+02	2.24E+01	-8.72E+01	4.91E-02	1.10E+00	-2.24E+01	4.91E+00	1.10E+00	-2.73E+01	1.47E-03	7.60E-03	-3.49E-02	1.11E+01
33	6.00E+02	1.70E+01	-8.70E+01	5.24E-02	8.92E-01	-1.70E+01	5.90E+00	8.92E-01	-2.29E+01	1.70E-03	8.75E-03	-4.15E-02	1.10E+01
34	7.00E+02	1.29E+01	-8.66E+01	5.98E-02	7.70E-01	-1.29E+01	6.88E+00	7.70E-01	-1.98E+01	1.97E-03	1.02E-02	-4.81E-02	1.09E+01
35	8.00E+02	9.50E+00	-8.60E+01	7.06E-02	6.69E-01	-9.48E+00	7.86E+00	6.69E-01	-1.73E+01	2.22E-03	1.15E-02	-5.48E-02	1.09E+01
36	9.00E+02	6.51E+00	-8.47E+01	9.28E-02	6.02E-01	-6.48E+00	8.84E+00	6.02E-01	-1.53E+01	2.56E-03	1.32E-02	-6.20E-02	1.10E+01
37	9.50E+02	5.13E+00	-8.36E+01	1.12E-01	5.70E-01	-5.10E+00	9.34E+00	5.70E-01	-1.44E+01	2.73E-03	1.41E-02	-6.59E-02	1.10E+01
38	1.00E+03	3.79E+00	-8.19E+01	1.42E-01	5.33E-01	-3.75E+00	9.83E+00	5.33E-01	-1.36E+01	2.89E-03	1.49E-02	-7.01E-02	1.12E+01

	f, MHz	ε'	tanδcorr	φcorr	ε''	Zcorr	Rcorr	ρ', Ohm-m	ρ''	losses,$\varepsilon''c$	losses	σ'', S/m
			G/B		ε'tanδ	Ohm	Ohm	(Rcorr)		$\sigma_c/\omega\varepsilon_0$	$\varepsilon''-\varepsilon''c$	
13	1.00E+00	9.83E+00	4.44E+00	-1.27E+01	4.36E+01	2.07E+03	2.02E+03	3.92E+02	8.83E+01	4.18E+01	1.89E+00	5.47E-04
14	1.00E+01	6.94E+00	7.97E-01	-5.14E+01	5.53E+00	1.05E+03	6.51E+02	1.26E+02	1.58E+02	4.18E+00	1.36E+00	3.86E-03
15	2.00E+01	6.95E+00	4.14E-01	-6.75E+01	2.88E+00	6.17E+02	2.36E+02	4.57E+01	1.10E+02	2.09E+00	7.89E-01	7.73E-03
16	3.00E+01	6.83E+00	2.88E-01	-7.39E+01	1.97E+00	4.35E+02	1.20E+02	2.33E+01	8.10E+01	1.39E+00	5.78E-01	1.14E-02
17	4.00E+01	6.76E+00	2.24E-01	-7.74E+01	1.51E+00	3.35E+02	7.32E+01	1.42E+01	6.33E+01	1.04E+00	4.69E-01	1.50E-02
18	5.00E+01	6.72E+00	1.83E-01	-7.96E+01	1.23E+00	2.72E+02	4.90E+01	9.50E+00	5.18E+01	8.35E-01	3.98E-01	1.87E-02
19	6.00E+01	6.70E+00	1.56E-01	-8.12E+01	1.04E+00	2.28E+02	3.51E+01	6.80E+00	4.37E+01	6.96E-01	3.46E-01	2.24E-02
20	7.00E+01	6.68E+00	1.36E-01	-8.23E+01	9.06E-01	1.97E+02	2.64E+01	5.12E+00	3.77E+01	5.97E-01	3.10E-01	2.60E-02
21	8.00E+01	6.67E+00	1.21E-01	-8.31E+01	8.05E-01	1.73E+02	2.07E+01	4.01E+00	3.32E+01	5.22E-01	2.83E-01	2.97E-02
22	9.00E+01	6.66E+00	1.09E-01	-8.38E+01	7.25E-01	1.54E+02	1.67E+01	3.23E+00	2.96E+01	4.64E-01	2.61E-01	3.33E-02
23	1.00E+02	6.65E+00	1.00E-01	-8.43E+01	6.65E-01	1.39E+02	1.38E+01	2.68E+00	2.68E+01	4.18E-01	2.48E-01	3.70E-02
24	1.20E+02	6.64E+00	8.62E-02	-8.51E+01	5.72E-01	1.16E+02	9.96E+00	1.93E+00	2.24E+01	3.48E-01	2.24E-01	4.43E-02
25	1.50E+02	6.62E+00	7.28E-02	-8.58E+01	4.82E-01	9.32E+01	6.77E+00	1.31E+00	1.80E+01	2.78E-01	2.04E-01	5.52E-02
26	2.00E+02	6.60E+00	5.99E-02	-8.66E+01	3.96E-01	7.01E+01	4.19E+00	8.13E-01	1.36E+01	2.09E-01	1.87E-01	7.35E-02
27	2.50E+02	6.58E+00	5.33E-02	-8.69E+01	3.51E-01	5.63E+01	3.00E+00	5.81E-01	1.09E+01	1.67E-01	1.84E-01	9.16E-02
28	3.00E+02	6.57E+00	4.81E-02	-8.72E+01	3.16E-01	4.70E+01	2.26E+00	4.37E-01	9.10E+00	1.39E-01	1.77E-01	1.10E-01
29	3.50E+02	6.55E+00	4.59E-02	-8.74E+01	3.01E-01	4.04E+01	1.85E+00	3.59E-01	7.83E+00	1.19E-01	1.81E-01	1.28E-01
30	4.00E+02	6.53E+00	4.44E-02	-8.75E+01	2.90E-01	3.55E+01	1.57E+00	3.05E-01	6.87E+00	1.04E-01	1.85E-01	1.45E-01
31	4.50E+02	6.50E+00	4.27E-02	-8.76E+01	2.77E-01	3.17E+01	1.35E+00	2.62E-01	6.14E+00	9.28E-02	1.85E-01	1.63E-01
32	5.00E+02	6.47E+00	4.22E-02	-8.76E+01	2.73E-01	2.86E+01	1.21E+00	2.34E-01	5.55E+00	8.35E-02	1.90E-01	1.80E-01
33	6.00E+02	6.42E+00	4.09E-02	-8.77E+01	2.62E-01	2.41E+01	9.83E-01	1.91E-01	4.66E+00	6.96E-02	1.93E-01	2.14E-01
34	7.00E+02	6.37E+00	4.09E-02	-8.77E+01	2.61E-01	2.08E+01	8.50E-01	1.65E-01	4.02E+00	5.97E-02	2.01E-01	2.48E-01
35	8.00E+02	6.35E+00	4.06E-02	-8.77E+01	2.58E-01	1.82E+01	7.39E-01	1.43E-01	3.53E+00	5.22E-02	2.06E-01	2.83E-01
36	9.00E+02	6.39E+00	4.12E-02	-8.76E+01	2.63E-01	1.61E+01	6.63E-01	1.29E-01	3.12E+00	4.64E-02	2.17E-01	3.20E-01
37	9.50E+02	6.43E+00	4.15E-02	-8.76E+01	2.67E-01	1.52E+01	6.28E-01	1.22E-01	2.94E+00	4.40E-02	2.23E-01	3.40E-01
38	1.00E+03	6.50E+00	4.12E-02	-8.76E+01	2.68E-01	1.43E+01	5.87E-01	1.14E-01	2.76E+00	4.18E-02	2.26E-01	3.62E-01

Row	A	B	C	D	E	F	G	H	I	J	K	L
1	Columbia - 14.2 %											
2	Low Frequency											
3												
4	Sands from Columbia University. Seaved, 0.6 mm.				Distilled water is added.				Data from "Sands", sheet 3, 4			
5	HP 4194A	Z-Probe	Combination of a new teflon spacer			(0.22025) and old one (0.20675). Sample: 2.573 cm					From	
6	26-07-1998	(measured)	Without upper pin.				Updated 2.7.08; 5.2.09				LFR-meth.	
7	Length,cm	R, cm		Weight	Dens.(wet)	Drying, weight in g.		Water cont.		Bulk dens.	σ_c, S/m	
8	2.573	0.7144	0.3102	M, g	γ, g/cm^3	Wet	Dry	W, %	Wv, %	g/cm^3	0.002323	
9	Subtracting the Electrode Polar.			5.24	1.56528	27.96	25.42	9.992	14.220	1.42308		
10	L, nH	Cs, pF	V, cm^3	Cs from Air meas. with combination of Tef.spacers (8.2.98).								
11	4.74	0.45600	3.3476521	L - from coaxial shorted (with brass disk).					Not using Rms for coaxial.			
12	f. MHz	Z, Ohm	ϕ, deg	tanδ	Res.meas.	Xmeas.	$X_L=\omega L$	Rm-Rms-	Xm-wL-	G, S	σ', S/m	B, S
13					R(s), Ohm	Zsinϕ	Ohm	Rel	Xel			
14	1.00E-02	2.30E+03	-1.18E+00	4.88E+01	2.30E+03	-4.72E+01	2.98E-04	2.23E+03	-4.72E+01	4.49E-04	2.32E-03	-9.50E-06
15	2.00E-02	2.30E+03	-1.75E+00	3.27E+01	2.29E+03	-7.03E+01	5.96E-04	2.24E+03	-3.17E+01	4.46E-04	2.30E-03	-6.26E-06
16	3.00E-02	2.27E+03	-1.75E+00	3.28E+01	2.27E+03	-6.92E+01	8.93E-04	2.22E+03	-3.77E+01	4.50E-04	2.32E-03	-7.55E-06
17	5.00E-02	2.25E+03	-1.43E+00	4.01E+01	2.25E+03	-5.61E+01	1.49E-03	2.22E+03	-3.17E+01	4.51E-04	2.33E-03	-6.31E-06
18	1.00E-01	2.25E+03	-1.82E+00	3.14E+01	2.25E+03	-7.15E+01	2.98E-03	2.22E+03	-5.43E+01	4.49E-04	2.32E-03	-1.07E-05
19	2.00E-01	2.24E+03	-3.19E+00	1.80E+01	2.24E+03	-1.25E+02	5.96E-03	2.22E+03	-1.12E+02	4.49E-04	2.32E-03	-2.22E-05
20	3.00E-01	2.23E+03	-4.58E+00	1.25E+01	2.22E+03	-1.78E+02	8.93E-03	2.21E+03	-1.68E+02	4.50E-04	2.32E-03	-3.34E-05
21	5.00E-01	2.21E+03	-7.30E+00	7.81E+00	2.19E+03	-2.80E+02	1.49E-02	2.18E+03	-2.73E+02	4.52E-04	2.33E-03	-5.51E-05
22	1.00E+00	2.13E+03	-1.34E+01	4.19E+00	2.07E+03	-4.94E+02	2.98E-02	2.06E+03	-4.88E+02	4.60E-04	2.37E-03	-1.06E-04
23	2.00E+00	1.93E+03	-2.32E+01	2.33E+00	1.78E+03	-7.61E+02	5.96E-02	1.77E+03	-7.57E+02	4.77E-04	2.46E-03	-1.98E-04
24	3.00E+00	1.75E+03	-3.05E+01	1.70E+00	1.51E+03	-8.87E+02	8.93E-02	1.50E+03	-8.84E+02	4.95E-04	2.55E-03	-2.82E-04
25	5.00E+00	1.45E+03	-4.07E+01	1.16E+00	1.10E+03	-9.45E+02	1.49E-01	1.09E+03	-9.43E+02	5.24E-04	2.71E-03	-4.38E-04
26	1.00E+01	9.99E+02	-5.53E+01	6.92E-01	5.68E+02	-8.21E+02	2.98E-01	5.66E+02	-8.20E+02	5.70E-04	2.94E-03	-7.98E-04
27	2.00E+01	5.99E+02	-6.86E+01	3.92E-01	2.19E+02	-5.57E+02	5.96E-01	2.17E+02	-5.57E+02	6.07E-04	3.13E-03	-1.50E-03
28	3.00E+01	4.21E+02	-7.47E+01	2.74E-01	1.11E+02	-4.06E+02	8.93E-01	1.10E+02	-4.06E+02	6.21E-04	3.21E-03	-2.21E-03
29	4.00E+01	3.23E+02	-7.81E+01	2.11E-01	6.68E+01	-3.16E+02	1.19E+00	6.56E+01	-3.16E+02	6.29E-04	3.24E-03	-2.92E-03
30	5.00E+01	2.61E+02	-8.02E+01	1.72E-01	4.43E+01	-2.57E+02	1.49E+00	4.33E+01	-2.58E+02	6.32E-04	3.26E-03	-3.63E-03
31	6.00E+01	2.19E+02	-8.17E+01	1.45E-01	3.14E+01	-2.17E+02	1.79E+00	3.05E+01	-2.18E+02	6.31E-04	3.25E-03	-4.33E-03
32	7.00E+01	1.88E+02	-8.28E+01	1.26E-01	2.35E+01	-1.87E+02	2.08E+00	2.26E+01	-1.88E+02	6.30E-04	3.25E-03	-5.04E-03
33	8.00E+01	1.65E+02	-8.36E+01	1.11E-01	1.82E+01	-1.64E+02	2.38E+00	1.74E+01	-1.65E+02	6.28E-04	3.24E-03	-5.75E-03
34	9.00E+01	1.46E+02	-8.43E+01	1.00E-01	1.46E+01	-1.46E+02	2.68E+00	1.37E+01	-1.48E+02	6.25E-04	3.22E-03	-6.45E-03
35	1.00E+02	1.32E+02	-8.48E+01	9.08E-02	1.19E+01	-1.31E+02	2.98E+00	1.11E+01	-1.33E+02	6.20E-04	3.20E-03	-7.16E-03

	M	N	O	P	Q	R	S	T	U	V	W	X
1	Columbia - 14.2 %											
2	Low Frequency											
3												
4												
5												
6												
7												
8												
9												
10												
11												
12	f, MHz	C, pF	ε'	tandcorr	φcorr	ε''	Zcorr	Rcorr	ρ', Ohm-m	ρ''	losses,ε''c	losses
13		B/w		G/B		ε'tanδ	Ohm	Ohm	(Rcorr)		$\sigma_c/\omega\varepsilon_0$	$\varepsilon''-\varepsilon''$c
14	1.00E-02	1.51E+02	8.81E+01	4.73E+01	-1.21E+00	4.16E+01	2.23E+03	2.23E+03	4.31E+02	9.13E+02	4.18E+03	-1.10E+01
15	2.00E-02	4.98E+01	2.90E+01	7.13E+01	-8.04E-01	2.07E+01	2.24E+03	2.24E+03	4.34E+02	6.09E+02	2.09E+03	-1.78E+01
16	3.00E-02	4.00E+01	2.33E+01	5.96E+01	-9.61E-01	1.39E+01	2.22E+03	2.22E+03	4.31E+02	7.23E+02	1.39E+03	-1.19E+01
17	5.00E-02	2.01E+01	1.17E+01	7.15E+01	-8.01E-01	8.36E+00	2.22E+03	2.22E+03	4.30E+02	6.01E+02	8.35E+02	1.22E+00
18	1.00E-01	1.70E+01	9.91E+00	4.21E+01	-1.36E+00	4.17E+00	2.22E+03	2.22E+03	4.31E+02	1.02E+02	4.18E+02	-7.04E-01
19	2.00E-01	1.76E+01	1.03E+01	2.03E+01	-2.82E+00	2.08E+00	2.22E+03	2.22E+03	4.30E+02	2.12E+01	2.09E+02	-4.46E-01
20	3.00E-01	1.77E+01	1.03E+01	1.35E+01	-4.24E+00	1.39E+00	2.22E+03	2.21E+03	4.28E+02	3.18E+01	1.39E+02	-1.07E-01
21	5.00E-01	1.76E+01	1.02E+01	8.20E+00	-6.96E+00	8.39E-01	2.20E+03	2.18E+03	4.22E+02	5.15E+01	8.35E+01	3.52E-01
22	1.00E+00	1.69E+01	9.84E+00	4.33E+00	-1.30E+01	4.26E-01	2.12E+03	2.07E+03	4.00E+02	9.24E+01	4.18E+01	8.71E-01
23	2.00E+00	1.58E+01	9.20E+00	2.41E+00	-2.26E+01	2.21E-01	1.93E+03	1.79E+03	3.46E+02	1.44E+02	2.09E+01	1.26E+00
24	3.00E+00	1.50E+01	8.73E+00	1.75E+00	-2.97E+01	1.53E-01	1.76E+03	1.52E+03	2.95E+02	1.69E+02	1.39E+01	1.38E+00
25	5.00E+00	1.39E+01	8.12E+00	1.20E+00	-3.98E+01	9.73E-02	1.46E+03	1.12E+03	2.18E+02	1.82E+02	8.35E+00	1.38E+00
26	1.00E+01	1.27E+01	7.40E+00	7.15E-01	-5.45E+01	5.29E-02	1.02E+03	5.93E+02	1.15E+02	1.61E+02	4.18E+00	1.11E+00
27	2.00E+01	1.20E+01	6.97E+00	4.04E-01	-6.80E+01	2.82E-02	6.17E+02	2.31E+02	4.48E+01	1.11E+02	2.09E+00	7.29E-01
28	3.00E+01	1.17E+01	6.83E+00	2.81E-01	-7.43E+01	1.92E-02	4.36E+02	1.18E+02	2.29E+01	8.13E+01	1.39E+00	5.29E-01
29	4.00E+01	1.16E+01	6.76E+00	2.16E-01	-7.78E+01	1.46E-02	3.35E+02	7.06E+01	1.37E+01	6.35E+01	1.04E+00	4.14E-01
30	5.00E+01	1.15E+01	6.73E+00	1.74E-01	-8.01E+01	1.17E-02	2.72E+02	4.66E+01	9.04E+00	5.19E+01	8.35E-01	3.37E-01
31	6.00E+01	1.15E+01	6.70E+00	1.45E-01	-8.17E+01	9.75E-03	2.28E+02	3.29E+01	6.37E+00	4.38E+01	6.96E-01	2.79E-01
32	7.00E+01	1.15E+01	6.68E+00	1.25E-01	-8.29E+01	8.34E-03	1.97E+02	2.44E+01	4.73E+00	3.78E+01	5.97E-01	2.38E-01
33	8.00E+01	1.14E+01	6.67E+00	1.09E-01	-8.38E+01	7.28E-03	1.73E+02	1.88E+01	3.64E+00	3.33E+01	5.22E-01	2.06E-01
34	9.00E+01	1.14E+01	6.65E+00	9.68E-02	-8.45E+01	6.44E-03	1.54E+02	1.49E+01	2.88E+00	2.97E+01	4.64E-01	1.80E-01
35	1.00E+02	1.14E+01	6.64E+00	8.67E-02	-8.50E+01	5.75E-03	1.39E+02	1.20E+01	2.33E+00	2.69E+01	4.18E-01	1.58E-01

	Y	Z	AA	AB	AC	AD	AE
1	Columbia - 14.2 %						
2	Low Frequency						
3							
4							
5							
6							
7							
8							
9							
10							
11							
12	f, MHz	$100/\omega^{0.5}$	R_{el}	Xel	σ"		Rms
13			y=192.62*x	y=-136.61*x	σ'/tanδ	C(s), pF	Ohm
14	1.00E-02	3.99E-01	7.68E+01	-5.45E+01	4.90E-05	3.37E+05	1.00E-03
15	2.00E-02	2.82E-01	5.43E+01	-3.85E+01	3.23E-05	1.13E+05	1.00E-03
16	3.00E-02	2.30E-01	4.44E+01	-3.15E+01	3.89E-05	7.67E+04	1.00E-03
17	5.00E-02	1.78E-01	3.44E+01	-2.44E+01	3.25E-05	5.67E+04	1.00E-03
18	1.00E-01	1.26E-01	2.43E+01	-1.72E+01	5.51E-05	2.23E+04	9.99E-04
19	2.00E-01	8.92E-02	1.72E+01	-1.22E+01	1.14E-04	6.39E+03	9.98E-04
20	3.00E-01	7.28E-02	1.40E+01	-9.95E+00	1.72E-04	2.98E+03	9.97E-04
21	5.00E-01	5.64E-02	1.09E+01	-7.71E+00	2.85E-04	1.14E+03	9.96E-04
22	1.00E+00	3.99E-02	7.68E+00	-5.45E+00	5.47E-04	3.22E+02	9.94E-04
23	2.00E+00	2.82E-02	5.43E+00	-3.85E+00	1.02E-03	1.05E+02	9.96E-04
24	3.00E+00	2.30E-02	4.44E+00	-3.15E+00	1.46E-03	5.98E+01	1.01E-03
25	5.00E+00	1.78E-02	3.44E+00	-2.44E+00	2.26E-03	3.37E+01	1.05E-03
26	1.00E+01	1.26E-02	2.43E+00	-1.72E+00	4.12E-03	1.94E+01	1.30E-03
27	2.00E+01	8.92E-03	1.72E+00	-1.22E+00	7.75E-03	1.43E+01	2.40E-03
28	3.00E+01	7.28E-03	1.40E+00	-9.95E-01	1.14E-02	1.31E+01	4.30E-03
29	4.00E+01	6.31E-03	1.22E+00	-8.62E-01	1.51E-02	1.26E+01	7.00E-03
30	5.00E+01	5.64E-03	1.09E+00	-7.71E-01	1.87E-02	1.24E+01	1.05E-02
31	6.00E+01	5.15E-03	9.92E-01	-7.04E-01	2.24E-02	1.22E+01	1.48E-02
32	7.00E+01	4.77E-03	9.18E-01	-6.51E-01	2.60E-02	1.22E+01	1.99E-02
33	8.00E+01	4.46E-03	8.59E-01	-6.09E-01	2.97E-02	1.22E+01	2.58E-02
34	9.00E+01	4.21E-03	8.10E-01	-5.74E-01	3.33E-02	1.21E+01	3.25E-02
35	1.00E+02	3.99E-03	7.68E-01	-5.45E-01	3.69E-02	1.21E+01	4.00E-02

Columbia (21.28%)

	A	B	C	D	E	F	G	H	I	J	K	L	M
1	Columbia - 21.28 %												
2	High Frequency												
3													
4	Sands from Columbia University. Seaved, 0.6 mm.				Distilled water is added.				fres = 990 MHz		Lres, H		
5	HP 4191A	Combination of a new teflon spacer (0.22025) and old one (0.20675). Sample: 2.573 cm									1.43E-09		
6	26-07-1998 (measured)			Without upper pin.				Updated 2.14.08; 5.7.09			From		
7	Length,cm	R, cm	r, cm	Weight	Dens.(wet)	Drying. weight in g.		Water cont.		Bulk dens.	Argand		
8	2.573	0.7144	0.3102	M, g	γ, g/cm^3	Wet	Dry	W, %	Wv, %	g/cm^3	sc, S/m		
9	L, nH	Cs, pF	V, cm^3	5.51	1.64593	29.78	25.93	14.848	21.279	1.43314	0.00424		
10	1.432	0.55200	3.347652		L from resonance.	Cs from air meas. with same Teflon spacers (AVG).							
11	f. MHz	Z, Ohm	φ, deg	tanδ	Res.meas.	Xmeas.	X_L=ωL	Rm-Rms	Xm-wL	G, S	σ', S/m	B, S	C, pF
12					R(s), Ohm	Zsinφ	Ohm	R'=Rm	X'				B/w
13	1.00E+00	1.15E+03	-9.30E+00	6.11E+00	1.14E+03	-1.86E+02	9.00E-03	1.14E+03	-1.86E+02	8.57E-04	4.42E-03	-1.37E-04	2.18E+01
14	1.00E+01	6.71E+02	-5.01E+01	8.36E-01	4.30E+02	-5.15E+02	9.00E-02	4.30E+02	-5.15E+02	9.56E-04	4.93E-03	-1.11E-03	1.76E+01
15	2.00E+01	3.99E+02	-6.71E+01	4.22E-01	1.55E+02	-3.67E+02	1.80E-01	1.55E+02	-3.68E+02	9.74E-04	5.03E-03	-2.24E-03	1.78E+01
16	3.00E+01	2.80E+02	-7.39E+01	2.88E-01	7.74E+01	-2.69E+02	2.70E-01	7.74E+01	-2.69E+02	9.89E-04	5.10E-03	-3.33E-03	1.77E+01
17	4.00E+01	2.14E+02	-7.75E+01	2.22E-01	4.64E+01	-2.09E+02	3.60E-01	4.64E+01	-2.10E+02	1.01E-03	5.19E-03	-4.41E-03	1.75E+01
18	5.00E+01	1.73E+02	-7.97E+01	1.81E-01	3.09E+01	-1.70E+02	4.50E-01	3.09E+01	-1.71E+02	1.02E-03	5.28E-03	-5.49E-03	1.75E+01
19	6.00E+01	1.45E+02	-8.13E+01	1.54E-01	2.20E+01	-1.44E+02	5.40E-01	2.20E+01	-1.44E+02	1.04E-03	5.35E-03	-6.57E-03	1.74E+01
20	7.00E+01	1.25E+02	-8.24E+01	1.34E-01	1.65E+01	-1.24E+02	6.30E-01	1.65E+01	-1.24E+02	1.05E-03	5.42E-03	-7.66E-03	1.74E+01
21	8.00E+01	1.09E+02	-8.32E+01	1.19E-01	1.29E+01	-1.09E+02	7.20E-01	1.29E+01	-1.09E+02	1.06E-03	5.49E-03	-8.74E-03	1.74E+01
22	9.00E+01	9.73E+01	-8.39E+01	1.08E-01	1.04E+01	-9.68E+01	8.10E-01	1.04E+01	-9.76E+01	1.08E-03	5.58E-03	-9.82E-03	1.74E+01
23	1.00E+02	8.76E+01	-8.44E+01	9.86E-02	8.59E+00	-8.72E+01	9.00E-01	8.59E+00	-8.81E+01	1.10E-03	5.66E-03	-1.09E-02	1.73E+01
24	1.20E+02	7.29E+01	-8.51E+01	8.54E-02	6.20E+00	-7.26E+01	1.08E+00	6.20E+00	-7.37E+01	1.13E-03	5.85E-03	-1.31E-02	1.73E+01
25	1.50E+02	5.81E+01	-8.59E+01	7.24E-02	4.19E+00	-5.79E+01	1.35E+00	4.19E+00	-5.93E+01	1.19E-03	6.13E-03	-1.63E-02	1.73E+01
26	2.00E+02	4.30E+01	-8.66E+01	6.01E-02	2.58E+00	-4.29E+01	1.80E+00	2.58E+00	-4.47E+01	1.29E-03	6.63E-03	-2.16E-02	1.72E+01
27	2.50E+02	3.38E+01	-8.69E+01	5.40E-02	1.82E+00	-3.37E+01	2.25E+00	1.82E+00	-3.60E+01	1.40E-03	7.24E-03	-2.69E-02	1.71E+01
28	3.00E+02	2.75E+01	-8.72E+01	4.93E-02	1.35E+00	-2.75E+01	2.70E+00	1.35E+00	-3.02E+01	1.48E-03	7.66E-03	-3.21E-02	1.70E+01
29	3.50E+02	2.29E+01	-8.72E+01	4.82E-02	1.10E+00	-2.29E+01	3.15E+00	1.10E+00	-2.60E+01	1.63E-03	8.39E-03	-3.71E-02	1.69E+01
30	4.00E+02	1.94E+01	-8.73E+01	4.80E-02	9.29E-01	-1.93E+01	3.60E+00	9.29E-01	-2.29E+01	1.76E-03	9.10E-03	-4.21E-02	1.68E+01
31	4.50E+02	1.65E+01	-8.72E+01	4.84E-02	7.98E-01	-1.65E+01	4.05E+00	7.98E-01	-2.05E+01	1.89E-03	9.74E-03	-4.70E-02	1.66E+01
32	5.00E+02	1.41E+01	-8.71E+01	5.05E-02	7.13E-01	-1.41E+01	4.50E+00	7.13E-01	-1.86E+01	2.05E-03	1.06E-02	-5.19E-02	1.65E+01
33	6.00E+02	1.03E+01	-8.68E+01	5.61E-02	5.79E-01	-1.03E+01	5.40E+00	5.79E-01	-1.57E+01	2.34E-03	1.21E-02	-6.14E-02	1.63E+01
34	7.00E+02	7.30E+00	-8.61E+01	6.90E-02	5.03E-01	-7.28E+00	6.30E+00	5.03E-01	-1.36E+01	2.72E-03	1.40E-02	-7.11E-02	1.62E+01
35	8.00E+02	4.68E+00	-8.46E+01	9.45E-02	4.40E-01	-4.65E+00	7.20E+00	4.40E-01	-1.19E+01	3.13E-03	1.61E-02	-8.15E-02	1.62E+01
36	9.00E+02	2.22E+00	-7.95E+01	1.85E-01	4.04E-01	-2.18E+00	8.10E+00	4.04E-01	-1.03E+01	3.82E-03	1.97E-02	-9.40E-02	1.66E+01
37	9.50E+02	1.04E+00	-6.80E+01	4.04E-01	3.88E-01	-9.61E-01	8.55E+00	3.88E-01	-9.51E+00	4.29E-03	2.21E-02	-1.02E-01	1.70E+01
38	1.00E+03	4.64E-01	3.65E+01	-1.35E+00	3.73E-01	2.76E-01	9.00E+00	3.73E-01	-8.72E+00	4.89E-03	2.53E-02	-1.11E-01	1.77E+01

	N	O	P	Q	R	S	T	U	V	W	X	Y	Z
1	Columbia - 21.28 %												
2	High Frequency												
3													
4													
5													
6													
7													
8													
9													
10													
11	f, MHz	ε'	tanδcorr	φcorr	ε''	Zcorr	Rcorr	ρ', Ohm-m	ρ''	losses,$\varepsilon''c$	ε''-$\varepsilon''c$ losses	Cmeas	σ'', S/m
12			G/B		ε'tanδ	Ohm	Ohm	(Rcorr)		$\sigma_c/\omega\varepsilon_0$	ε''-$\varepsilon''c$	par.	
13	1.00E+00	1.27E+01	6.26E+00	-9.07E+00	7.95E+01	1.15E+03	1.14E+03	2.20E+02	3.52E+01	7.62E+01	3.31E+00	2.23E+01	7.07E-04
14	1.00E+01	1.03E+01	8.62E-01	-4.92E+01	8.87E+00	6.83E+02	4.46E+02	8.65E+01	1.00E+02	7.62E+02	1.24E+00	1.82E+01	5.72E-03
15	2.00E+01	1.04E+01	4.35E-01	-6.65E+01	4.52E+00	4.09E+02	1.63E+02	3.17E+01	7.28E+01	3.81E+01	7.08E-01	1.84E+01	1.16E-02
16	3.00E+01	1.03E+01	2.97E-01	-7.35E+01	3.06E+00	2.88E+02	8.19E+01	1.59E+01	5.35E+01	2.54E+00	5.18E-01	1.82E+01	1.72E-02
17	4.00E+01	1.02E+01	2.28E-01	-7.71E+01	2.33E+00	2.21E+02	4.92E+01	9.53E+00	4.18E+01	1.91E+00	4.29E-01	1.81E+01	2.28E-02
18	5.00E+01	1.02E+01	1.86E-01	-7.94E+01	1.90E+00	1.79E+02	3.28E+01	6.36E+00	3.41E+01	1.52E+00	3.76E-01	1.81E+01	2.83E-02
19	6.00E+01	1.02E+01	1.58E-01	-8.10E+01	1.60E+00	1.50E+02	2.34E+01	4.54E+00	2.88E+01	1.27E+00	3.34E-01	1.81E+01	3.39E-02
20	7.00E+01	1.01E+01	1.37E-01	-8.22E+01	1.39E+00	1.29E+02	1.76E+01	3.41E+00	2.48E+01	1.09E+00	3.03E-01	1.81E+01	3.95E-02
21	8.00E+01	1.01E+01	1.22E-01	-8.31E+01	1.23E+00	1.14E+02	1.37E+01	2.66E+00	2.19E+01	9.53E-01	2.82E-01	1.81E+01	4.51E-02
22	9.00E+01	1.01E+01	1.10E-01	-8.37E+01	1.11E+00	1.01E+02	1.11E+01	2.15E+00	1.95E+01	8.47E-01	2.67E-01	1.81E+01	5.07E-02
23	1.00E+02	1.01E+01	1.01E-01	-8.43E+01	1.02E+00	9.13E+01	9.15E+00	1.77E+00	1.76E+01	7.62E-01	2.56E-01	1.81E+01	5.62E-02
24	1.20E+02	1.01E+01	8.68E-02	-8.50E+01	8.76E-01	7.63E+01	6.60E+00	1.28E+00	1.47E+01	6.35E-01	2.41E-01	1.81E+01	6.74E-02
25	1.50E+02	1.01E+01	7.30E-02	-8.58E+01	7.34E-01	6.13E+01	4.46E+00	8.65E-01	1.19E+01	5.08E-01	2.26E-01	1.82E+01	8.39E-02
26	2.00E+02	1.00E+01	5.95E-02	-8.66E+01	5.96E-01	4.62E+01	2.75E+00	5.33E-01	8.94E+00	3.81E-01	2.15E-01	1.85E+01	1.11E-01
27	2.50E+02	9.96E+00	5.22E-02	-8.70E+01	5.21E-01	3.72E+01	1.94E+00	3.76E-01	7.20E+00	3.05E-01	2.16E-01	1.88E+01	1.39E-01
28	3.00E+02	9.91E+00	4.63E-02	-8.73E+01	4.59E-01	3.12E+01	1.44E+00	2.79E-01	6.03E+00	2.54E-01	2.05E-01	1.93E+01	1.65E-01
29	3.50E+02	9.84E+00	4.38E-02	-8.75E+01	4.31E-01	2.69E+01	1.18E+00	2.28E-01	5.21E+00	2.18E-01	2.13E-01	1.98E+01	1.92E-01
30	4.00E+02	9.77E+00	4.18E-02	-8.76E+01	4.09E-01	2.37E+01	9.91E-01	1.92E-01	4.59E+00	1.91E-01	2.18E-01	2.05E+01	2.17E-01
31	4.50E+02	9.69E+00	4.01E-02	-8.77E+01	3.89E-01	2.12E+01	8.52E-01	1.65E-01	4.11E+00	1.69E-01	2.20E-01	2.14E+01	2.43E-01
32	5.00E+02	9.63E+00	3.96E-02	-8.77E+01	3.81E-01	1.93E+01	7.61E-01	1.48E-01	3.73E+00	1.52E-01	2.28E-01	2.25E+01	2.68E-01
33	6.00E+02	9.49E+00	3.81E-02	-8.78E+01	3.62E-01	1.63E+01	6.19E-01	1.20E-01	3.15E+00	1.27E-01	2.35E-01	2.56E+01	3.17E-01
34	7.00E+02	9.42E+00	3.83E-02	-8.78E+01	3.61E-01	1.41E+01	5.38E-01	1.04E-01	2.72E+00	1.09E-01	2.52E-01	3.11E+01	3.67E-01
35	8.00E+02	9.45E+00	3.84E-02	-8.78E+01	3.63E-01	1.23E+01	4.70E-01	9.12E-02	2.37E+00	9.53E-02	2.67E-01	4.24E+01	4.20E-01
36	9.00E+02	9.69E+00	4.06E-02	-8.77E+01	3.93E-01	1.06E+01	4.31E-01	8.35E-02	2.06E+00	8.47E-02	3.09E-01	7.85E+01	4.85E-01
37	9.50E+02	9.93E+00	4.22E-02	-8.76E+01	4.19E-01	9.82E+00	4.14E-01	8.02E-02	1.90E+00	8.02E-02	3.39E-01	1.50E+02	5.25E-01
38	1.00E+03	1.03E+01	4.41E-02	-8.75E+01	4.54E-01	9.00E+00	3.97E-01	7.69E-02	1.74E+00	7.62E-02	3.78E-01	-2.04E+02	5.73E-01

	A	B	C	D	E	F	G	H	I	J	K	L
1	Columbia - 21.28 %											
2	Low Frequency											
3												
4	Sands from Columbia University. Seaved, 0.6 mm.					Distilled water is added.						
5	HP 4194A	Z-Probe	Combination of a new teflon spacer (0.22025) and old one (0.20675). Sample: 2.573 cm									
6	26-07-1998	(measured)	Without upper pin.				Updated 2.13.08; 5.2.09			From		
7	Length,cm	R, cm	r, cm	Weight	Dens.(wet)	Drying, weight in g.		Water cont.		Bulk dens.	PL method	
8	2.573	0.7144	0.3102	M, g	γ, g/cm^3	Wet	Dry	W, %	Wv, %	g/cm^3	sc, S/m	
9	Subtracting the Electrode Polar.			5.51	1.64593	29.78	25.93	14.848	21.279	1.43314	0.00427	
10	L, nH		Cs, pF	Vol.cm^3	Cs from Air meas. with combination of Tef.spacers (8.2.98)				Not using Rms for coaxial.			
11	4.74	0.45600		3.3476521	L - from coaxial shorted (with brass disk).							
12	f. MHz	Z, Ohm	ϕ, deg	tanδ	Res.meas.	Xmeas.	$X_L=\omega L$	Rm-Rms-	Xm-wL-	G, S	σ', S/m	B, S
13					R(s), Ohm	Zsinϕ	Ohm	Rel	Xel			
14	1.00E-02	1.25E+03	-1.02E+00	5.61E+01	1.25E+03	-2.22E+01	2.98E-04	1.21E+03	-2.22E+01	8.29E-04	4.28E-03	-1.52E-05
15	2.00E-02	1.24E+03	-1.54E+00	3.71E+01	1.24E+03	-3.35E+01	5.96E-04	1.24E+03	-5.36E+00	8.04E-04	4.15E-03	-3.41E-06
16	3.00E-02	1.23E+03	-1.52E+00	3.78E+01	1.23E+03	-3.25E+01	8.93E-04	1.23E+03	-9.56E+00	8.13E-04	4.20E-03	-6.24E-06
17	5.00E-02	1.22E+03	-1.09E+00	5.28E+01	1.22E+03	-2.32E+01	1.49E-03	1.22E+03	-5.37E+00	8.18E-04	4.22E-03	-3.45E-06
18	1.00E-01	1.22E+03	-1.25E+00	4.60E+01	1.22E+03	-2.65E+01	2.98E-03	1.22E+03	-1.40E+01	8.19E-04	4.22E-03	-9.07E-06
19	2.00E-01	1.22E+03	-2.13E+00	2.69E+01	1.22E+03	-4.54E+01	5.96E-03	1.22E+03	-3.65E+01	8.20E-04	4.23E-03	-2.40E-05
20	3.00E-01	1.22E+03	-3.07E+00	1.86E+01	1.22E+03	-6.52E+01	8.93E-03	1.22E+03	-5.80E+01	8.21E-04	4.24E-03	-3.83E-05
21	5.00E-01	1.21E+03	-4.95E+00	1.15E+01	1.21E+03	-1.05E+02	1.49E-02	1.21E+03	-9.89E+01	8.23E-04	4.25E-03	-6.61E-05
22	1.00E+00	1.19E+03	-9.42E+00	6.03E+00	1.17E+03	-1.95E+02	2.98E-02	1.17E+03	-1.91E+02	8.29E-04	4.28E-03	-1.32E-04
23	2.00E+00	1.13E+03	-1.75E+01	3.18E+00	1.08E+03	-3.40E+02	5.96E-02	1.08E+03	-3.38E+02	8.42E-04	4.34E-03	-2.57E-04
24	3.00E+00	1.07E+03	-2.44E+01	2.21E+00	9.73E+02	-4.40E+02	8.93E-02	9.73E+02	-4.38E+02	8.55E-04	4.41E-03	-3.76E-04
25	5.00E+00	9.32E+02	-3.55E+01	1.40E+00	7.58E+02	-5.41E+02	1.49E-01	7.58E+02	-5.40E+02	8.75E-04	4.52E-03	-6.09E-04
26	1.00E+01	6.69E+02	-5.24E+01	7.71E-01	4.08E+02	-5.30E+02	2.98E-01	4.08E+02	-5.29E+02	9.14E-04	4.72E-03	-1.16E-03
27	2.00E+01	4.02E+02	-6.76E+01	4.12E-01	1.53E+02	-3.72E+02	5.96E-01	1.53E+02	-3.72E+02	9.49E-04	4.90E-03	-2.24E-03
28	3.00E+01	2.82E+02	-7.42E+01	2.84E-01	7.69E+01	-2.71E+02	8.93E-01	7.69E+01	-2.71E+02	9.68E-04	4.99E-03	-3.33E-03
29	4.00E+01	2.15E+02	-7.77E+01	2.18E-01	4.59E+01	-2.10E+02	1.19E+00	4.59E+01	-2.11E+02	9.83E-04	5.07E-03	-4.41E-03
30	5.00E+01	1.74E+02	-7.99E+01	1.78E-01	3.04E+01	-1.71E+02	1.49E+00	3.04E+01	-1.72E+02	9.95E-04	5.13E-03	-5.49E-03
31	6.00E+01	1.46E+02	-8.15E+01	1.50E-01	2.16E+01	-1.44E+02	1.79E+00	2.16E+01	-1.45E+02	1.00E-03	5.17E-03	-6.56E-03
32	7.00E+01	1.25E+02	-8.26E+01	1.31E-01	1.62E+01	-1.24E+02	2.08E+00	1.62E+01	-1.26E+02	1.01E-03	5.21E-03	-7.63E-03
33	8.00E+01	1.09E+02	-8.34E+01	1.16E-01	1.26E+01	-1.09E+02	2.38E+00	1.26E+01	-1.11E+02	1.02E-03	5.24E-03	-8.70E-03
34	9.00E+01	9.71E+01	-8.40E+01	1.04E-01	1.01E+01	-9.66E+01	2.68E+00	1.01E+01	-9.89E+01	1.02E-03	5.27E-03	-9.75E-03
35	1.00E+02	8.73E+01	-8.46E+01	9.53E-02	8.28E+00	-8.69E+01	2.98E+00	8.28E+00	-8.94E+01	1.03E-03	5.29E-03	-1.08E-02

	M	N	O	P	Q	R	S	T	U	V	W	X
1	Columbia - 21.28 %											
2	Low Frequency											
3												
4												
5												
6												
7												
8												
9												
10												
11												
12	f, MHz	C, pF	ε'	tanδcorr	φcorr	ε''	Zcorr	Rcorr	ρ', Ohm-m	ρ"	losses,$\varepsilon''c$	losses
13		B/w		G/B		ε'tanδ	Ohm	Ohm	(Rcorr)		$\sigma_c/\omega\varepsilon_0$	$\varepsilon''-\varepsilon''c$
14	1.00E-02	2.43E+02	1.41E+02	5.44E+01	-1.05E+02	7.69E+03	1.21E+03	1.21E+03	2.34E+02	4.30E+02	7.68E+03	1.28E+01
15	2.00E-02	2.71E+01	1.58E+01	2.36E+02	-2.43E-01	3.73E+03	1.24E+03	1.24E+03	2.41E+02	1.02E+02	3.84E+03	-1.08E+02
16	3.00E-02	3.31E+01	1.93E+01	1.30E+02	-4.39E-01	2.51E+03	1.23E+03	1.23E+03	2.38E+02	1.83E+02	2.56E+03	-4.45E+01
17	5.00E-02	1.10E+01	6.40E+00	2.37E+02	-2.42E-01	1.52E+03	1.22E+03	1.22E+03	2.37E+02	9.99E-01	1.54E+03	-1.74E+01
18	1.00E-01	1.44E+01	8.41E+00	9.03E+01	-6.35E-01	7.59E+02	1.22E+03	1.22E+03	2.37E+02	2.62E+02	7.68E+02	-8.11E+00
19	2.00E-01	1.91E+01	1.11E+01	3.42E+01	-1.67E+00	3.80E+02	1.22E+03	1.22E+03	2.36E+02	6.90E+02	3.84E+02	-3.56E+00
20	3.00E-01	2.03E+01	1.18E+01	2.14E+01	-2.67E+00	2.54E+02	1.22E+03	1.22E+03	2.36E+02	1.10E+01	2.56E+02	-2.02E+00
21	5.00E-01	2.10E+01	1.23E+01	1.25E+01	-4.59E+00	1.53E+02	1.21E+03	1.21E+03	2.34E+02	1.88E+01	1.54E+02	-7.85E-01
22	1.00E+00	2.10E+01	1.22E+01	6.29E+00	-9.04E+00	7.69E+01	1.19E+03	1.18E+03	2.28E+02	3.62E+01	7.68E+01	1.70E-01
23	2.00E+00	2.04E+01	1.19E+01	3.28E+00	-1.70E+01	3.91E+01	1.14E+03	1.09E+03	2.11E+02	6.42E+01	3.84E+01	6.73E-01
24	3.00E+00	2.00E+01	1.16E+01	2.27E+00	-2.38E+01	2.64E+01	1.07E+03	9.80E+02	1.90E+02	8.36E+01	2.56E+01	8.43E-01
25	5.00E+00	1.94E+01	1.13E+01	1.44E+00	-3.48E+01	1.62E+01	9.38E+02	7.70E+02	1.49E+02	1.04E+02	1.54E+01	8.86E-01
26	1.00E+01	1.84E+01	1.07E+01	7.91E-01	-5.16E+01	8.48E+00	6.79E+02	4.21E+02	8.16E+01	1.03E+02	7.68E+00	8.07E-01
27	2.00E+01	1.78E+01	1.04E+01	4.23E-01	-6.71E+01	4.40E+00	4.11E+02	1.60E+02	3.10E+01	7.33E+01	3.84E+00	5.64E-01
28	3.00E+01	1.77E+01	1.03E+01	2.91E-01	-7.38E+01	2.99E+00	2.89E+02	8.06E+01	1.56E+01	5.37E+01	2.56E+00	4.34E-01
29	4.00E+01	1.75E+01	1.02E+01	2.23E-01	-7.74E+01	2.28E+00	2.21E+02	4.82E+01	9.33E+00	4.19E+01	1.92E+00	3.61E-01
30	5.00E+01	1.75E+01	1.02E+01	1.81E-01	-7.97E+01	1.85E+00	1.79E+02	3.20E+01	6.20E+00	3.42E+01	1.54E+00	3.10E-01
31	6.00E+01	1.74E+01	1.01E+01	1.53E-01	-8.13E+01	1.55E+00	1.51E+02	2.27E+01	4.40E+00	2.89E+01	1.28E+00	2.69E-01
32	7.00E+01	1.74E+01	1.01E+01	1.32E-01	-8.25E+01	1.34E+00	1.30E+02	1.70E+01	3.30E+00	2.49E+01	1.10E+00	2.42E-01
33	8.00E+01	1.73E+01	1.01E+01	1.17E-01	-8.33E+01	1.18E+00	1.14E+02	1.33E+01	2.57E+00	2.20E+01	9.59E-01	2.18E-01
34	9.00E+01	1.72E+01	1.01E+01	1.05E-01	-8.40E+01	1.05E+00	1.02E+02	1.06E+01	2.06E+00	1.97E+01	8.53E-01	1.99E-01
35	1.00E+02	1.72E+01	1.00E+01	9.50E-02	-8.46E+01	9.52E-01	9.22E+01	8.72E+00	1.69E+00	1.78E+01	7.68E-01	1.84E-01

	Y	Z	AA	AB	AC	AD	AE
1	Columbia - 21.28 %						
2	Low Frequency						
3							
4							
5							
6							
7							
8							
9							
10							
11							
12	f, MHz	$100/\omega^{0.5}$	R_{el}	Xel	σ''	C(s), pF	Rms
13			$y=103.05*x$	$y=-99.795*x$	$\sigma'/\tan\delta$		Ohm
14	1.00E-02	3.99E-01	4.11E+01	-3.98E+01	7.87E-05	7.16E+05	1.00E-03
15	2.00E-02	2.82E-01	2.91E+01	-2.82E+01	1.76E-05	2.37E+05	1.00E-03
16	3.00E-02	2.30E-01	2.37E+01	-2.30E+01	3.22E-05	1.63E+05	1.00E-03
17	5.00E-02	1.78E-01	1.84E+01	-1.78E+01	1.78E-05	1.37E+05	1.00E-03
18	1.00E-01	1.26E-01	1.30E+01	-1.26E+01	4.68E-05	6.00E+04	9.99E-04
19	2.00E-01	8.92E-02	9.19E+00	-8.90E+00	1.24E-04	1.75E+04	9.98E-04
20	3.00E-01	7.28E-02	7.51E+00	-7.27E+00	1.98E-04	8.13E+03	9.97E-04
21	5.00E-01	5.64E-02	5.81E+00	-5.63E+00	3.41E-04	3.04E+03	9.96E-04
22	1.00E+00	3.99E-02	4.11E+00	-3.98E+00	6.81E-04	8.17E+02	9.94E-04
23	2.00E+00	2.82E-02	2.91E+00	-2.82E+00	1.33E-03	2.34E+02	9.96E-04
24	3.00E+00	2.30E-02	2.37E+00	-2.30E+00	1.94E-03	1.20E+02	1.01E-03
25	5.00E+00	1.78E-02	1.84E+00	-1.78E+00	3.14E-03	5.88E+01	1.05E-03
26	1.00E+01	1.26E-02	1.30E+00	-1.26E+00	5.96E-03	3.00E+01	1.30E-03
27	2.00E+01	8.92E-03	9.19E-01	-8.90E-01	1.16E-02	2.14E+01	2.40E-03
28	3.00E+01	7.28E-03	7.51E-01	-7.27E-01	1.72E-02	1.96E+01	4.30E-03
29	4.00E+01	6.31E-03	6.50E-01	-6.29E-01	2.28E-02	1.89E+01	7.00E-03
30	5.00E+01	5.64E-03	5.81E-01	-5.63E-01	2.83E-02	1.86E+01	1.05E-02
31	6.00E+01	5.15E-03	5.31E-01	-5.14E-01	3.39E-02	1.84E+01	1.48E-02
32	7.00E+01	4.77E-03	4.91E-01	-4.76E-01	3.94E-02	1.83E+01	1.99E-02
33	8.00E+01	4.46E-03	4.60E-01	-4.45E-01	4.49E-02	1.83E+01	2.58E-02
34	9.00E+01	4.21E-03	4.33E-01	-4.20E-01	5.03E-02	1.83E+01	3.25E-02
35	1.00E+02	3.99E-03	4.11E-01	-3.98E-01	5.57E-02	1.83E+01	4.00E-02

Fort Huachuca (7.18%)

Fort Huachuca, Arizona - 7.18 %

Low Frequency

Parallel-plate sample holder (Disk electrodes, d=5 cm) — L & Cs-from Air — ε_0, F/m = 8.85E-12 — μ_0, H/m = 1.26E-06

HP 4194A, Z-Probe, 51 points (Labview program) — Calibrating Z-Probe, in the shield box.

Measuring soil from ATF-Subsurface (Allison), sample # 11- sieved. With **5%** distilled water (by weight)

Copy from "ATFSbsr5-03"

Diameter	Thickness	Volume	Weight	Dens. (wet)	Drying, weight in g.		Water cont.		Bulk density	
D, cm	h, cm	V, cm^3	M, g	γ, g/cm^3	Wet	Dry	W, %	Wv, %	g/cm^3	σ_c, S/m
5	0.7067	13.87602	20.96	1.51052	32.78	31.22	4.997	7.189	1.44	0.00599

L, nH	Cs	From File: ImpedAn94-DiskEl-08, Air		Cs	L, nH
29.15735	4.19484	(using polynoms for repeated data,12.10.02):		4.194843	29.15735

Sample # 11, sieved, 5 % water

f, MHz	Z, Ohm	ϕ, deg.	tanδ	Res.meas. R(s), Ohm	Xmeas. Zsinϕ	$X_L=\omega L$ Ohm	R_{el} y=54.49x	Xel y=-44.932x	R' / Rm-Rms- / Rel	X' / Xm-wL- / Xel	G, S	σ', S/m	B, S	f, MHz
1.00E-02	6.24E+02	-2.28E+00	2.51E+01	6.23E+02	-2.49E+01	1.83E-03	2.17E+01	-1.79E+01	6.01E+02	-6.94E+00	1.66E-03	5.98E-03	-1.89E-05	1.00E-02
1.20E-02	6.19E+02	-2.08E+00	2.75E+01		-2.25E+01	2.20E-03	1.98E+01	-1.63E+01	-1.98E+01	-6.14E+00	-4.60E-02	-1.66E-01	-1.42E-02	1.20E-02
1.45E-02	6.19E+02	-2.07E+00	2.77E+01	6.18E+02	-2.24E+01	2.65E-03	1.81E+01	-1.49E+01	6.00E+02	-7.45E+00	1.67E-03	5.99E-03	-2.03E-05	1.45E-02
1.74E-02	6.08E+02	-1.42E+00	4.05E+01			3.18E-03	1.65E+01	-1.36E+01	-1.65E+01	1.36E+01	-3.61E-02	-1.30E-01	2.98E-02	1.74E-02
2.09E-02	6.14E+02	-1.74E+00	3.29E+01			3.83E-03	1.50E+01	-1.24E+01	-1.50E+01	1.24E+01	-3.96E-02	-1.42E-01	3.26E-02	2.09E-02
2.51E-02	6.15E+02	-1.90E+00	3.02E+01	6.15E+02	-2.04E+01	4.60E-03	1.37E+01	-1.13E+01	6.01E+02	-9.06E+00	1.66E-03	5.99E-03	-2.44E-05	2.51E-02
3.02E-02	6.13E+02	-1.69E+00	3.39E+01	6.13E+02	-1.81E+01	5.53E-03	1.25E+01	-1.03E+01	6.00E+02	-7.79E+00	1.67E-03	5.99E-03	-2.08E-05	3.02E-02
3.63E-02	6.10E+02	-7.44E-01	7.70E+01			6.65E-03	1.14E+01	-9.41E+00	-1.14E+01	9.40E+00	-5.22E-02	-1.88E-01	4.30E-02	3.63E-02
4.37E-02	6.11E+02	-1.09E+00	5.27E+01	6.11E+02	-1.36E+01	8.00E-03	1.04E+01	-8.58E+00	6.00E+02	8.57E+00	1.66E-03	5.99E-03	2.49E-05	4.37E-02
5.25E-02	6.11E+02	-1.28E+00	4.48E+01	6.11E+02	-1.35E+01	9.61E-03	9.49E+00	-7.82E+00	6.01E+02	-5.80E+00	1.66E-03	5.99E-03	-1.47E-05	5.25E-02
6.31E-02	6.10E+02	-1.26E+00	4.53E+01	6.10E+02	-1.41E+01	1.16E-02	8.65E+00	-7.14E+00	6.01E+02	-6.34E+00	1.66E-03	5.99E-03	-1.59E-05	6.31E-02
7.59E-02	6.09E+02	-1.33E+00	4.32E+01	6.09E+02	-1.64E+01	1.39E-02	7.89E+00	-6.51E+00	6.01E+02	-7.60E+00	1.66E-03	5.99E-03	-1.90E-05	7.59E-02
9.12E-02	6.08E+02	-1.55E+00	3.70E+01	6.08E+02	-1.77E+01	1.67E-02	7.20E+00	-5.94E+00	6.01E+02	-1.05E+01	1.66E-03	5.99E-03	-2.67E-05	9.12E-02
1.10E-01	6.07E+02	-1.68E+00	3.42E+01	6.06E+02	-1.97E+01	2.01E-02	6.56E+00	-5.41E+00	6.00E+02	-1.24E+01	1.67E-03	6.00E-03	-3.14E-05	1.10E-01
1.32E-01	6.05E+02	-1.87E+00	3.07E+01	6.05E+02	-2.17E+01	2.42E-02	5.99E+00	-4.94E+00	5.99E+02	-1.48E+01	1.67E-03	6.01E-03	-3.78E-05	1.32E-01
1.58E-01	6.04E+02	-2.06E+00	2.78E+01	6.03E+02	-2.45E+01	2.90E-02	5.46E+00	-4.50E+00	5.98E+02	-1.72E+01	1.67E-03	6.01E-03	-4.40E-05	1.58E-01
1.91E-01	6.02E+02	-2.33E+00	2.46E+01	6.02E+02	-2.71E+01	3.49E-02	4.98E+00	-4.11E+00	5.97E+02	-2.04E+01	1.67E-03	6.02E-03	-5.23E-05	1.91E-01
2.29E-01	6.01E+02	-2.59E+00	2.21E+01	6.00E+02	-3.05E+01	4.20E-02	4.54E+00	-3.75E+00	5.95E+02	-2.34E+01	1.68E-03	6.04E-03	-5.99E-05	2.29E-01
2.75E-01	5.99E+02	-2.92E+00	1.96E+01	5.98E+02	-3.43E+01	5.05E-02	4.14E+00	-3.42E+00	5.94E+02	-2.71E+01	1.68E-03	6.05E-03	-6.96E-05	2.75E-01
3.31E-01	5.97E+02	-3.30E+00	1.74E+01	5.96E+02	-3.83E+01	6.07E-02	3.78E+00	-3.12E+00	5.92E+02	-3.12E+01	1.68E-03	6.06E-03	-8.02E-05	3.31E-01
3.98E-01	5.95E+02	-3.70E+00	1.55E+01	5.93E+02		7.29E-02	3.45E+00	-2.84E+00	5.90E+02	-3.56E+01	1.69E-03	6.08E-03	-9.13E-05	3.98E-01

Fort Huachuca, Arizona - 7.18 %

Low Frequency

	C, pF	ε'	tanδcorr	φcorr	ε"	Zcorr	Rcorr	ρ', Ohm-m	ρ"	Conduct. losses,e"c	Diel. losses	100/ω^0.5	Rms	C(s), pF	σ", S/m
	B/w		G/B		ε'*tanδ	Ohm	Ohm	(Rcorr)	ρ'/tanδ	$\sigma_d/\omega\varepsilon_0$	ε"-ε"c		Ohm		
16	3.01E+02	1.22E+02	8.79E+01	-6.52E-01	1.08E+04	6.01E+02	6.01E+02	1.67E+02	1.90E+00	1.08E+04	-1.10E+01	3.99E-01	5.51E-03	6.40E+05	6.81E-05
17	1.89E+05	7.67E+04	-3.23E+00	1.72E+01	-2.48E+05	2.08E+01	1.98E+01	5.51E+00	-1.71E+00	8.96E+03	-2.57E+05	3.64E-01	5.51E-03	5.89E+05	5.13E-02
18	2.23E+02	9.08E+01	8.21E+01	-6.98E-01	7.45E+03	6.00E+02	6.00E+02	1.67E+02	2.03E+00	7.45E+03	4.79E+00	3.32E-01	5.51E-03	4.93E+05	7.30E-05
19	-2.72E+05	-1.11E+05	1.21E+00	-3.95E+00	-1.34E+05	2.14E+01	1.65E+02	4.58E+00	3.78E+00	6.20E+03	-1.41E+05	3.03E-01	5.51E-03	6.10E+05	-1.07E-01
20	-2.49E+05	-1.01E+05	1.21E+00	-3.95E+00	-1.23E+05	1.95E+01	1.50E+02	4.18E+00	3.44E+00	5.15E+03	-1.28E+05	2.76E-01	5.51E-03	4.08E+05	-1.17E-01
21	1.55E+02	6.29E+01	6.81E+01	-8.41E-01	4.28E+03	6.01E+02	6.01E+02	1.67E+02	2.45E+00	4.29E+03	-3.48E+00	2.52E-01	5.51E-03	3.11E+05	8.78E-05
22	1.10E+02	4.46E+01	8.00E+01	-7.16E-01	3.57E+03	6.00E+02	6.00E+02	1.67E+02	2.08E+00	3.57E+03	2.02E+00	2.30E-01	5.51E-03	2.91E+05	7.49E-05
23	-1.88E+05	-7.66E+04	1.21E+00	-3.95E+00	-9.30E+04	1.48E+01	1.14E+02	3.17E+00	2.61E+00	2.97E+03	-9.60E+04	2.09E-01	5.51E-03	5.54E+05	-1.55E-01
24	-9.08E+01	-3.69E+01	-6.68E+01	8.57E-01	2.47E+03	6.01E+02	6.00E+02	1.67E+02	-2.50E+00	2.47E+03	1.07E+00	1.91E-01	5.51E-03	3.14E+05	-8.97E-05
25	4.45E+01	1.81E+01	1.13E+02	-5.05E-01	2.05E+03	6.01E+02	6.01E+02	1.67E+02	1.47E+00	2.05E+03	-4.66E-01	1.74E-01	5.51E-03	2.23E+05	5.28E-05
26	4.01E+01	1.63E+01	1.05E+02	-5.47E-01	1.71E+03	6.01E+02	6.01E+02	1.67E+02	1.60E+00	1.71E+03	-1.26E+00	1.59E-01	5.51E-03	1.87E+05	5.72E-05
27	3.99E+01	1.62E+01	8.74E+01	-6.56E-01	1.42E+03	6.01E+02	6.01E+02	1.67E+02	1.91E+00	1.42E+03	-1.09E+00	1.45E-01	5.51E-03	1.49E+05	6.85E-05
28	4.66E+01	1.89E+01	6.24E+01	-9.18E-01	1.18E+03	6.01E+02	6.01E+02	1.67E+02	2.67E+00	1.18E+03	3.03E-01	1.32E-01	5.51E-03	1.06E+05	9.61E-05
29	4.56E+01	1.86E+01	5.30E+01	-1.08E+00	9.84E+02	6.00E+02	6.00E+02	1.67E+02	3.14E+00	9.82E+02	1.54E+00	1.20E-01	5.18E-03	8.18E+04	1.13E-04
30	4.56E+01	1.86E+01	4.41E+01	-1.30E+00	8.19E+02	5.99E+02	5.99E+02	1.66E+02	3.77E+00	8.17E+02	2.14E+00	1.10E-01	5.28E-03	6.12E+04	1.36E-04
31	4.42E+01	1.80E+01	3.80E+01	-1.51E+00	6.82E+02	5.98E+02	5.98E+02	1.66E+02	4.37E+00	6.79E+02	2.81E+00	1.00E-01	5.41E-03	4.63E+04	1.58E-04
32	4.37E+01	1.78E+01	3.20E+01	-1.79E+00	5.68E+02	5.97E+02	5.97E+02	1.66E+02	5.18E+00	5.65E+02	3.25E+00	9.14E-02	5.55E-03	3.41E+04	1.88E-04
33	4.16E+01	1.69E+01	2.80E+01	-2.04E+00	4.74E+02	5.96E+02	5.96E+02	1.65E+02	5.91E+00	4.70E+02	3.57E+00	8.34E-02	5.73E-03	2.56E+04	2.15E-04
34	4.02E+01	1.63E+01	2.42E+01	-2.37E+00	3.95E+02	5.94E+02	5.94E+02	1.65E+02	6.83E+00	3.91E+02	3.87E+00	7.60E-02	5.94E-03	1.89E+04	2.50E-04
35	3.85E+01	1.57E+01	2.10E+01	-2.73E+00	3.29E+02	5.93E+02	5.92E+02	1.65E+02	7.83E+00	3.25E+02	4.02E+00	6.93E-02	6.20E-03	1.40E+04	2.89E-04
36	3.65E+01	1.48E+01	1.85E+01	-3.09E+00	2.74E+02	5.91E+02	5.90E+02	1.64E+02	8.87E+00	2.70E+02	4.00E+00	6.32E-02	6.50E-03	1.04E+04	3.29E-04

	A	B	C	D	E	F	G	H	I	J	K	L	M	N	O
37															
38	f, MHz	Z, Ohm	φ, deg.	tanδ	Res.meas.	Xmeas.	$X_L=\omega L$	R_{el}	Xel	Rm-Rms-	Xm-wL-	G, S	σ', S/m	B, S	f, MHz
39					R(s), Ohm	Zsinφ	Ohm	y=54.49x	y=-44.932x	R' Rel	X' Xel				
40	4.79E-01	5.92E+02	-4.16E+00	1.37E+01	5.91E+02	-4.30E+01	8.77E-02	3.14E+00	-2.59E+00	5.87E+02	-4.05E+01	1.69E-03	6.10E-03	-1.04E-04	4.79E-01
41	5.75E-01	5.89E+02	-4.78E+00	1.20E+01	5.87E+02	-4.91E+01	1.05E-01	2.87E+00	-2.36E+00	5.84E+02	-4.68E+01	1.70E-03	6.12E-03	-1.21E-04	5.75E-01
42	6.92E-01	5.86E+02	-5.47E+00	1.04E+01	5.84E+02	-5.58E+01	1.27E-01	2.61E+00	-2.16E+00	5.81E+02	-5.38E+01	1.71E-03	6.14E-03	-1.40E-04	6.92E-01
43	8.32E-01	5.83E+02	-6.25E+00	9.13E+00	5.80E+02	-6.35E+01	1.52E-01	2.38E+00	-1.97E+00	5.77E+02	-6.16E+01	1.71E-03	6.17E-03	-1.61E-04	8.32E-01
44	1.00E+00	5.79E+02	-7.16E+00	7.96E+00	5.74E+02	-7.22E+01	1.83E-01	2.17E+00	-1.79E+00	5.72E+02	-7.06E+01	1.72E-03	6.20E-03	-1.86E-04	1.00E+00
45	1.20E+00	5.74E+02	-8.23E+00	6.91E+00	5.68E+02	-8.22E+01	2.20E-01	1.98E+00	-1.63E+00	5.66E+02	-8.08E+01	1.73E-03	6.23E-03	-2.15E-04	1.20E+00
46	1.45E+00	5.69E+02	-9.45E+00	6.01E+00	5.61E+02	-9.34E+01	2.65E-01	1.81E+00	-1.49E+00	5.59E+02	-9.22E+01	1.74E-03	6.26E-03	-2.49E-04	1.45E+00
47	1.74E+00	5.63E+02	-1.09E+01	5.21E+00	5.53E+02	-1.06E+02	3.18E-01	1.65E+00	-1.36E+00	5.51E+02	-1.05E+02	1.75E-03	6.30E-03	-2.88E-04	1.74E+00
48	2.09E+00	5.55E+02	-1.25E+01	4.51E+00	5.42E+02	-1.20E+02	3.83E-01	1.50E+00	-1.24E+00	5.40E+02	-1.19E+02	1.76E-03	6.35E-03	-3.34E-04	2.09E+00
49	2.51E+00	5.46E+02	-1.44E+01	3.90E+00	5.29E+02	-1.36E+02	4.60E-01	1.37E+00	-1.13E+00	5.28E+02	-1.35E+02	1.78E-03	6.40E-03	-3.88E-04	2.51E+00
50	3.02E+00	5.35E+02	-1.65E+01	3.38E+00	5.13E+02	-1.52E+02	5.53E-01	1.25E+00	-1.03E+00	5.12E+02	-1.52E+02	1.80E-03	6.46E-03	-4.52E-04	3.02E+00
51	3.63E+00	5.22E+02	-1.89E+01	2.92E+00	4.94E+02	-1.69E+02	6.65E-01	1.14E+00	-9.41E-01	4.92E+02	-1.69E+02	1.82E-03	6.54E-03	-5.28E-04	3.63E+00
52	4.37E+00	5.06E+02	-2.16E+01	2.53E+00	4.71E+02	-1.86E+02	8.00E-01	1.04E+00	-8.58E-01	4.70E+02	-1.86E+02	1.84E-03	6.62E-03	-6.15E-04	4.37E+00
53	5.25E+00	4.88E+02	-2.45E+01	2.19E+00	4.44E+02	-2.03E+02	9.61E-01	9.49E-01	-7.82E-01	4.43E+02	-2.03E+02	1.87E-03	6.72E-03	-7.15E-04	5.25E+00
54	6.31E+00	4.67E+02	-2.77E+01	1.90E+00	4.13E+02	-2.17E+02	1.16E+00	8.65E-01	-7.14E-01	4.12E+02	-2.17E+02	1.90E-03	6.83E-03	-8.35E-04	6.31E+00
55	7.59E+00	4.42E+02	-3.11E+01	1.66E+00	3.79E+02	-2.28E+02	1.39E+00	7.89E-01	-6.51E-01	3.78E+02	-2.29E+02	1.94E-03	6.97E-03	-9.73E-04	7.59E+00
56	9.12E+00	4.15E+02	-3.46E+01	1.45E+00	3.41E+02	-2.35E+02	1.67E+00	7.20E-01	-5.94E-01	3.41E+02	-2.36E+02	1.98E-03	7.13E-03	-1.13E-03	9.12E+00
57	1.10E+01	3.85E+02	-3.81E+01	1.27E+00	3.03E+02	-2.38E+02	2.01E+00	6.56E-01	-5.41E-01	3.02E+02	-2.39E+02	2.04E-03	7.33E-03	-1.32E-03	1.10E+01
58	1.32E+01	3.53E+02	-4.17E+01	1.12E+00	2.64E+02	-2.35E+02	2.42E+00	5.99E-01	-4.94E-01	2.63E+02	-2.37E+02	2.10E-03	7.55E-03	-1.54E-03	1.32E+01
59	1.58E+01	3.22E+02	-4.52E+01	9.92E-01	2.27E+02	-2.28E+02	2.90E+00	5.46E-01	-4.50E-01	2.26E+02	-2.31E+02	2.17E-03	7.79E-03	-1.79E-03	1.58E+01
60	1.91E+01	2.89E+02	-4.86E+01	8.82E-01	1.91E+02	-2.17E+02	3.49E+00	4.98E-01	-4.11E-01	1.91E+02	-2.20E+02	2.25E-03	8.10E-03	-2.09E-03	1.91E+01
61	2.29E+01	2.58E+02	-5.18E+01	7.86E-01	1.59E+02	-2.03E+02	4.20E+00	4.54E-01	-3.75E-01	1.59E+02	-2.07E+02	2.34E-03	8.42E-03	-2.44E-03	2.29E+01
62	2.75E+01	2.27E+02	-5.48E+01	7.05E-01	1.31E+02	-1.85E+02	5.05E+00	4.14E-01	-3.42E-01	1.30E+02	-1.90E+02	2.45E-03	8.83E-03	-2.85E-03	2.75E+01
63	3.31E+01	1.98E+02	-5.74E+01	6.40E-01	1.07E+02	-1.66E+02	6.07E+00	3.78E-01	-3.12E-01	1.06E+02	-1.72E+02	2.59E-03	9.33E-03	-3.34E-03	3.31E+01
64	3.98E+01	1.71E+02	-5.95E+01	5.89E-01	8.68E+01	-1.47E+02	7.29E+00	3.45E-01	-2.84E-01	8.63E+01	-1.54E+02	2.76E-03	9.93E-03	-3.89E-03	3.98E+01
65	4.79E+01	1.46E+02	-6.14E+01	5.46E-01	7.01E+01	-1.28E+02	8.77E+00	3.14E-01	-2.59E-01	6.97E+01	-1.37E+02	2.95E-03	1.06E-02	-4.54E-03	4.79E+01
66	5.75E+01	1.23E+02	-6.30E+01	5.09E-01	5.60E+01	-1.10E+02	1.05E+01	2.87E-01	-2.36E-01	5.55E+01	-1.20E+02	3.16E-03	1.14E-02	-5.34E-03	5.75E+01
67	6.92E+01	1.02E+02	-6.39E+01	4.89E-01	4.50E+01	-9.19E+01	1.27E+01	2.61E-01	-2.16E-01	4.45E+01	-1.04E+02	3.45E-03	1.24E-02	-6.28E-03	6.92E+01
68	8.32E+01	8.32E+01	-6.43E+01	4.80E-01	3.60E+01	-7.50E+01	1.52E+01	2.38E-01	-1.97E-01	3.55E+01	-9.00E+01	3.80E-03	1.37E-02	-7.42E-03	8.32E+01
69	1.00E+02	6.51E+01	-6.37E+01	4.94E-01	2.89E+01	-5.84E+01	1.83E+01	2.17E-01	-1.79E-01	2.84E+01	-7.65E+01	4.26E-03	1.53E-02	-8.85E-03	1.00E+02

	P	Q	R	S	T	U	V	W	X	Y	Z	AA	AB	AC	AD
37										Conduct.	Diel.				
38	C, pF	ε'	tanδcorr	φcorr	$\varepsilon"$	Zcorr	Rcorr	ρ', Ohm-m	$\rho"$	losses,e"c	losses	$100/\omega^{0.5}$	Rms	C(s), pF	σ", S/m
39	B/w		G/B		$\varepsilon'*$tanδ	Ohm	Ohm	(Rcorr)	ρ'/tanδ	$\sigma_c/\omega\varepsilon_0$	$\varepsilon"-\varepsilon"c$		Ohm		
40	3.46E+01	1.41E+01	1.63E+01	-3.52E+00	2.29E+02	5.89E+02	5.88E+02	1.63E+02	1.00E+01	2.25E+02	4.06E+00	5.77E-02	6.87E-03	7.74E+03	3.75E-04
41	3.35E+01	1.36E+01	1.40E+01	-4.07E+00	1.91E+02	5.87E+02	5.85E+02	1.63E+02	1.16E+01	1.87E+02	4.03E+00	5.26E-02	7.31E-03	5.63E+03	4.36E-04
42	3.22E+01	1.31E+01	1.22E+01	-4.68E+00	1.60E+02	5.84E+02	5.82E+02	1.62E+02	1.33E+01	1.56E+02	3.96E+00	4.80E-02	7.84E-03	4.12E+03	5.03E-04
43	3.08E+01	1.25E+01	1.06E+01	-5.37E+00	1.33E+02	5.81E+02	5.79E+02	1.61E+02	1.51E+01	1.29E+02	3.81E+00	4.37E-02	8.48E-03	3.02E+03	5.80E-04
44	2.96E+01	1.20E+01	9.26E+00	-6.17E+00	1.11E+02	5.78E+02	5.74E+02	1.60E+02	1.72E+01	1.08E+02	3.71E+00	3.99E-02	9.24E-03	2.20E+03	6.69E-04
45	2.85E+01	1.16E+01	8.04E+00	-7.09E+00	9.31E+01	5.74E+02	5.69E+02	1.58E+02	1.97E+01	8.96E+01	3.54E+00	3.64E-02	1.02E-02	1.61E+03	7.74E-04
46	2.74E+01	1.11E+01	7.00E+00	-8.13E+00	7.79E+01	5.69E+02	5.63E+02	1.56E+02	2.24E+01	7.45E+01	3.40E+00	3.32E-02	1.13E-02	1.18E+03	8.95E-04
47	2.64E+01	1.07E+01	6.07E+00	-9.35E+00	6.52E+01	5.63E+02	5.56E+02	1.54E+02	2.54E+01	6.20E+01	3.25E+00	3.03E-02	1.26E-02	8.63E+02	1.04E-03
48	2.55E+01	1.04E+01	5.28E+00	-1.07E+01	5.46E+01	5.57E+02	5.47E+02	1.52E+02	2.88E+01	5.15E+01	3.10E+00	2.76E-02	1.41E-02	6.34E+02	1.20E-03
49	2.46E+01	1.00E+01	4.58E+00	-1.23E+01	4.58E+01	5.49E+02	5.37E+02	1.49E+02	3.25E+01	4.29E+01	2.95E+00	2.52E-02	1.60E-02	4.68E+02	1.40E-03
50	2.38E+01	9.68E+00	3.97E+00	-1.41E+01	3.85E+01	5.40E+02	5.24E+02	1.45E+02	3.66E+01	3.57E+01	2.82E+00	2.30E-02	1.83E-02	3.47E+02	1.63E-03
51	2.31E+01	9.40E+00	3.44E+00	-1.62E+01	3.24E+01	5.29E+02	5.08E+02	1.41E+02	4.10E+01	2.97E+01	2.72E+00	2.09E-02	2.10E-02	2.59E+02	1.90E-03
52	2.24E+01	9.11E+00	2.99E+00	-1.85E+01	2.73E+01	5.16E+02	4.89E+02	1.36E+02	4.54E+01	2.47E+01	2.60E+00	1.91E-02	2.42E-02	1.96E+02	2.21E-03
53	2.17E+01	8.82E+00	2.61E+00	-2.10E+01	2.30E+01	5.00E+02	4.67E+02	1.30E+02	4.98E+01	2.05E+01	2.49E+00	1.74E-02	2.79E-02	1.50E+02	2.57E-03
54	2.11E+01	8.56E+00	2.27E+00	-2.37E+01	1.95E+01	4.82E+02	4.41E+02	1.23E+02	5.39E+01	1.71E+01	2.40E+00	1.59E-02	3.25E-02	1.16E+02	3.00E-03
55	2.04E+01	8.30E+00	1.99E+00	-2.67E+01	1.65E+01	4.61E+02	4.12E+02	1.15E+02	5.76E+01	1.42E+01	2.32E+00	1.45E-02	3.78E-02	9.20E+01	3.50E-03
56	1.98E+01	8.05E+00	1.75E+00	-2.98E+01	1.41E+01	4.38E+02	3.80E+02	1.06E+02	6.05E+01	1.18E+01	2.25E+00	1.32E-02	4.41E-02	7.42E+01	4.08E-03
57	1.92E+01	7.80E+00	1.54E+00	-3.30E+01	1.20E+01	4.12E+02	3.46E+02	9.60E+01	6.23E+01	9.82E+00	2.19E+00	1.20E-02	5.14E-02	6.11E+01	4.76E-03
58	1.86E+01	7.56E+00	1.36E+00	-3.63E+01	1.03E+01	3.84E+02	3.10E+02	8.61E+01	6.31E+01	8.17E+00	2.13E+00	1.10E-02	6.00E-02	5.14E+01	5.54E-03
59	1.80E+01	7.32E+00	1.21E+00	-3.96E+01	8.84E+00	3.56E+02	2.74E+02	7.61E+01	6.30E+01	6.79E+00	2.05E+00	1.00E-02	7.00E-02	4.40E+01	6.46E-03
60	1.75E+01	7.10E+00	1.08E+00	-4.29E+01	7.64E+00	3.26E+02	2.38E+02	6.62E+01	6.16E+01	5.65E+00	1.99E+00	9.14E-03	8.16E-02	3.85E+01	7.53E-03
61	1.69E+01	6.89E+00	9.59E-01	-4.62E+01	6.61E+00	2.96E+02	2.05E+02	5.69E+01	5.93E+01	4.70E+00	1.91E+00	8.34E-03	9.47E-02	3.43E+01	8.78E-03
62	1.65E+01	6.70E+00	8.60E-01	-4.93E+01	5.76E+00	2.66E+02	1.73E+02	4.81E+01	5.60E+01	3.91E+00	1.85E+00	7.60E-03	1.10E-01	3.12E+01	1.03E-02
63	1.60E+01	6.52E+00	7.77E-01	-5.22E+01	5.07E+00	2.37E+02	1.45E+02	4.03E+01	5.19E+01	3.25E+00	1.81E+00	6.93E-03	1.26E-01	2.89E+01	1.20E-02
64	1.55E+01	6.32E+00	7.10E-01	-5.46E+01	4.49E+00	2.10E+02	1.21E+02	3.37E+01	4.75E+01	2.70E+00	1.78E+00	6.32E-03	1.45E-01	2.71E+01	1.40E-02
65	1.51E+01	6.14E+00	6.50E-01	-5.70E+01	3.99E+00	1.85E+02	1.01E+02	2.80E+01	4.30E+01	2.25E+00	1.74E+00	5.77E-03	1.64E-01	2.59E+01	1.63E-02
66	1.48E+01	6.00E+00	5.93E-01	-5.93E+01	3.56E+00	1.61E+02	8.22E+01	2.28E+01	3.85E+01	1.87E+00	1.69E+00	5.26E-03	1.86E-01	2.51E+01	1.92E-02
67	1.45E+01	5.87E+00	5.50E-01	-6.12E+01	3.23E+00	1.39E+02	6.72E+01	1.87E+01	3.40E+01	1.56E+00	1.67E+00	4.80E-03	2.08E-01	2.50E+01	2.26E-02
68	1.42E+01	5.77E+00	5.12E-01	-6.29E+01	2.95E+00	1.20E+02	5.47E+01	1.52E+01	2.97E+01	1.29E+00	1.66E+00	4.37E-03	2.32E-01	2.55E+01	2.67E-02
69	1.41E+01	5.73E+00	4.81E-01	-6.43E+01	2.75E+00	1.02E+02	4.41E+01	1.23E+01	2.55E+01	1.08E+00	1.68E+00	3.99E-03	2.60E-01	2.72E+01	3.19E-02

Fort Huachuca (14.6%)

Fort Huachuca, Arizona - 14.6 %
Low Frequency

Parallel-plate sample holder (Disk electrodes, d=5 cm)	L & Cs-from Air	ε_0, F/m	μ_0, H/m
HP 4194A, Z-Probe, 51 points (Labview program) — Calibrating Z-Probe		8.85E-12	1.26E-06

Measuring soil from ATF-Subsurface (Allison) in the shield box. # 11- sieved. With about 10% distilled water by weight

Copy from "ATFSbsr10-03"

Diameter	Thickness	Volume	Weight	Dens.(wet)	Drying, weight in g.		Water cont.		Bulk density	
D, cm	h, cm	V, cm^3	M, g	γ, g/cm^3	Wet	Dry	W, %	Wv, %	g/cm^3	σst, S/m
5	0.7289	14.311918	25.12	1.75518	25.05	22.96	9.103	14.644	1.61	0.01103
L, nH	Cs						Cs		L, nH	
29	4.18160	From File: ImpedAn94-DiskEl-08, Air					4.1816555	28.99754		

(using polynoms for repeated data, 12.10.02):   R'   X'

Sample # 11, sieved, 10 % water

f, MHz	Z, Ohm	φ, deg.	tanδ	Res.meas. R(s), Ohm	Xmeas. Zsinφ	$X_L=\omega L$ Ohm	R_{el} (y=52.317x)	Xel (y=-24.761x)	Rm-Rms- (R')	Xm-wL- (X')	G, S	σ', S/m	B, S	C, pF (B/w)
1.00E-02	3.58E+02	-2.08E+00	2.75E+01	3.57E+02	-1.30E+01	1.82E-03	2.09E+01	-9.88E+00	3.36E+02	-3.13E+00	2.97E-03	1.10E-02	-2.74E-05	4.35E+02
1.20E-02	3.55E+02	-1.99E+00	2.88E+01	3.55E+02	-1.23E+01	2.19E-03	1.90E+01	-9.01E+00	3.36E+02	-3.30E+00	2.98E-03	1.11E-02	-2.90E-05	3.84E+02
1.45E-02	3.55E+02	-2.13E+00	2.69E+01			2.63E-03	1.74E+01	-8.22E+00	3.38E+02	8.21E+00	2.96E-03	1.10E-02	7.24E-05	-7.97E+02
1.74E-02	3.49E+02	-1.65E+00	3.46E+01	3.52E+02	-1.01E+01	3.17E-03	1.58E+01	-7.49E+00	3.37E+02	-2.58E+00	-6.15E-02	-2.28E-01	-1.00E-02	9.18E+04
2.09E-02	3.52E+02	-2.11E+00	2.71E+01			3.81E-03	1.44E+01	-6.83E+00	3.37E+02	6.83E+00	2.97E-03	1.10E-02	6.06E-05	-4.62E+02
2.51E-02	3.52E+02	-2.40E+00	2.38E+01			4.58E-03	1.32E+01	-6.23E+00	3.37E+02	6.23E+00	-6.20E-02	-2.30E-01	2.93E-02	-1.86E+05
3.02E-02	3.49E+02	-2.31E+00	2.48E+01			5.50E-03	1.20E+01	-5.68E+00	3.37E+02	5.68E+00	2.97E-03	1.10E-02	5.07E-05	-2.67E+02
3.63E-02	3.47E+02	-1.40E+00	4.10E+01	3.49E+02	-8.46E+00	6.62E-03	1.10E+01	-5.18E+00	3.36E+02	-3.28E+00	-8.37E-02	-3.11E-01	-2.51E-02	1.10E+05
4.37E-02	3.47E+02	-1.77E+00	3.24E+01			7.95E-03	9.99E+00	-4.73E+00	3.36E+02	4.72E+00	2.97E-03	1.10E-02	4.28E-05	-1.56E+02
5.25E-02	3.46E+02	-1.95E+00	2.93E+01	3.46E+02	-1.18E+01	9.56E-03	9.11E+00	-4.31E+00	3.36E+02	-7.49E+00	2.97E-03	1.10E-02	-6.47E-05	1.96E+02
6.31E-02	3.45E+02	-1.94E+00	2.95E+01	3.45E+02	-1.17E+01	1.15E-02	8.31E+00	-3.93E+00	3.36E+02	-7.76E+00	2.97E-03	1.10E-02	-6.68E-05	1.69E+02
7.59E-02	3.44E+02	-2.00E+00	2.86E+01	3.44E+02	-1.20E+01	1.38E-02	7.58E+00	-3.59E+00	3.36E+02	-8.45E+00	2.97E-03	1.10E-02	-7.27E-05	1.53E+02
9.12E-02	3.43E+02	-2.23E+00	2.57E+01	3.43E+02	-1.33E+01	1.66E-02	6.91E+00	-3.27E+00	3.36E+02	-1.01E+01	2.98E-03	1.11E-02	-8.69E-05	1.52E+02
1.10E-01	3.41E+02	-2.37E+00	2.42E+01	3.41E+02	-1.41E+01	2.00E-02	6.30E+00	-2.98E+00	3.35E+02	-1.11E+01	2.98E-03	1.11E-02	-9.63E-05	1.40E+02
1.32E-01	3.40E+02	-2.57E+00	2.23E+01	3.40E+02	-1.52E+01	2.40E-02	5.75E+00	-2.72E+00	3.34E+02	-1.25E+01	2.99E-03	1.11E-02	-1.09E-04	1.31E+02
1.58E-01	3.39E+02	-2.76E+00	2.07E+01	3.38E+02	-1.63E+01	2.89E-02	5.24E+00	-2.48E+00	3.33E+02	-1.39E+01	3.00E-03	1.11E-02	-1.21E-04	1.21E+02
1.91E-01	3.37E+02	-3.03E+00	1.89E+01	3.37E+02	-1.78E+01	3.47E-02	4.78E+00	-2.26E+00	3.32E+02	-1.56E+01	3.01E-03	1.12E-02	-1.36E-04	1.14E+02
2.29E-01	3.36E+02	-3.27E+00	1.75E+01	3.35E+02	-1.92E+01	4.17E-02	4.36E+00	-2.06E+00	3.31E+02	-1.72E+01	3.01E-03	1.12E-02	-1.50E-04	1.04E+02
2.75E-01	3.34E+02	-3.59E+00	1.60E+01	3.33E+02	-2.09E+01	5.02E-02	3.98E+00	-1.88E+00	3.29E+02	-1.91E+01	3.03E-03	1.13E-02	-1.68E-04	9.70E+01
3.31E-01	3.32E+02	-3.93E+00	1.45E+01	3.32E+02	-2.28E+01	6.03E-02	3.63E+00	-1.72E+00	3.28E+02	-2.11E+01	3.04E-03	1.13E-02	-1.87E-04	8.99E+01
3.98E-01	3.31E+02	-4.30E+00	1.33E+01	3.30E+02	-2.48E+01	7.25E-02	3.31E+00	-1.57E+00	3.26E+02	-2.33E+01	3.05E-03	1.14E-02	-2.07E-04	8.29E+01
4.79E-01	3.29E+02	-4.72E+00	1.21E+01	3.27E+02	-2.71E+01	8.72E-02	3.02E+00	-1.43E+00	3.24E+02	-2.57E+01	3.06E-03	1.14E-02	-2.30E-04	7.65E+01
5.75E-01	3.26E+02	-5.27E+00	1.08E+01	3.25E+02	-3.00E+01	1.05E-01	2.75E+00	-1.30E+00	3.22E+02	-2.88E+01	3.08E-03	1.14E-02	-2.60E-04	7.19E+01
6.92E-01	3.24E+02	-5.87E+00	9.72E+00	3.22E+02	-3.32E+01	1.26E-01	2.51E+00	-1.19E+00	3.20E+02	-3.21E+01	3.09E-03	1.15E-02	-2.92E-04	6.73E+01
8.32E-01	3.22E+02	-6.55E+00	8.70E+00	3.20E+02	-3.67E+01	1.52E-01	2.29E+00	-1.08E+00	3.17E+02	-3.58E+01	3.11E-03	1.16E-02	-3.29E-04	6.30E+01

Fort Huachuca, Arizona - 14.6 %

Low Frequency

	P	Q	R	S	T	U	V	W	X	Y	Z	AA	AB	AC	AD
13										Conduct.	Diel.				
14	f, MHz	ε'	tanδcorr	φcorr	ε''	Zcorr	Rcorr	ρ', Ohm-m	ρ''	losses,ε''c	losses	$100/\omega^{0.5}$	Rms	C(s), pF	σ'', S/m
15			G/B		ε'*tanδ	Ohm	Ohm	(Rcorr)	ρ'/tanδ	$\sigma_d/\omega\varepsilon_0$	$\varepsilon''-\varepsilon''c$		Ohm		
16	1.00E-02	1.83E+02	1.09E+02	-5.27E-01	1.98E+04	3.36E+02	3.36E+02	9.06E+01	8.34E-01	1.98E+04	7.87E+00	3.99E-01	5.51E-03	1.22E+06	1.02E-04
17	1.20E-02	1.61E+02	1.03E+02	-5.57E-01	1.65E+04	3.36E+02	3.36E+02	9.04E+01	8.79E-01	1.65E+04	4.57E+01	3.64E-01	5.51E-03	1.08E+06	1.08E-04
18	1.45E-02	-3.34E+02	-4.09E+01	1.40E+00	1.37E+04	3.38E+02	3.38E+02	9.09E+01	-2.22E+00	1.37E+04	-4.81E+01	3.32E-01	5.51E-03	8.36E+05	-2.69E-04
19	1.74E-02	3.85E+04	-6.13E+00	9.26E+00	-2.36E+05	1.60E+01	1.58E+01	4.27E+00	-6.95E-01	1.14E+04	-2.48E+05	3.03E-01	5.51E-03	9.09E+05	3.72E-02
20	2.09E-02	-1.94E+02	-4.89E+01	1.17E+00	9.47E+03	3.37E+02	3.37E+02	9.08E+01	-1.86E+00	9.49E+03	-1.81E+01	2.76E-01	5.51E-03	5.88E+05	-2.25E-04
21	2.51E-02	-7.79E+04	2.12E+00	-2.53E+01	-1.65E+05	1.46E+01	1.32E+01	3.55E+00	1.68E+00	7.89E+03	-1.73E+05	2.52E-01	5.51E-03	4.30E+05	-1.09E-01
22	3.02E-02	-1.12E+02	-5.84E+01	9.80E-01	6.55E+03	3.37E+02	3.37E+02	9.08E+01	-1.55E+00	6.57E+03	-1.37E+01	2.30E-01	5.51E-03	3.74E+05	-1.88E-04
23	3.63E-02	4.61E+02	-3.34E+00	1.67E+01	-1.54E+05	1.14E+01	1.10E+01	2.95E+00	-8.84E-01	5.46E+03	-1.59E+05	2.09E-01	5.51E-03	5.18E+05	9.31E-02
24	4.37E-02	-6.55E+01	-6.94E+01	8.26E-01	4.54E+03	3.37E+02	3.36E+02	9.06E+01	-1.31E+00	4.54E+03	4.35E-02	1.91E-01	5.51E-03	3.41E+05	-1.59E-04
25	5.25E-02	8.23E+01	4.59E+01	-1.25E+00	3.78E+03	3.37E+02	3.36E+02	9.06E+01	1.97E+00	3.78E+03	-8.90E-01	1.74E-01	5.51E-03	2.57E+05	2.40E-04
26	6.31E-02	7.07E+01	4.44E+01	-1.29E+00	3.14E+03	3.37E+02	3.36E+02	9.06E+01	2.04E+00	3.14E+03	-8.36E-01	1.59E-01	5.51E-03	2.16E+05	2.48E-04
27	7.59E-02	6.40E+01	4.09E+01	-1.40E+00	2.61E+03	3.36E+02	3.36E+02	9.06E+01	2.22E+00	2.61E+03	8.95E-01	1.45E-01	5.51E-03	1.74E+05	2.70E-04
28	9.12E-02	6.36E+01	3.43E+01	-1.67E+00	2.18E+03	3.36E+02	3.36E+02	9.04E+01	2.64E+00	2.17E+03	4.28E+00	1.32E-01	5.51E-03	1.31E+05	3.22E-04
29	1.10E-01	5.86E+01	3.10E+01	-1.85E+00	1.82E+03	3.35E+02	3.35E+02	9.02E+01	2.91E+00	1.81E+03	7.35E+00	1.20E-01	5.18E-03	1.03E+05	3.57E-04
30	1.32E-01	5.50E+01	2.75E+01	-2.08E+00	1.51E+03	3.34E+02	3.34E+02	9.00E+01	3.27E+00	1.50E+03	9.08E+00	1.10E-01	5.28E-03	7.93E+04	4.04E-04
31	1.58E-01	5.08E+01	2.48E+01	-2.31E+00	1.26E+03	3.33E+02	3.33E+02	8.97E+01	3.61E+00	1.25E+03	1.07E+01	1.00E-01	5.41E-03	6.15E+04	4.48E-04
32	1.91E-01	4.77E+01	2.21E+01	-2.60E+00	1.05E+03	3.32E+02	3.32E+02	8.95E+01	4.06E+00	1.04E+03	1.19E+01	9.14E-02	5.55E-03	4.68E+04	5.06E-04
33	2.29E-01	4.38E+01	2.01E+01	-2.85E+00	8.78E+02	3.31E+02	3.31E+02	8.91E+01	4.44E+00	8.65E+02	1.26E+01	8.34E-02	5.73E-03	3.62E+04	5.58E-04
34	2.75E-01	4.07E+01	1.80E+01	-3.17E+00	7.33E+02	3.30E+02	3.30E+02	8.88E+01	4.92E+00	7.20E+02	1.30E+01	7.60E-02	5.94E-03	2.77E+04	6.23E-04
35	3.31E-01	3.77E+01	1.62E+01	-3.53E+00	6.12E+02	3.29E+02	3.28E+02	8.84E+01	5.45E+00	5.99E+02	1.32E+01	6.93E-02	6.20E-03	2.11E+04	6.95E-04
36	3.98E-01	3.47E+01	1.47E+01	-3.89E+00	5.11E+02	3.27E+02	3.27E+02	8.80E+01	5.98E+00	4.98E+02	1.29E+01	6.32E-02	6.50E-03	1.61E+04	7.69E-04
37	4.79E-01	3.21E+01	1.33E+01	-4.30E+00	4.27E+02	3.26E+02	3.25E+02	8.75E+01	6.57E+00	4.14E+02	1.28E+01	5.77E-02	6.87E-03	1.23E+04	8.54E-04
38	5.75E-01	3.01E+01	1.18E+01	-4.83E+00	3.57E+02	3.24E+02	3.23E+02	8.69E+01	7.34E+00	3.45E+02	1.24E+01	5.26E-02	7.31E-03	9.22E+03	9.65E-04
39	6.92E-01	2.82E+01	1.06E+01	-5.40E+00	2.98E+02	3.22E+02	3.20E+02	8.63E+01	8.15E+00	2.87E+02	1.19E+01	4.80E-02	7.84E-03	6.93E+03	1.09E-03
40	8.32E-01	2.64E+01	9.46E+00	-6.04E+00	2.50E+02	3.20E+02	3.18E+02	8.56E+01	9.05E+00	2.38E+02	1.13E+01	4.37E-02	8.48E-03	5.21E+03	1.22E-03

	A	B	C	D	E	F	G	H	I	J	K	L	M	N	O
41										R'	X'				
42	f, MHz	Z, Ohm	φ, deg.	tanδ	Res.meas.	Xmeas.	$X_L=\omega L$	R_{el}	Xel	Rm-Rms-	Xm-wL-	G, S	σ', S/m	B, S	C, pF
43	Sample # 11, sieved, 10 % water				R(s), Ohm	Zsinφ	Ohm	y=52.317x Rel	y=-24.761x Xel	Rel	Xel				B/w
44	1.00E+00	3.19E+02	-7.34E+00	7.76E+00	3.16E+02	-4.08E+01	1.82E-01	2.09E+00	-9.88E-01	3.14E+02	-3.99E+01	3.13E-03	1.16E-02	-3.72E-04	5.92E+01
45	1.20E+00	3.16E+02	-8.26E+00	6.89E+00	3.13E+02	-4.54E+01	2.19E-01	1.90E+00	-9.01E-01	3.11E+02	-4.47E+01	3.15E-03	1.17E-02	-4.22E-04	5.58E+01
46	1.45E+00	3.13E+02	-9.30E+00	6.11E+00	3.09E+02	-5.05E+01	2.63E-01	1.74E+00	-8.22E-01	3.07E+02	-5.00E+01	3.17E-03	1.18E-02	-4.79E-04	5.27E+01
47	1.74E+00	3.09E+02	-1.05E+01	5.39E+00	3.04E+02	-5.64E+01	3.17E-01	1.58E+00	-7.49E-01	3.02E+02	-5.59E+01	3.20E-03	1.19E-02	-5.46E-04	5.00E+01
48	2.09E+00	3.05E+02	-1.19E+01	4.75E+00	2.98E+02	-6.28E+01	3.81E-01	1.44E+00	-6.83E-01	2.97E+02	-6.25E+01	3.23E-03	1.20E-02	-6.25E-04	4.76E+01
49	2.51E+00	3.00E+02	-1.35E+01	4.17E+00	2.92E+02	-7.00E+01	4.58E-01	1.32E+00	-6.23E-01	2.90E+02	-6.98E+01	3.26E-03	1.21E-02	-7.17E-04	4.54E+01
50	3.02E+00	2.94E+02	-1.53E+01	3.65E+00	2.84E+02	-7.78E+01	5.50E-01	1.20E+00	-5.68E-01	2.83E+02	-7.78E+01	3.29E-03	1.22E-02	-8.26E-04	4.35E+01
51	3.63E+00	2.88E+02	-1.74E+01	3.18E+00	2.74E+02	-8.61E+01	6.62E-01	1.10E+00	-5.18E-01	2.73E+02	-8.63E+01	3.33E-03	1.24E-02	-9.56E-04	4.19E+01
52	4.37E+00	2.80E+02	-1.98E+01	2.78E+00	2.63E+02	-9.46E+01	7.95E-01	9.99E-01	-4.73E-01	2.62E+02	-9.50E+01	3.37E-03	1.25E-02	-1.11E-03	4.03E+01
53	5.25E+00	2.71E+02	-2.24E+01	2.43E+00	2.50E+02	-1.03E+02	9.56E-01	9.11E-01	-4.31E-01	2.49E+02	-1.04E+02	3.42E-03	1.27E-02	-1.28E-03	3.89E+01
54	6.31E+00	2.60E+02	-2.52E+01	2.12E+00	2.35E+02	-1.11E+02	1.15E+00	8.31E-01	-3.93E-01	2.34E+02	-1.12E+02	3.48E-03	1.29E-02	-1.49E-03	3.76E+01
55	7.59E+00	2.48E+02	-2.83E+01	1.86E+00	2.18E+02	-1.17E+02	1.38E+00	7.58E-01	-3.59E-01	2.17E+02	-1.18E+02	3.55E-03	1.32E-02	-1.73E-03	3.64E+01
56	9.12E+00	2.34E+02	-3.15E+01	1.63E+00	1.99E+02	-1.22E+02	1.66E+00	6.91E-01	-3.27E-01	1.99E+02	-1.24E+02	3.63E-03	1.35E-02	-2.02E-03	3.52E+01
57	1.10E+01	2.19E+02	-3.48E+01	1.44E+00	1.79E+02	-1.25E+02	2.00E+00	6.30E-01	-2.98E-01	1.79E+02	-1.26E+02	3.73E-03	1.38E-02	-2.35E-03	3.41E+01
58	1.32E+01	2.02E+02	-3.81E+01	1.27E+00	1.59E+02	-1.25E+02	2.40E+00	5.75E-01	-2.72E-01	1.58E+02	-1.27E+02	3.85E-03	1.43E-02	-2.73E-03	3.30E+01
59	1.58E+01	1.85E+02	-4.15E+01	1.13E+00	1.39E+02	-1.23E+02	2.89E+00	5.24E-01	-2.48E-01	1.38E+02	-1.25E+02	3.97E-03	1.48E-02	-3.19E-03	3.20E+01
60	1.91E+01	1.67E+02	-4.46E+01	1.01E+00	1.19E+02	-1.18E+02	3.47E+00	4.78E-01	-2.26E-01	1.19E+02	-1.21E+02	4.14E-03	1.54E-02	-3.71E-03	3.10E+01
61	2.29E+01	1.50E+02	-4.76E+01	9.12E-01	1.01E+02	-1.11E+02	4.17E+00	4.36E-01	-2.06E-01	1.01E+02	-1.15E+02	4.32E-03	1.60E-02	-4.33E-03	3.01E+01
62	2.75E+01	1.32E+02	-5.03E+01	8.29E-01	8.45E+01	-1.02E+02	5.02E+00	3.98E-01	-1.88E-01	8.40E+01	-1.07E+02	4.55E-03	1.69E-02	-5.06E-03	2.92E+01
63	3.31E+01	1.16E+02	-5.26E+01	7.64E-01	7.02E+01	-9.19E+01	6.03E+00	3.63E-01	-1.72E-01	6.97E+01	-9.77E+01	4.84E-03	1.80E-02	-5.91E-03	2.84E+01
64	3.98E+01	9.99E+01	-5.44E+01	7.17E-01	5.82E+01	-8.12E+01	7.25E+00	3.31E-01	-1.57E-01	5.77E+01	-8.83E+01	5.19E-03	1.93E-02	-6.89E-03	2.75E+01
65	4.79E+01	8.50E+01	-5.57E+01	6.81E-01	4.79E+01	-7.03E+01	8.72E+00	3.02E-01	-1.43E-01	4.74E+01	-7.88E+01	5.60E-03	2.08E-02	-8.06E-03	2.68E+01
66	5.75E+01	7.08E+01	-5.66E+01	6.58E-01	3.89E+01	-5.91E+01	1.05E+01	2.75E-01	-1.30E-01	3.85E+01	-6.95E+01	6.10E-03	2.26E-02	-9.50E-03	2.63E+01
67	6.92E+01	5.74E+01	-5.64E+01	6.65E-01	3.17E+01	-4.78E+01	1.26E+01	2.51E-01	-1.19E-01	3.13E+01	-6.03E+01	6.79E-03	2.52E-02	-1.13E-02	2.59E+01
68	8.32E+01	4.47E+01	-5.47E+01	7.07E-01	2.58E+01	-3.65E+01	1.52E+01	2.29E-01	-1.08E-01	2.54E+01	-5.16E+01	7.68E-03	2.85E-02	-1.34E-02	2.57E+01
69	1.00E+02	3.26E+01	-4.99E+01	8.41E-01	2.10E+01	-2.49E+01	1.82E+01	2.09E-01	-9.88E-02	2.05E+01	-4.31E+01	9.02E-03	3.35E-02	-1.63E-02	2.59E+01

	P	Q	R	S	T	U	V	W	X	Y	Z	AA	AB	AC	AD
41															
42	f, MHz	ε'	tanδcorr	φcorr	ε"	Zcorr	Rcorr	ρ', Ohm-m	ρ"	Conduct. losses,ε"c	Diel. losses ε"-ε"c	100/ω^0.5	Rms	C(s), pF	σ", S/m
43			G/B		ε'*tanδ	Ohm	Ohm	(Rcorr)	ρ'/tanδ	$\sigma_c/\omega\varepsilon_0$	ε"-ε"c		Ohm		
44	1.00E+00	2.48E+01	8.42E+00	-6.77E+00	2.09E+02	3.17E+02	3.15E+02	8.48E+01	1.01E+01	1.98E+02	1.07E+01	3.99E-02	9.24E-03	3.91E+03	1.38E-03
45	1.20E+00	2.34E+01	7.47E+00	-7.62E+00	1.75E+02	3.14E+02	3.12E+02	8.40E+01	1.12E+01	1.65E+02	1.00E+01	3.64E-02	1.02E-02	2.92E+03	1.57E-03
46	1.45E+00	2.21E+01	6.63E+00	-8.58E+00	1.47E+02	3.11E+02	3.08E+02	8.30E+01	1.25E+01	1.37E+02	9.39E+00	3.32E-02	1.13E-02	2.18E+03	1.78E-03
47	1.74E+00	2.10E+01	5.85E+00	-9.69E+00	1.23E+02	3.08E+02	3.04E+02	8.18E+01	1.40E+01	1.14E+02	8.76E+00	3.03E-02	1.26E-02	1.62E+03	2.03E-03
48	2.09E+00	2.00E+01	5.16E+00	-1.10E+01	1.03E+02	3.04E+02	2.99E+02	8.05E+01	1.56E+01	9.49E+01	8.14E+00	2.76E-02	1.41E-02	1.21E+03	2.32E-03
49	2.51E+00	1.90E+01	4.54E+00	-1.24E+01	8.65E+01	3.00E+02	2.93E+02	7.89E+01	1.74E+01	7.89E+01	7.56E+00	2.52E-02	1.60E-02	9.05E+02	2.66E-03
50	3.02E+00	1.82E+01	3.98E+00	-1.41E+01	7.27E+01	2.95E+02	2.86E+02	7.70E+01	1.93E+01	6.57E+01	7.02E+00	2.30E-02	1.83E-02	6.77E+02	3.07E-03
51	3.63E+00	1.76E+01	3.48E+00	-1.60E+01	6.12E+01	2.89E+02	2.78E+02	7.48E+01	2.15E+01	5.46E+01	6.57E+00	2.09E-02	2.10E-02	5.09E+02	3.55E-03
52	4.37E+00	1.69E+01	3.05E+00	-1.82E+01	5.15E+01	2.82E+02	2.68E+02	7.21E+01	2.37E+01	4.54E+01	6.11E+00	1.91E-02	2.42E-02	3.85E+02	4.11E-03
53	5.25E+00	1.63E+01	2.67E+00	-2.05E+01	4.35E+01	2.74E+02	2.56E+02	6.91E+01	2.59E+01	3.78E+01	5.69E+00	1.74E-02	2.79E-02	2.94E+02	4.76E-03
54	6.31E+00	1.58E+01	2.33E+00	-2.32E+01	3.68E+01	2.64E+02	2.43E+02	6.55E+01	2.80E+01	3.14E+01	5.35E+00	1.59E-02	3.25E-02	2.28E+02	5.53E-03
55	7.59E+00	1.52E+01	2.05E+00	-2.60E+01	3.12E+01	2.53E+02	2.28E+02	6.13E+01	3.00E+01	2.61E+01	5.06E+00	1.45E-02	3.78E-02	1.79E+02	6.43E-03
56	9.12E+00	1.48E+01	1.80E+00	-2.91E+01	2.65E+01	2.41E+02	2.11E+02	5.67E+01	3.15E+01	2.17E+01	4.80E+00	1.32E-02	4.41E-02	1.43E+02	7.49E-03
57	1.10E+01	1.43E+01	1.59E+00	-3.22E+01	2.27E+01	2.27E+02	1.92E+02	5.17E+01	3.26E+01	1.81E+01	4.60E+00	1.20E-02	5.14E-02	1.16E+02	8.72E-03
58	1.32E+01	1.38E+01	1.41E+00	-3.54E+01	1.95E+01	2.12E+02	1.73E+02	4.65E+01	3.31E+01	1.50E+01	4.43E+00	1.10E-02	6.00E-02	9.68E+01	1.02E-02
59	1.58E+01	1.34E+01	1.25E+00	-3.87E+01	1.67E+01	1.96E+02	1.53E+02	4.13E+01	3.31E+01	1.25E+01	4.22E+00	1.00E-02	7.00E-02	8.19E+01	1.18E-02
60	1.91E+01	1.30E+01	1.11E+00	-4.19E+01	1.45E+01	1.80E+02	1.34E+02	3.61E+01	3.24E+01	1.04E+01	4.08E+00	9.14E-03	8.16E-02	7.10E+01	1.38E-02
61	2.29E+01	1.26E+01	9.98E-01	-4.51E+01	1.26E+01	1.64E+02	1.16E+02	3.11E+01	3.12E+01	8.65E+00	3.93E+00	8.34E-03	9.47E-02	6.27E+01	1.61E-02
62	2.75E+01	1.23E+01	8.99E-01	-4.80E+01	1.10E+01	1.47E+02	9.82E+01	2.65E+01	2.94E+01	7.20E+00	3.83E+00	7.60E-03	1.10E-01	5.67E+01	1.88E-02
63	3.31E+01	1.19E+01	8.18E-01	-5.07E+01	9.75E+00	1.31E+02	8.29E+01	2.23E+01	2.73E+01	5.99E+00	3.76E+00	6.93E-03	1.26E-01	5.23E+01	2.20E-02
64	3.98E+01	1.16E+01	7.53E-01	-5.30E+01	8.70E+00	1.16E+02	6.97E+01	1.88E+01	2.49E+01	4.98E+00	3.72E+00	6.32E-03	1.45E-01	4.93E+01	2.56E-02
65	4.79E+01	1.12E+01	6.95E-01	-5.52E+01	7.81E+00	1.02E+02	5.81E+01	1.57E+01	2.25E+01	4.14E+00	3.67E+00	5.77E-03	1.64E-01	4.73E+01	2.99E-02
66	5.75E+01	1.10E+01	6.42E-01	-5.73E+01	7.07E+00	8.86E+01	4.78E+01	1.29E+01	2.01E+01	3.45E+00	3.63E+00	5.26E-03	1.86E-01	4.68E+01	3.53E-02
67	6.92E+01	1.09E+01	6.03E-01	-5.89E+01	6.55E+00	7.61E+01	3.93E+01	1.06E+01	1.76E+01	2.87E+00	3.68E+00	4.80E-03	2.08E-01	4.82E+01	4.18E-02
68	8.32E+01	1.08E+01	5.72E-01	-6.02E+01	6.16E+00	6.46E+01	3.21E+01	8.64E+00	1.51E+01	2.38E+00	3.78E+00	4.37E-03	2.32E-01	5.24E+01	4.99E-02
69	1.00E+02	1.09E+01	5.53E-01	-6.10E+01	6.02E+00	5.37E+01	2.60E+01	7.00E+00	1.27E+01	1.98E+00	4.03E+00	3.99E-03	2.60E-01	6.38E+01	6.05E-02

Fort Huachuca (25%)

Fort Huachuca, Arizona - 25 %

Low Frequency

Parallel-plate sample holder (Disk electrodes, d=5 cm) — L & Cs-from Air

HP 4194A, Z-Probe, 51 points (Labview program) — Calibrating Z-Probe — ε_0, F/m = 8.85E-12 — μ_0, H/m = 1.26E-06

Measuring soil from ATF-Subsurface (Allison) in the shield box, # 11- sieved. With 15% distilled water

Subsurface — Copy from "ATFSbsr15-03"

	Diameter	Thickness	Volume	Weight	Dens.(wet)	Drying, weight in g.		Water cont.		Bulk density	
	D, cm	h, cm	V, cm^3	M. g	γ, g/cm^3	Wet	Dry	W, %	Wv, %	g/cm^3	σ_{st}, S/m
	5	0.7100	13.940817	27.38	1.96402	27.26	23.79	14.586	25.001	1.71	0.01331

L, nH	Cs	From File:	ImpedAn94-DiskEl-08, Air				Cs	L, nH
29.1332	4.19290	(using polynoms for repeated data,12.10.02):					4.1928997	29.13316

Sample # 11, sieved, 15 % water

Measurement data table. Columns H (R_{el}) and I (X_{el}) carry the fits $y=58.771x$ and $y=-13.435x$ respectively. Column O subtitle "B/w". Columns J, K additionally labelled R', X'.

f, MHz	Z, Ohm	ϕ, deg.	$\tan\delta$	Res.meas. R(s), Ohm	Xmeas. Zsinϕ	$X_L=\omega_0 L$ Ohm	R_{el}	X_{el}	Rm-Rms- R_{el}	Xm-wL- X_{el}	G, S	σ', S/m	B, S	C, pF
1.00E-02	2.95E+02	-2.38E+00	2.40E+01	2.95E+02	-1.23E+01	1.83E-03	2.34E+01	-5.36E+00	2.71E+02	-6.91E+00	3.68E-03	1.33E-02	-9.35E-05	1.49E+03
3.02E-02	2.86E+02	-2.88E+00	1.99E+01	2.86E+02		5.53E-03	1.35E+01	-3.08E+00	2.73E+02	3.08E+00	3.67E-03	1.33E-02	4.22E-05	-2.22E+02
4.37E-02	2.83E+02	-2.37E+00	2.41E+01	2.83E+02	-1.17E+01	7.99E-03	1.12E+01	-2.57E+00	2.72E+02	-9.18E+00	3.67E-03	1.33E-02	-1.23E-04	4.48E+02
5.25E-02	2.82E+02	-2.58E+00	2.22E+01	2.82E+02	-1.27E+01	9.61E-03	1.02E+01	-2.34E+00	2.72E+02	-1.04E+01	3.68E-03	1.33E-02	-1.39E-04	4.21E+02
6.31E-02	2.81E+02	-2.58E+00	2.22E+01	2.81E+02	-1.27E+01	1.15E-02	9.33E+00	-2.13E+00	2.72E+02	-1.05E+01	3.68E-03	1.33E-02	-1.41E-04	3.56E+02
7.59E-02	2.80E+02	-2.66E+00	2.15E+01	2.80E+02	-1.30E+01	1.39E-02	8.51E+00	-1.95E+00	2.71E+02	-1.11E+01	3.68E-03	1.33E-02	-1.48E-04	3.11E+02
9.12E-02	2.79E+02	-2.89E+00	1.98E+01	2.78E+02	-1.41E+01	1.67E-02	7.76E+00	-1.77E+00	2.70E+02	-1.23E+01	3.69E-03	1.33E-02	-1.66E-04	2.89E+02
1.10E-01	2.77E+02	-3.05E+00	1.88E+01	2.77E+02	-1.47E+01	2.01E-02	7.08E+00	-1.62E+00	2.70E+02	-1.31E+01	3.70E-03	1.34E-02	-1.77E-04	2.57E+02
3.31E-01	2.68E+02	-4.70E+00	1.22E+01	2.67E+02	-2.19E+01	6.06E-02	4.07E+00	-9.31E-01	2.60E+02	-2.11E+01	3.78E-03	1.37E-02	-2.94E-04	1.42E+02
4.79E-01	2.64E+02	-5.57E+00	1.03E+01	2.63E+02	-2.57E+01	8.76E-02	3.39E+00	-7.75E-01	2.60E+02	-2.50E+01	3.82E-03	1.38E-02	-3.54E-04	1.18E+02
6.92E-01	2.60E+02	-6.87E+00	8.30E+00	2.58E+02	-3.11E+01	1.27E-01	2.82E+00	-6.44E-01	2.55E+02	-3.06E+01	3.86E-03	1.40E-02	-4.44E-04	1.02E+02
1.00E+00	2.56E+02	-8.59E+00	6.62E+00	2.53E+02	-3.82E+01	1.83E-01	2.34E+00	-5.36E-01	2.50E+02	-3.78E+01	3.91E-03	1.41E-02	-5.64E-04	8.97E+01
4.37E+00	2.18E+02	-2.36E+01	2.29E+00	1.99E+02	-8.72E+01	7.99E-01	1.12E+00	-2.57E-01	1.98E+02	-8.78E+01	4.22E-03	1.52E-02	-1.75E-03	6.39E+01
5.25E+00	2.09E+02	-2.66E+01	1.99E+00	1.86E+02	-9.36E+01	9.61E-01	1.02E+00	-2.34E-01	1.85E+02	-9.43E+01	4.29E-03	1.55E-02	-2.04E-03	6.19E+01
6.31E+00	1.98E+02	-2.99E+01	1.74E+00	1.72E+02	-9.87E+01	1.15E+00	9.33E-01	-2.13E-01	1.71E+02	-9.96E+01	4.37E-03	1.58E-02	-2.39E-03	6.02E+01
7.59E+00	1.86E+02	-3.33E+01	1.52E+00	1.55E+02	-1.02E+02	1.39E+00	8.51E-01	-1.95E-01	1.54E+02	-1.03E+02	4.48E-03	1.62E-02	-2.79E-03	5.85E+01
9.12E+00	1.73E+02	-3.67E+01	1.34E+00	1.38E+02	-1.03E+02	1.67E+00	7.76E-01	-1.77E-01	1.38E+02	-1.05E+02	4.60E-03	1.66E-02	-3.26E-03	5.69E+01
1.10E+01	1.59E+02	-4.00E+01	1.19E+00	1.21E+02	-1.02E+02	2.01E+00	7.08E-01	-1.62E-01	1.21E+02	-1.04E+02	4.76E-03	1.72E-02	-3.81E-03	5.53E+01
2.29E+01	1.02E+02	-5.17E+01	7.89E-01	6.29E+01	-7.97E+01	4.19E+00	4.90E-01	-1.12E-01	6.23E+01	-8.38E+01	5.72E-03	2.07E-02	-7.08E-03	4.92E+01
3.31E+01	7.54E+01	-5.56E+01	6.85E-01	4.26E+01	-6.22E+01	6.06E+00	4.07E-01	-9.31E-02	4.21E+01	-6.82E+01	6.55E-03	2.37E-02	-9.75E-03	4.68E+01
4.79E+01	5.26E+01	-5.71E+01	6.47E-01	2.86E+01	-4.41E+01	8.76E+00	3.39E-01	-7.75E-02	2.81E+01	-5.28E+01	7.84E-03	2.84E-02	-1.35E-02	4.49E+01
5.75E+01	4.20E+01	-5.66E+01	6.58E-01	2.31E+01	-3.51E+01	1.05E+01	3.09E-01	-7.07E-02	2.26E+01	-4.55E+01	8.75E-03	3.16E-02	-1.61E-02	4.46E+01
6.92E+01	3.19E+01	-5.41E+01	7.24E-01	1.87E+01	-2.58E+01	1.27E+01	2.82E-01	-6.44E-02	1.82E+01	-3.84E+01	1.01E-02	3.64E-02	-1.94E-02	4.47E+01
8.32E+01	2.25E+01	-4.76E+01	9.14E-01	1.52E+01	-1.66E+01	1.52E+01	2.57E-01	-5.88E-02	1.47E+01	-3.17E+01	1.20E-02	4.33E-02	-2.38E-02	4.55E+01
1.00E+02	1.41E+01	-2.95E+01	1.77E+00	1.23E+01	-6.94E+00	1.83E+01	2.34E-01	-5.36E-01	1.18E+01	-2.52E+01	1.52E-02	5.50E-02	-3.00E-02	4.77E+01

Fort Huachuca, Arizona - 25 %

Low Frequency

#	f, MHz	ε'	tanδcorr (G/B)	φcorr	ε" (ε'*tanδ)	Zcorr (Ohm)	Rcorr (Ohm)	ρ', Ohm-m (Rcorr)	ρ" (ρ'/tanδ)	losses,ε"c ($\sigma_c/\omega\varepsilon_0$)	losses (ε"-ε"c)	$100/\omega^{0.5}$	Rms (Ohm)	C(s), pF	σ", S/m
16	1.00E-02	6.08E+02	3.94E+01	-1.46E+00	2.39E+04	2.71E+02	2.71E+02	7.51E+01	1.91E+00	2.39E+04	1.03E+00	3.99E-01	5.51E-03	1.30E+06	3.38E-04
17	3.02E-02	-9.09E+01	-8.69E+01	6.59E-01	7.89E+03	2.73E+02	2.73E+02	7.54E+01	-8.68E-01	7.92E+03	-2.87E+01	2.30E-01	5.51E-03	3.66E+05	-1.53E-04
18	4.37E-02	1.83E+02	2.99E+01	-1.92E+00	5.47E+03	2.72E+02	2.72E+02	7.52E+01	2.52E+00	5.48E+03	-9.96E+00	1.91E-01	5.51E-03	3.11E+05	4.45E-04
19	5.25E-02	1.72E+02	2.64E+01	-2.17E+00	4.55E+03	2.72E+02	2.72E+02	7.51E+01	2.84E+00	4.56E+03	-6.86E+00	1.74E-01	5.51E-03	2.39E+05	5.03E-04
20	6.31E-02	1.45E+02	2.61E+01	-2.20E+00	3.79E+03	2.72E+02	2.72E+02	7.51E+01	2.88E+00	3.79E+03	-3.97E+00	1.59E-01	5.51E-03	1.99E+05	5.10E-04
21	7.59E-02	1.27E+02	2.48E+01	-2.30E+00	3.15E+03	2.71E+02	2.71E+02	7.50E+01	3.02E+00	3.15E+03	7.11E-01	1.45E-01	5.51E-03	1.61E+05	5.36E-04
22	9.12E-02	1.18E+02	2.23E+01	-2.57E+00	2.63E+03	2.71E+02	2.70E+02	7.48E+01	3.35E+00	2.62E+03	6.53E+00	1.32E-01	5.51E-03	1.24E+05	5.98E-04
23	1.10E-01	1.05E+02	2.09E+01	-2.74E+00	2.19E+03	2.70E+02	2.70E+02	7.46E+01	3.57E+00	2.18E+03	1.14E+01	1.20E-01	5.18E-03	9.86E+04	6.41E-04
24	3.31E-01	5.78E+01	1.28E+01	-4.45E+00	7.42E+02	2.64E+02	2.63E+02	7.27E+01	5.66E+00	7.23E+02	1.97E+01	6.93E-02	6.20E-03	2.19E+04	1.06E-03
25	4.79E-01	4.81E+01	1.08E+01	-5.31E+00	5.18E+02	2.61E+02	2.60E+02	7.18E+01	6.67E+00	5.00E+02	1.85E+01	5.77E-02	6.87E-03	1.30E+04	1.28E-03
26	6.92E-01	4.17E+01	8.69E+00	-6.57E+00	3.63E+02	2.57E+02	2.56E+02	7.07E+01	8.14E+00	3.46E+02	1.67E+01	4.80E-02	7.84E-03	7.39E+03	1.61E-03
27	1.00E+00	3.66E+01	6.93E+00	-8.21E+00	2.54E+02	2.53E+02	2.51E+02	6.94E+01	1.00E+01	2.39E+02	1.46E+01	3.99E-02	9.24E-03	4.17E+03	2.04E-03
28	4.37E+00	2.61E+01	2.41E+00	-2.26E+01	6.28E+01	2.19E+02	2.02E+02	5.59E+01	2.32E+01	5.48E+01	7.98E+01	1.91E-02	2.42E-02	4.18E+02	6.33E-03
29	5.25E+00	2.53E+01	2.10E+00	-2.55E+01	5.31E+01	2.11E+02	1.90E+02	5.26E+01	2.51E+01	4.56E+01	7.50E+01	1.74E-02	2.79E-02	3.24E+02	7.38E-03
30	6.31E+00	2.46E+01	1.83E+00	-2.86E+01	4.50E+01	2.01E+02	1.76E+02	4.87E+01	2.66E+01	3.79E+01	7.12E+01	1.59E-02	3.25E-02	2.56E+02	8.63E-03
31	7.59E+00	2.39E+01	1.60E+00	-3.19E+01	3.84E+01	1.90E+02	1.61E+02	4.45E+01	2.77E+01	3.15E+01	6.83E+01	1.45E-02	3.78E-02	2.06E+02	1.01E-02
32	9.12E+00	2.32E+01	1.41E+00	-3.53E+01	3.28E+01	1.77E+02	1.45E+02	4.00E+01	2.83E+01	2.62E+01	6.58E+01	1.32E-02	4.41E-02	1.69E+02	1.18E-02
33	1.10E+01	2.26E+01	1.25E+00	-3.87E+01	2.82E+01	1.64E+02	1.28E+02	3.54E+01	2.83E+01	2.18E+01	6.41E+01	1.20E-02	5.14E-02	1.42E+02	1.38E-02
34	2.29E+01	2.01E+01	8.07E-01	-5.11E+01	1.62E+01	1.10E+02	6.90E+01	1.91E+01	2.36E+01	1.04E+01	5.77E+00	8.34E-03	9.47E-02	8.72E+01	2.56E-02
35	3.31E+01	1.91E+01	6.72E-01	-5.61E+01	1.29E+01	8.51E+01	4.75E+01	1.31E+01	1.95E+01	7.23E+00	5.63E+00	6.93E-03	1.26E-01	7.72E+01	3.52E-02
36	4.79E+01	1.83E+01	5.81E-01	-5.98E+01	1.07E+01	6.40E+01	3.22E+01	8.90E+00	1.53E+01	5.00E+00	5.65E+00	5.77E-03	1.64E-01	7.53E+01	4.88E-02
37	5.75E+01	1.82E+01	5.43E-01	-6.15E+01	9.88E+00	5.46E+01	2.60E+01	7.20E+00	1.33E+01	4.16E+00	5.72E+00	5.26E-03	1.86E-01	7.89E+01	5.82E-02
38	6.92E+01	1.82E+01	5.19E-01	-6.26E+01	9.46E+00	4.57E+01	2.11E+01	5.82E+00	1.12E+01	3.46E+00	6.00E+00	4.80E-03	2.08E-01	8.90E+01	7.02E-02
39	8.32E+01	1.86E+01	5.04E-01	-6.32E+01	9.37E+00	3.76E+01	1.69E+01	4.68E+00	9.28E+00	2.88E+00	6.49E+00	4.37E-03	2.32E-01	1.15E+02	8.60E-02
40	1.00E+02	1.95E+01	5.08E-01	-6.31E+01	9.89E+00	2.98E+01	1.35E+01	3.73E+00	7.34E+00	2.39E+00	7.50E+00	3.99E-03	2.60E-01	2.29E+02	1.08E-01

Attenuation and Velocity Data

Wv=0%			Wv=5.8%			Wv=10.6%		
Frequency	Attenuation	Velocity	Frequency	Attenuation	Velocity	Frequency	Attenuation	Velocity
MHz	dB/m	m/s	MHz	dB/m	m/s	MHz	dB/m	m/s
0.001	9.02E-06	1.62E+08	0.001	3.26E-02	1.41E+06	0.001	8.51E-02	5.87E+05
0.002	1.64E-05	1.63E+08	0.002	4.85E-02	1.97E+06	0.002	1.24E-01	8.30E+05
0.003	2.28E-05	1.64E+08	0.003	6.06E-02	2.41E+06	0.003	1.53E-01	1.02E+06
0.005	3.45E-05	1.66E+08	0.005	8.02E-02	3.10E+06	0.005	2.00E-01	1.31E+06
0.01	6.11E-05	1.67E+08	0.01	1.17E-01	4.38E+06	0.01	2.85E-01	1.85E+06
0.02	1.16E-04	1.68E+08	0.02	1.67E-01	6.17E+06	0.02	4.07E-01	2.61E+06
0.03	1.64E-04	1.69E+08	0.03	2.07E-01	7.52E+06	0.03	5.00E-01	3.19E+06
0.05	2.70E-04	1.70E+08	0.05	2.69E-01	9.66E+06	0.05	6.47E-01	4.11E+06
0.1	5.09E-04	1.71E+08	0.1	3.82E-01	1.35E+07	0.1	9.19E-01	5.78E+06
0.2	9.40E-04	1.72E+08	0.2	5.41E-01	1.88E+07	0.2	1.30E+00	8.12E+06
0.3	1.39E-03	1.73E+08	0.3	6.61E-01	2.27E+07	0.3	1.60E+00	9.89E+06
0.5	2.23E-03	1.74E+08	0.5	8.48E-01	2.87E+07	0.5	2.06E+00	1.27E+07
1	4.17E-03	1.75E+08	1	1.18E+00	3.87E+07	1	2.91E+00	1.76E+07
2	7.89E-03	1.76E+08	2	1.62E+00	5.09E+07	2	4.10E+00	2.44E+07
3	1.12E-02	1.76E+08	3	1.94E+00	5.86E+07	3	4.99E+00	2.93E+07
5	1.69E-02	1.77E+08	5	2.41E+00	6.85E+07	5	6.36E+00	3.66E+07
10	3.49E-02	1.78E+08	10	3.26E+00	8.13E+07	10	8.72E+00	4.84E+07
20	5.85E-02	1.79E+08	20	4.49E+00	9.31E+07	20	1.17E+01	6.17E+07
30	8.71E-02	1.79E+08	30	5.49E+00	9.97E+07	30	1.38E+01	6.96E+07
40	1.12E-01	1.80E+08	40	6.37E+00	1.04E+08	40	1.54E+01	7.52E+07
59	1.47E-01	1.79E+08	30	5.27E+00	9.95E+07	10	8.06E+00	5.11E+07
79	2.00E-01	1.80E+08	40	6.18E+00	1.04E+08	20	1.09E+01	6.41E+07
98	2.52E-01	1.80E+08	50	7.00E+00	1.07E+08	30	1.30E+01	7.20E+07
118	3.15E-01	1.80E+08	100	1.03E+01	1.17E+08	40	1.47E+01	7.75E+07
138	3.71E-01	1.80E+08	150	1.30E+01	1.23E+08	50	1.62E+01	8.17E+07
157	4.36E-01	1.80E+08	200	1.52E+01	1.27E+08	100	2.18E+01	9.37E+07
177	5.00E-01	1.80E+08	250	1.72E+01	1.29E+08	150	2.61E+01	1.00E+08
196	5.65E-01	1.80E+08	300	1.88E+01	1.31E+08	200	2.95E+01	1.04E+08
255	7.93E-01	1.81E+08	350	2.04E+01	1.33E+08	250	3.25E+01	1.07E+08
314	1.06E+00	1.81E+08	400	2.18E+01	1.35E+08	300	3.50E+01	1.10E+08
412	1.58E+00	1.81E+08	450	2.32E+01	1.36E+08	350	3.73E+01	1.12E+08
510	2.10E+00	1.81E+08	500	2.45E+01	1.37E+08	400	3.94E+01	1.13E+08
647	2.81E+00	1.81E+08	600	2.67E+01	1.39E+08	500	4.33E+01	1.16E+08
745	3.46E+00	1.82E+08	700	2.88E+01	1.41E+08	600	4.68E+01	1.18E+08
843	4.01E+00	1.82E+08	800	3.08E+01	1.42E+08	700	5.04E+01	1.19E+08
902	4.44E+00	1.82E+08	900	3.31E+01	1.42E+08	800	5.46E+01	1.20E+08
1000	4.99E+00	1.81E+08	1000	3.59E+01	1.42E+08	900	6.01E+01	1.21E+08

Wv=16.4 %			Wv=25 %			Wv=45 %		
Frequency	Attenuation	Velocity	Frequency	Attenuation	Velocity	Frequency	Attenuation	Velocity
MHz	dB/m	m/s	MHz	dB/m	m/s	MHz	dB/m	m/s
0.001	1.05E-01	4.66E+05	0.001	1.52E-01	3.29E+05	0.01	4.96E-01	1.06E+06
0.002	1.54E-01	6.57E+05	0.002	2.20E-01	4.66E+05	0.02	7.06E-01	1.49E+06
0.003	1.91E-01	8.05E+05	0.003	2.73E-01	5.71E+05	0.03	8.69E-01	1.82E+06
0.005	2.50E-01	1.04E+06	0.005	3.56E-01	7.37E+05	0.05	1.13E+00	2.34E+06
0.01	3.58E-01	1.46E+06	0.01	5.08E-01	1.04E+06	0.1	1.60E+00	3.29E+06
0.02	5.11E-01	2.06E+06	0.02	7.24E-01	1.46E+06	0.2	2.28E+00	4.61E+06
0.03	6.29E-01	2.52E+06	0.03	8.90E-01	1.79E+06	0.3	2.80E+00	5.61E+06
0.05	8.16E-01	3.24E+06	0.05	1.15E+00	2.30E+06	0.5	3.62E+00	7.19E+06
0.1	1.16E+00	4.55E+06	0.1	1.64E+00	3.22E+06	1	5.13E+00	1.00E+07
0.2	1.65E+00	6.38E+06	0.2	2.33E+00	4.52E+06	2	7.24E+00	1.39E+07
0.3	2.02E+00	7.76E+06	0.3	2.86E+00	5.50E+06	3	8.82E+00	1.67E+07
0.5	2.62E+00	9.93E+06	0.5	3.70E+00	7.04E+06	5	1.12E+01	2.10E+07
1	3.71E+00	1.38E+07	1	5.25E+00	9.81E+06	10	1.54E+01	2.80E+07
2	5.23E+00	1.91E+07	2	7.41E+00	1.36E+07	20	2.05E+01	3.58E+07
3	6.37E+00	2.30E+07	3	9.04E+00	1.64E+07	30	2.39E+01	4.02E+07
5	8.07E+00	2.87E+07	5	1.15E+01	2.07E+07	40	2.65E+01	4.32E+07
10	1.11E+01	3.84E+07	10	1.58E+01	2.78E+07	50	2.87E+01	4.53E+07
20	1.48E+01	4.92E+07	20	2.09E+01	3.62E+07	60	3.05E+01	4.68E+07
30	1.72E+01	5.57E+07	30	2.38E+01	4.17E+07	1	5.27E+00	9.79E+06
40	1.88E+01	6.03E+07	40	2.54E+01	4.58E+07	10	1.59E+01	2.75E+07
50	2.28E+01	6.62E+07	50	3.29E+01	4.96E+07	20	2.12E+01	3.53E+07
60	2.45E+01	6.92E+07	100	4.24E+01	5.87E+07	30	2.47E+01	3.99E+07
70	2.61E+01	7.17E+07	200	5.29E+01	6.66E+07	40	2.75E+01	4.29E+07
80	2.74E+01	7.38E+07	250	5.61E+01	6.90E+07	50	2.97E+01	4.51E+07
90	2.87E+01	7.55E+07	300	5.86E+01	7.09E+07	60	3.17E+01	4.68E+07
100	2.98E+01	7.72E+07	350	6.05E+01	7.26E+07	90	3.64E+01	5.01E+07
150	3.45E+01	8.29E+07	400	6.21E+01	7.41E+07	100	3.78E+01	5.09E+07
200	3.79E+01	8.66E+07	500	6.49E+01	7.67E+07	200	4.86E+01	5.55E+07
300	4.25E+01	9.15E+07	600	6.85E+01	7.89E+07	300	5.75E+01	5.78E+07
400	4.61E+01	9.49E+07	700	7.54E+01	8.04E+07	400	6.78E+01	5.93E+07
500	4.76E+01	9.75E+07	800	9.02E+01	8.09E+07	500	8.35E+01	5.99E+07
600	4.98E+01	9.99E+07	900	1.23E+02	7.99E+07	600	1.15E+02	5.93E+07
700	5.17E+01	1.02E+08	950	1.54E+02	7.89E+07	700	1.97E+02	5.71E+07
800	5.38E+01	1.03E+08	1000	2.05E+02	7.80E+07	800	4.47E+02	5.96E+07
900	6.26E+01	1.04E+08						
1000	7.49E+01	1.04E+08						

Montmorillonite 70.6%			Avra 10.6%			Brookhaven 9.75%			Brookhaven 0%		
Frequency MHz	Attenuation dB/m	Velocity m/s	Frequency MHz	Attenuation dB/m	Velocity m/s	Frequency MHz	Attenuation dB/m	Velocity m/s	Frequency MHz	Attenuation dB/m	Velocity m/s
0.001	2.60E-01	1.70E+05	0.001	8.51E-02	5.87E-05	0.001	2.52E-02	2.13E+06	0.001	3.17E-06	1.82E+08
0.002	3.93E-01	2.41E+05	0.002	1.24E-01	8.30E-05	0.002	3.58E-02	3.02E+06	0.002	3.02E-06	1.83E+08
0.003	4.94E-01	2.96E+05	0.003	1.53E-01	1.02E-06	0.003	4.39E-02	3.70E+06	0.003	4.02E-06	1.82E+08
0.005	6.53E-01	3.82E+05	0.005	2.00E-01	1.31E-06	0.005	5.67E-02	4.77E+06	0.005	5.18E-06	1.83E+08
0.010	9.45E-01	5.38E+05	0.010	2.85E-01	1.85E-06	0.01	8.04E-02	6.74E+06	0.01	6.54E-05	1.83E+08
0.020	1.36E+00	7.56E+05	0.020	4.07E-01	2.61E-06	0.02	1.14E-01	9.52E+06	0.02	1.63E-05	1.83E+08
0.030	1.68E+00	9.21E+05	0.030	5.00E-01	3.19E-06	0.03	1.39E-01	1.16E+07	0.03	1.78E-05	1.83E+08
0.050	2.18E+00	1.18E+06	0.050	6.47E-01	4.11E-06	0.05	1.79E-01	1.50E+07	0.05	3.78E-05	1.83E+08
0.100	3.12E+00	1.65E+06	0.100	9.19E-01	5.78E-06	0.1	2.53E-01	2.11E+07	0.1	6.45E-05	1.84E+08
0.200	4.47E+00	2.30E+06	0.200	1.30E+00	8.12E-06	0.2	3.55E-01	2.95E+07	0.2	7.78E-05	1.84E+08
0.300	5.51E+00	2.80E+06	0.300	1.60E+00	9.89E-06	0.3	4.31E-01	3.58E+07	0.3	1.20E-04	1.84E+08
0.500	7.17E+00	3.57E+06	0.500	2.06E+00	1.27E-07	0.5	5.47E-01	4.53E+07	0.5	1.83E-04	1.84E+08
1.000	1.02E+01	4.97E+06	1.000	2.91E+00	1.76E-07	1	7.45E-01	6.11E+07	1	2.99E-04	1.84E+08
2.000	1.44E+01	6.89E+06	2.000	4.10E+00	2.44E-07	2	1.02E+00	7.80E+07	2	5.94E-04	1.84E+08
3.000	1.76E+01	8.30E+06	3.000	4.99E+00	2.93E-07	3	1.18E+00	8.86E+07	3	7.98E-04	1.84E+08
5.000	2.24E+01	1.04E+07	5.000	6.36E+00	3.66E-07	7	1.54E+00	1.08E+08	7	2.03E-03	1.84E+08
10.000	3.03E+01	1.39E+07	10	8.06E+00	5.11E-07	10	1.70E+00	1.14E+08	10	2.90E-03	1.84E+08
20.000	3.86E+01	1.81E+07	20	1.09E+01	6.41E-07	20	1.96E+00	1.23E+08	15	4.34E-03	1.84E+08
30.000	4.16E+01	2.12E+07	30	1.30E+01	7.20E-07	30	2.08E+00	1.26E+08	20	5.79E-03	1.84E+08
40.000	4.13E+01	2.39E+07	40	1.47E+01	7.75E-07	40	2.15E+00	1.27E+08	30	8.68E-03	1.84E+08
			50	1.62E+01	8.17E-07	50	2.20E+00	1.28E+08	40	1.16E-02	1.84E+08
			100	2.18E+01	9.37E-07	50	2.13E+00	1.28E+08	50	1.44E-02	1.84E+08
			150	2.61E+01	1.00E-08	88	2.27E+00	1.29E+08	50	1.44E-02	1.84E+08
			200	2.95E+01	1.04E-08	107	2.34E+00	1.29E+08	69	2.59E-02	1.85E+08
			300	3.50E+01	1.10E-08	145	2.48E+00	1.29E+08	88	4.06E-02	1.85E+08
			400	3.94E+01	1.13E-08	202	2.74E+00	1.30E+08	107	5.24E-02	1.85E+08
			500	4.33E+01	1.16E-08	259	3.01E+00	1.30E+08	145	9.99E-02	1.85E+08
			600	4.68E+01	1.18E-08	297	3.21E+00	1.30E+08	202	1.84E-01	1.85E+08
			700	5.04E+01	1.19E-08	411	3.95E+00	1.31E+08	297	3.84E-01	1.85E+08
			800	5.46E+01	1.20E-08	506	4.58E+00	1.31E+08	411	7.65E-01	1.85E+08
			900	6.01E+01	1.21E-08	601	5.26E+00	1.32E+08	506	1.09E+00	1.85E+08
			1000	6.82E+01	1.20E-08	696	6.08E+00	1.32E+08	601	1.39E+00	1.85E+08
						810	7.13E+00	1.32E+08	696	1.58E+00	1.85E+08
						924	8.29E+00	1.32E+08	810	2.14E+00	1.85E+08
						1000	9.12E+00	1.32E+08	924	2.57E+00	1.85E+08
									1000	2.85E+00	1.85E+08

Spreadsheet formulas for processing laboratory-sample electromagnetic measurements

Chapter Outline

E.1 High Frequency—Coaxial Air Line

High Frequency Spreadsheet

	A	B	C	D
1	Avra - 6.15% Sample			
2	High Frequency - Coaxial Air Line			
3				
4	Sample holder, 3 cm,#154, spacers-1.5 mm (flat, h=0.324cm)			
5	Length	R, cm	r, cm	Weight
6	cm			M, g
7	2.676	0.7144	0.3102	5.28
8	L, nH	C_s, pF		Volume
9	1.856	0.705		V, cm^3
10				=(PI()*B7^2-PI()*C7^2)*$A$7
11	f, MHz	Z, Ohm	ϕ, deg	tanδ
12				
13	1	989	-12.8	=-1/TAN(C13*PI()/180)

	E	F	G
1			
2			
3			
4			
5	Dens.(wet)		Drying, weight in g.
6	g, g/cm^3	Wet	Dry
7	=D7/D10	5.21	5
8			
9			
10			
11	Res.meas.	Xmeas.	$X_L=\omega L$
12	R(s), Ohm	Zsinϕ	Ohm
13	=B13*COS(C13*PI()/180)	=B13*SIN(C13*PI()/180)	=2*PI()*A13*10^6*$A$9*10^-9

	H	I	J
1			
2			
3			
4			
5	Water cont.		Bulk density
6	W , %	Wv, %	g/cm^3
7	=(F7-G7)*100/G7	6.15	=G7*E7/F7
8			sc (from PL method)
9			0.005186
10			
11	Rm-Rms	Xm-ωL	G, S
12	R' = Rm	X'	
13	=E13	=F13-G13	=H13/(H13^2+I13^2)

High Frequency Spreadsheet

	K	L	M
1			
2			
3			
4			
5			
6			
7			
8			
9			
10			
11	σ', S/m	B, S	tanδcorr
12			G/B
13	=100*LN($B$7/$C$7)*J13/(2*PI()*$A$7)	=(I13/(H13^2+I13^2))+2*PI()*A13*10^6*$B$9*10^-12	=-J13/L13

	N	O	P
1	Avra - 6.15% Sample		
2	High Frequency - Coaxial Air Line		
3			
4			
5			
6			
7			
8			
9			
10			ε'
11	f, MHz	C, pF	
12		B/w	
13	1	=-(10^12*L13/(2*PI()*A13*10^6))	=LN($B$7/$C$7)*O13/(2*PI()*$A$7*0.08854)

High Frequency Spreadsheet

	Q	R	S
1			
2			
3			
4			
5			
6			
7			
8			
9			
10			
11	ϕcorr	ε''	Zcorr
12		ε'tanδ	Ohm
13	=-ATAN(1/M13)*180/PI()	=P13*M13	=1/SQRT(J13^2+L13^2)

High Frequency Spreadsheet

	T	U
1		
2		
3		
4		
5		
6		
7		
8		
9		
10		
11	Rcorr	ρ', Ohm-m
12	Ohm	(Rcorr)
13	=S13*COS(Q13*PI()/180)	=2*PI()*$A$7*T13/(100*LN($B$7/$C$7))

	V
1	
2	
3	
4	
5	
6	
7	
8	
9	
10	
11	ρ"
12	
13	=2*PI()*A13*10^6*P13*8.854*10^-12/(K13^2+(2*PI()*A13*10^6*P13*8.854*10^-12)^2)

High Frequency Spreadsheet

	W	X	Y
1			
2			
3			
4			
5			
6			
7			
8			
9			
10	Conductive	Dielectric	
11	losses ε''_c	losses $\varepsilon''-\varepsilon''_c$	σ'', S/m
12	sc/$\omega\varepsilon_0$		
13	=\$J\$9/(2*PI()*N13*10^6*8.854*10^-12)	=R13-W13	=K13/M13

E.2 Low Frequency—Disk Electrodes

Low Frequency Spreadsheet

	A	B	C	D
1	Avra - 6.15% Sample			
2	Low Frequency - Disk Electrodes			
3				
4	Avra, # 16-2.5 (From the bag)			
5	HP 4194A, Fixture.			
6	Diameter	Thickness	Volume	Weight
7	D, cm	h, cm	V, cm^3	M, g
8	5	0.9233	=PI()*A8^2*B8/4	28.67
9	L, nH	Cs, pF		Cs, calc.
10	50	2.1813		=-0.8827*B8+2.9963
11				
12	f, MHz	Z, Ohm	ϕ, deg	ϵ'tanδ
13				
14	0.001	1091.06	-10.028	=-1/TAN(C14*PI()/180)

Low Frequency Spreadsheet

	E	F	G	H
1				
2				
3				
4				
5			σ_c, S/m (from FL - method)	0.0518577
6	Dens.(wet)		Drying, weight in g.	Water cont.
7	g, g/cm^3	Wet	Dry	W, %
8	=D8/C8	28.61	27.49	=(F8-G8)*100/G8
9			(Regression egn. for Air meas. File "Methods"	
10				
11				
12	Res.meas.	Xmeas.	$X_L = \omega L$	R_{el}
13	R(s), Ohm	Zsinϕ	Ohm	y=134.53x
14	=B14*COS(C14*PI()/180)	=B14*SIN(C14*PI()/180)	=2*PI()*A14*10^6*$A$10*10^-9	=134.5322*AB14

Low Frequency Spreadsheet

	I	J	K	L
1				
2				
3				
4				
5				
6			Bulk density	
7	Wv, %	g/cm^3		
8	6.15	=G8*E8/F8		
9				
10				
11			R'	
12	Xel	Rm-Rms-	$Xm-\omega L-$	G, S
13	y=-141.84x	Rel	Xel	
14	=-141.84*AB14	=E14-AD14-H14	=F14-G14-I14	=J14/(J14^2+K14^2)

	M	N	O
1	Avra - 6.15% Sample		
2	Low Frequency - Disk Electrodes		
3			
4			
5			
6			
7			
8			
9			
10			
11			
12	f, MHz	σ', S/m	B, S
13			
14	0.001	=400*$B$8*L14/(PI()*$A$8^2)	=(K14/(J14^2+K14^2))+2*PI()*A14*10^6*$B$10*10^-12

Low Frequency Spreadsheet

	P	Q	R	S
1				
2				
3				
4				
5				
6				
7				
8				
9				
10				
11				
12	C, pF	ε'	tanδcorr	φcorr
13	B/w		G/B	
14	=-(10^12*O14/(2*PI()*A14*10^6))	=400*$B$8*P14/(PI()*$A$8^2*8.854)	=-L14/O14	=-ATAN(1/R14)*180/PI()

Low Frequency Spreadsheet

	T	U	V	W
1				
2				
3				
4				
5				
6				
7				
8				
9				
10				
11				
12	ε''	Zcorr	Rcorr	ρ', Ohm-m
13	ε'tanδ	Ohm	Ohm	(Rcorr)
14	=Q14*R14	=1/SQRT(L14^2+O14^2)	=U14*COS(S14*PI()/180)	=PI()*$A$8^2*V14/(400*$B$8)

Low Frequency Spreadsheet

	X
1	
2	
3	
4	
5	
6	
7	
8	
9	
10	
11	
12	ρ"
13	
14	=2*PI()*A14*10^6*Q14*8.854*10^-12/(N14^2+(2*PI()*A14*10^6*Q14*8.854*10^-12)^2)

Low Frequency Spreadsheet

	Y	Z	AA	AB
1	Avra - 6.15% Sample			
2	Low Frequency - Disk Electrodes			
3				
4				
5				
6				
7				
8				
9				
10				
11	Conduct.		Diel.	
12	f, MHz	losses,$\varepsilon''c$	losses	$100/\omega^{0.5}$
13		sc/we0	$\varepsilon'' - \varepsilon''c$	
14	0.001	=$I$5/(2*PI()*M14*10^6*8.854*10^-12)	=T14-Z14	=100/SQRT(2*PI()*M14*10^6)

Low Frequency Spreadsheet

	AC	AD	AE	AF
1				
2				
3				
4				
5				
6				
7				
8				
9				
10				
11				
12	Rms-Avg	Rms # 1	C(s), pF	σ'', S/m
13	Ohm	Used		
14	0.0187	0.01508	=-(10^12/(2*PI()*A14*10^6*B14*SIN(C14*PI()/180)))	=N14/R14

Low Frequency Spreadsheet

	AG	AH	AI	AJ
1				
2				
3				
4				
5				
6				
7				
8				
9				
10				
11				
12	σ'diel	σ"diel	$1/\omega^{0.5}$	ρ"
13	σ'-σ'ohmic			
14	=N14-$I$5	=AG14/R14	=1/SQRT(2*PI()*M14*10^6)	=W14/R14

E.3 Attenuation and Velocity Equations

Attenuation and Velocity Equations

	A	B	C	D	E	F
1	Attenuation and velocity equations for Chapter 11.3.					Water cont.
2						W, %
3						6.495
4	4.13.99	Wet sample: Brookhaven, # I, 6-8'				
5	HP 4194A,	Fixture	Coaxial sample holder #154, new teflon spacers (without upper pin)			
6	Length,cm	R, cm	r, cm	Weight	Dens.(wet)	Drying, weight in g.
7	2.568	0.7144	0.3102	M, g	γ, g/cm^3	
8	L, nH	Cs, pF	V, cm^3	5.34	1.59825	Wet
9	40	0.43972	3.3411	Ct, pF		20.66
10				0.401450748		
11	f, MHz	Z, Ohm	ϕ, deg	tanδ	Res.meas.	Xmeas.
12					R(s), Ohm	Zsinϕ
13	40.000	390.665	-76.501	=-1/TAN(C16*PI()/180)	=B16*COS(C16*PI()/180)	=B16*SIN(C16*PI()/180)
14	30.000	509.269	-73.122	0.3034	147.86	-487.3312744

Attenuation and Velocity Equations

	G	H	I	J
1				
2	Wv, %			
3		9.747		
4	Bulk density g/cm^3			
5		1.50		
6				
7				
8	Dry			
9		19.4		
10				
11	$X_L = \omega L$	Rm-Rms	Xm-wL	G, S
12	Ohm	R'=Rm	X'	
13	=2*PI()*A16*10^6*$A$11*10^-9	=B16*COS(C16*PI()/180)	=F16-G16	=H16/(H16^2+I16^2)
14	7.5398	147.9	-494.87	0.00055

Attenuation and Velocity Equations

	K	L
1		
2		
3		
4		
5		
6		
7		
8		
9		
10		
11	σ', S/m	B, S
12		
13	=100*LN($B$9/$C$9)*J16/(2*PI()*$A$9)	=(I16/(H16^2+I16^2))+2*PI()*A16*10^6*$B$11*10^-12
14	2.9E-03	-0.001772228

Attenuation and Velocity Equations

	M	N	O
1			
2			
3			
4			
5			
6			
7			
8			
9			
10			
11	f, MHz	C, pF	ε'
12		B/w	
13	40.000	=-(10^12*L16/(2*PI()*A16*10^6))	=LN($B$9/$C$9)*N16/(2*PI()*$A$9*0.08854)
14	30.000	9.402	5.49

Attenuation and Velocity Equations

	P	Q	R	S
1				
2				
3				
4				
5				
6				
7				
8				
9				
10				
11	tandcorr	φcorr	ε''	Zcorr
12	G/B		e'tand	Ohm
13	=-J16/L16	=-ATAN(1/P16)*180/PI()	=O16*P16	=1/SQRT(J16^2+L16^2)
14	0.312762767	-72.632	1.72	538.5359066

Attenuation and Velocity Equations

	T	U
1		
2		
3		
4		
5		
6		
7		
8		
9		
10		
11	Rcorr	ρ', Ohm-m
12	Ohm	(Rcorr)
13	=S16*COS(Q16*PI()/180)	=2*PI()*$A$9*T16/(100*LN($B$9/$C$9))
14	160.7548313	31.09

Attenuation and Velocity Equations

	V
1	
2	
3	
4	
5	
6	
7	
8	
9	
10	
11	ρ''
12	
13	=2*PI()*A16*10^6*O16*8.854*10^-12/(K16^2+(2*PI()*A16*10^6*O16*8.854*10^-12)^2)
14	99.41

Attenuation and Velocity Equations

	W
1	
2	
3	
4	
5	
6	
7	
8	
9	
10	
11	Atten.
12	dB/m
13	=8.686*2*PI()*10^6*M16*SQRT(O16*8.854*1.257*10^-18/2*(SQRT(1+P16^2)-1))
14	1.98E+00

Attenuation and Velocity Equations

	X
1	
2	
3	
4	
5	
6	
7	
8	
9	
10	
11	Velocity
12	m/s
13	=1/SQRT(O16*8.854*10^-12*1.257*10^-6/2*(SQRT(1+P16^2)+1))
14	1.26E+08

Index

Note: Page numbers followed by "*f*" and "*t*" refer to figures and tables, respectively.